Interdisciplinary and Multidimensional Perspectives in Telecommunications and Networking:

Emerging Findings

Michael R. Bartolacci
Pennsylvania State University - Berks, USA

Steven R. Powell
California State Polytechnic University - Pomona, USA

Information Science REFERENCE

INFORMATION SCIENCE REFERENCE
Hershey · New York

Senior Editorial Director:	Kristin Klinger
Director of Book Publications:	Julia Mosemann
Editorial Director:	Lindsay Johnston
Acquisitions Editor:	Erika Carter
Development Editor:	Myla Harty
Production Coordinator:	Jamie Snavely
Typesetters:	Michael Brehm, Jennifer Romanchak, Milan Vracarich, Jr. & Deanna Zombro
Cover Design:	Nick Newcomer

Published in the United States of America by
Information Science Reference (an imprint of IGI Global)
701 E. Chocolate Avenue
Hershey PA 17033
Tel: 717-533-8845
Fax: 717-533-8661
E-mail: cust@igi-global.com
Web site: http://www.igi-global.com/reference

Library of Congress Cataloging-in-Publication Data

Interdisciplinary and multidimensional perspectives in telecommunications and networking : emerging findings / Michael R. Bartolacci and Steven R. Powell, editors.
p. cm.
Includes bibliographical references and index.
ISBN 978-1-60960-505-6 (hardcover) -- ISBN 978-1-60960-506-3 (ebook)
1. Telecommunication--Technological innovations. 2. Telecommunication--Social aspects. I. Bartolacci, Michael R., 1954- II. Powell, Steven.
TK5101.I555 2011
384--dc22

2011005745

British Cataloguing in Publication Data
A Cataloguing in Publication record for this book is available from the British Library.

All work contributed to this book is new, previously-unpublished material. The views expressed in this book are those of the authors, but not necessarily of the publisher.

Table of Contents

Detailed Table of Contents

The interdisciplinary approach can be beneficial to researchers, educators, and practitioners for solving complex problems and synthesizing knowledge. While a great deal has been written about interdisciplinary activity in relation to the humanities and the natural and social sciences, it has been important in the telecommunications and networking field as well. This article investigates the subject of interdisciplinarity, especially as it relates to telecommunications and networking research and education. Among the questions examined are: What exactly is meant by "interdisciplinary"? Why is the interdisciplinary approach important in telecommunications and networking research and education and what are some examples of how is it being utilized? How does the International Journal of Interdisciplinary Telecommunications & Networking differ from other telecommunications and networking publications in its treatment of interdisciplinary research and what does the journal hope to accomplish?

Much of the rationale for the launch of 3G networks from 2000 onwards was to promote the expected rise in wireless data usage. Consumers were predicted to make widespread use of mobile internet and mobile TV but it was felt that the data rates offered by 2G networks were insufficiently fast to give them the experience that they expected. But more recently success has been achieved with different types of phone such as the iPhone and with new applications rather than higher data rates. This article takes a look at predictions of the future, considers whether they need to be modified in the light of events and makes its own suggestions as to where things might go from here.

Internet services are gradually expanding to mobile devices. This is driven by the evolution towards Internet-based mobile networks (the ongoing mobile broadband diffusion), compelling value already embedded in existing Internet services (e.g. instant messaging and social networking services), and changing business logic in capitalizing on mobile services (e.g. mobile advertising; emergence of mobile software platforms; and device, service, subscription bundling). Given that the technical conditions are fulfilled, the uncertainty on the emergence of the mobile Internet is born in the changing dynamics of the mobile business landscape. This article studies the emergence of the mobile Internet business from six strategic perspectives, highlighting the ongoing transformations of the mobile industry. The study introduces recent case examples in elaborating each of the perspectives. The theoretical foundation of the article is based on a literature review on competitive market environments, ICT (information and communication technology) business dynamics and value network modeling. The research problem of the article is to assess the different future research domains of the mobile Internet business. The article suggests through a fact-based study that companies of the mobile industry can reconstruct their business models in light of the emerging mobile Internet with six different kinds of choices: 1. Positioning differently in the value network, involving both vertical (e.g. from hardware to services) and horizontal (e.g. provisioning new kinds of services) movements. 2. Acquiring and developing critical assets (content, network infrastructure, IPRs, brands) and competencies (e.g. creation of software platforms) to build sustaining and dynamic competitive strategies. 3. Inventing new mobile business models by utilizing the lessons from the Internet, e.g. bundling products, utilizing two-sided markets and advertising, using new revenue sharing schemes, dropping of old business models or radically restructuring them. 4. Integrating end-users in making and redefining services, optimizing pricing logic, introducing new business models with end-users content creation in mind, and making services easy to adopt by end-users. 5. Leveraging and focusing on disruptive, winning technologies that are changing the rules of the business. 6. Adapting to relevant policy schemes, and proactively planning and reconsidering business decisions under a dynamic regulatory environment.

The major US wireless operators already have announced their plans for the evolution of their networks towards 3G, but some uncertainties remain, such as emergence of new technologies (WiMAX and WLAN) and the consolidation among operators (AT&T Mobile and Sprint Nextel). The article discusses a real option based model for technology decisionsand applies it to the US wireless industry as a case study. We also discuss what decisions must be made, what the outcomes are, and how the options model is validated. The preliminary results show that the evolution of wireless network technologies between generations (inter-generations migration scenario) is desirable (a positive net option value), but not desirable (a negative net option value) within generations (intra-generation migration scenario), in the US.

Vehicular ad-hoc networks (VANETs) have been gained importance for the inter-vehicle communication that supports local communication between vehicles without any expensive infrastructure and considerable configuration efforts. As many value-added or customized applications can be provided in VANETs, geographic routing has received much attention to achieve efficient end-to-end data delivery. However, how to provide light-weight and scalable location management service which facilitates geographic routing in VANETs remains a fundamental issue. In this article we will present a novel peer-to-peer location management protocol, called PLM, to provide location management service in VANETs. PLM makes use of high mobility in VANETs to disseminate vehicles' historical location information over the network; and then a vehicle is able to predict current location of other vehicles with Kalman filtering technique. Our theoretical analysis shows that PLM is able to achieve high location information availability with a low protocol overhead and latency. We also evaluate the performance of PLM via extensive simulations. The simulation results indicate that PLM can provide fairly accurate location information with quite low communication overhead in VANETs. The performance of geographic routing with PLM is close to that with perfect location information.

The authors describe a methodology for evaluating the performability (combined performance and reliability) of large communications networks. Networks are represented by a 4-level hierarchical model, consisting of traffic matrix, network graph, "components" representing failure modes, and reliability information. Network states are assignments of modes to the network components, which usually represent network elements and their key modules, although they can be more abstract. The components can be binary or multi-modal, and each of their failure modes may change a set of attributes of the graph (e.g. the capacity or cost of a link). Their methodology also captures the effect of automatic restoration against network failures by including two common rerouting methods. To compute network performability measures, including upper and lower bounds on their cumulative distribution functions, we augment existing probabilistic state-space generation algorithms with our new "hybrid" algorithm. To characterize the network failures of highest impact, we compute the Pareto boundaries of the network's risk space. The authors have developed a network analysis tool called nperf that embodies this methodology. To illustrate the methodology and the practicality of the tool, they describe a performability analysis of three design alternatives for a large commercial IP backbone network.

The (information digital) network bandwidth and its usage through Internet subscription, by far, is perhaps the only uni-modal commodity provided to today's consumers at a flat rate. Regardless of his/her actual bandwidth usage, a consumer is charged a uniform subscription fee by the Internet Service

Providers (ISP). Meanwhile, the ISPs have a considerable stake in the overall infrastructure which supports the local perimeters of the Internet. This article presents a rigorous statistical analysis with the objective of determining the optimal billing or pricing policy for the ISPs of the U.S. on the basis of the proposed 'variable subscription-rate' business model. Two leading global consumer profiles were selected for data comparison on a cardinal scale; NYC, NY, U.S.A. and Seoul, South Korea.

The parametric cost estimation approach has proved to be an efficient method for analyzing complex systems such as spacecraft, missiles, ships, buildings, etc where cost varies according to a number of parameters. The cost to provision a telecom network also depends on a number of parameters; but little research effort has been applied to estimate cost using this approach. In estimating the cost of a telecom network, most published research has considered two parameters; distance and bandwidth of a link and ignored the effects of other parameters. We have modelled the cost based on distance, bandwidth, geographical terrain and technology simultaneously using a parametric cost estimation methodology applied to real data obtained from the Indian Telecom Company, BSNL. Using the model, we show how a cost optimized network can be designed given the real world constraints. The applicability of our model to determine revenue sharing mechanism for an international call is also demonstrated.

This article presents a new transmitted reference ultra-wideband receiver, which utilizes the orthogonal property of even and odd order derivatives of Gaussian pulses in neighboring chips for synchronization. This system, referred to as orthogonal TR-UWB (OTR-UWB), employs only a single spreading code, which results in much lower mean detection time compared to DS-UWB systems since smaller search space is involved for coarse acquisition. The hardware complexity for OTR-UWB receiver is significantly reduced against conventional TR-UWB systems. In addition, simulation results show that bit error rate (BER) performance is improved, while the new system is capable of supporting higher data rates. Also, this article presents the FPGA implementation OTR-UWB, with a bit-rate of 25Mb/s without using equalizer. It incorporates 4-bit of resolution with 8GHz sampling rate for analog to digital conversion which has been implemented by parallel low frequency ADC blocks. Nano second short pulses with giga samples can't be implemented on FPGA. So we present the DSP algorithm of baseband. Hardware of this system is implemented on two different FPGAs from ALTERA and XILINX, CycloneII (EP2C35F672C6) and Spartan 3 (3s4000fg676-5). Gate estimation and power analysis are performed by Quartus II 7.2 (ALTERA) and ISE 8.1 (XILINX) softwares.

The emergence of Web 2.0 and its rapid adoption by the corporate world, known as Enterprise 2.0, has radically modified access to tacit knowledge and significantly reduced business latency. Meanwhile, technology for wireless LAN and mobile communication, combined with advances in handsets, has allowed for superior quality of experience with acceptable productivity level. Combining mobility with Enterprise 2.0 is the next big step in evolution. This article presents the drivers for Enterprise 2.0 and the challenge of tapping tacit knowledge and, in parallel, the evolution of wireless and mobile technologies. The era of mobile broadband life is made possible, creating new ways of use and expectations for Millennials. The next business generation will be able to unleash the full potential of mobility and Enterprise 2.0. Three scenarios selected from different vertical domains—healthcare, education and emergency services—illustrate the benefits of mobility and Enterprise 2.0 in action. With the expected continued strong growth of wireless access, mobility support is one, if not the, key success factor for Enterprise 2.0.

Companies can benefit from diversifying internationally. This article analyzes the internationalization efforts of six major European wireless telecommunications service providers: Telefonica, Deutsche Telekom, France Telecom, Vodafone, Telenor and TeliaSonera. Although all six companies have invested heavily in foreign markets, their internationalization paths have not been the same, resulting in foreign market portfolios with different characteristics. Utilizing a proportionate customer weighting scheme based on the number of subscribers controlled by a company in each market, the article examines how the companies compared in 2007 versus 2002 in their extent and scope of internationalization and in the attractiveness of their foreign market portfolios with respect to some key growth, profitability, and risk market characteristics.

Mediation in news industries has received significant attention by researchers for more than half a century. Gatekeepers decide which information should be delivered to different audiences. The Shoemaker/Reese Gatekeeping Model identifies five different filters of content processing: individual influences, professional routines, the organization, extra-media influences and ideology. Journalism practices, intra-organization and extra-media-related procedures and strategic alliances, including culture and ideology, add more complexity in the contemporary globalized media landscape. Gatekeeping is being processed through out all the above mentioned pillars. ICT technologies related to the media

Preface

This book containing articles from the four issues of Volume 1 of the *International Journal of Interdisciplinary Telecommunications and Networking* (*IJITN*) reflects the journal's mission of publishing high-quality original interdisciplinary academic and practitioner research, surveys, and case studies that address telecommunications and networking issues, answer telecommunications and networking questions, or solve telecommunications and networking problems. The articles represented in the book also reflect the journal's objective of covering a wide variety of topics related to telecommunications and networking technology, management, policy, economics, and social impact from a diversity of disciplinary viewpoints, including electrical engineering, computer science, operations research, business, and law. The interdisciplinary approach can be beneficial to researchers, educators, and practitioners for solving complex problems and synthesizing knowledge.

Steven Powell investigates the subject of interdisciplinarity and its benefits, especially as they relate to telecommunications and networking research and education, in greater depth in his article, "Interdisciplinarity and Telecommunications and Networking," published in Issue 1 and included here. While a great deal has been written about interdisciplinary activity in relation to the humanities and the natural and social sciences, he observes that interdisciplinarity also has been important in telecommunications and networking research and education. In Bell Laboratories, to take an industrial example, some of the most innovative teams working on very challenging end-to-end communication issues are interdisciplinary. Some university degree programs and professional conferences related to telecommunications and networking also emphasize interdisciplinarity. Powell concludes his article by examining how *IJITN* can provide a home for interdisciplinary telecommunications and networking research, practice, and education.

An example of a professional conference that emphasizes interdisciplinarity in telecommunications and networking is the Wireless Telecommunications Symposium. With global mobile connections now topping five billion and expected to reach six billion by the first half of 2011 according to the research firm Wireless Intelligence, the key role that wireless communications plays in society today is undeniable. Issues Number 1 and Number 4 of *IJITN* are special issues containing articles from researchers who participated in the seventh annual Wireless Telecommunications Symposium, WTS 2008, and the eighth, WTS 2009. The Wireless Telecommunications Symposium is an interdisciplinary mobile communications and wireless communications conference which brings together leaders and experts from industry, government, and universities around the world to discuss advances in wireless technology, management, policy, applications, and security. WTS is hosted by the College of Business Administration and College of Engineering at the California State Polytechnic University, Pomona; the IEEE Communications Society is technical co-sponsor for WTS. WTS 2008 was held in Pomona, California USA on April 24-26, 2008; WTS 2009 was held in Prague, Czech Republic on April 22-24, 2009.

William Webb, who is Head of Research and Development at the UK Office of Communications (Ofcom) and a wireless communications visionary, was a guest speaker at WTS 2008. His article in Issue 1, "Wireless Communications: Is the Future Playing Out as Predicted," was based on his presentation at the conference and addresses the future of wireless communications from a technological, socio-economic, structural, and regulatory perspective. Webb investigates predictions of the future he made a few years ago, considers whether they need to be modified in the light of events, and makes his own suggestions as to where things might go from here. He concludes that much of what he had predicted is still correct.

In his article in Issue 1, "Analysis of the Forces Reshaping the Mobile Internet Business," Hannu Verkasalo describes the evolution toward Internet-based mobile networks and analyzes the mobile industry's transformation from different strategic perspectives. He suggests that there are six main ways companies in the mobile industry can change their business models and adapt to the changing business environment: (1) position the company differently in the value network; (2) acquire and develop critical sustainable assets and competencies; (3) invent new business models; (4) pay greater attention to end-users in defining, pricing, and delivering new services; (5) leverage and focus on new, disruptive, and winning technologies that may change the rules of business; and (6) adapt to new regulatory schemes and policy initiatives on the mobile Internet.

In their article in Issue 1,"Strategic Technology Options in the Wireless Industry: A Case Study for U.S. Wireless Carriers," Hak Kim, Martin Weiss, and Benoit Morel discuss a real option based model for technology decisions and apply it to the U.S. wireless industry as a case study. The authors also discuss what decisions are made, what the outcomes are, and how the options model is validated. Their preliminary results show that inter-generation migration is desirable, while intra-generation migration is undesirable.

In the final article in Issue 1, "Time-Based Confidentiality Enhancement Scheme for Mobile Wireless Networks," Qunwei Zheng, Xiaoyan Hong, Jun Liu, and Lei Tang propose a novel approach that exploits mobility to enhance confidentiality in mobile ad hoc wireless networks. The authors used simulation with different parameters and network scenarios to evaluate their scheme and found that it is particularly suitable for delay tolerant applications and networks where spatial multi-paths are not available.

The second volume of 2009 contained a variety of articles ranging from a security architecture for self-organizing networks to channel assignment for Wi-Fi networks; but all dealing with some aspect of wireless communications. The importance of wireless communications research is growing as the world becomes ever more dependent on networks from the "smart home" to the battlefield to health care. Wireless networks of one form or another are becoming ubiquitous in an effort to organize and control aspects of daily life on a real time basis from anywhere in the world. Leonard Kleinrock of UCLA, and a founding father of the Internet, in the 1990's coined the term "nomadic networks" to describe the future ability of mobile networks connected to the wired worldwide system to connect anyone to any piece of information or application from any point on the earth. Although his terminology has been substituted with other network names describing the type of wireless technology employed such as WiMAX or Mesh Networks for making the final link to the user from the wired network, the overall concept is steadily progressing towards reality.

In the first article of this volume, entitled "Hybrid Trust Structure in Self-Organizing Networks," authors Zhou and Harn deal with a topic of ever-increasing importance: trust in self-organizing networks. Ad hoc networks are those which arise "on the fly" with nodes joining and leaving in a usually unpredictable pattern. The ability to identify nodes of authority in this type of a network is key to its operation and efficiency. Without such a mechanism, security could be easily breached and the network rendered ineffective or worse yet destructive for the nodes that use it. This work examines a methodology for developing trust within such a network type.

The second article of the issue, "On Peer-to-Peer Location Management in Vehicular Ad Hoc Networks," deals with another aspect of wireless mobility: location management for mobile vehicular nodes in an ad hoc network. Vehicular nodes present a great challenge when put into an ad hoc networking context. Their mobility makes network management very challenging since the topology of the network can change so rapidly. One of the key issues in network management for MANETs such as these is the tracking of locations for all nodes in the network. Without careful tracking of locations for nodes, network routing becomes an exercise in futility and the network may become disconnected or painfully slow in its speed of transmission from source node to destination node. The authors of this article present a protocol that contains a two-step process for tracking nodal locations. The first step involved mobility-assisted location information dissemination (informing nodes of the locations of other nodes). The second step is a Kalman filter-based location information prediction methodology (using current information on nodal locations to predict future locations). The authors use simulation to justify the merits of their proposed protocol.

The third volume of 2009 contains four articles of greatly varying topics in telecommunications. All four articles do share one commonality: a mathematical or statistical approach to modeling a telecommunications problem versus the use of a qualitative or narrative model.

The first of these articles, entitled "Multi-Layer Network Performance and Reliability Analysis", details work conducted at AT&T Research Laboratories which is well known for its groundbreaking research in various aspects of telecommunications over the years. The authors present an operations research-based approach for evaluating both the reliability and the operating performance of a large telecommunications network. The approach is based on a the theoretical analysis and solution of a hierarchical network model which is in direct contrast to the most common approach of utilizing a simulation model for network evaluation. Their approach was able to find solutions for realistic-sized networks and provide their company with a tool for network evaluation and planning. Such a tool must prove invaluable to a company such as AT&T and its affiliated entities due to the sheer scale of the size of networks they build and operate. What might seem like a small cost savings realized in a planning model using a tool such as the one the authors describe can be magnified when actually implemented in a network's construction and operation.

The second article of this volume, "Next-Generation Strategic Business Model for the U.S. Internet Service Providers: Rate-Based Internet Subscription," involves a detailed analysis of the relevant factors in rate-based pricing for U.S.-based Internet service providers. Author Willow puts forth three hypotheses regarding which factors influence pricing and demand for Internet access. These hypotheses dealt with the issue of pricing Internet access strictly by access speeds. He then statistically analyzes these hypotheses using stratified cluster sampling for New York City, New York State, the United States, and Seoul, South Korea. Chi-Square goodness of fit tests were conducted and showed that strict rate-based pricing would not be appropriate for the American market.

The third article of this volume, "Application of Parametric Cost Estimation to Telecommunication Networks," proposes a parametric pricing model for provisioning telecommunication networks. Network provisioning is an important process where each component of a network must be specified in order to ascertain total cost of construction, purchase or construct required components, and develop a timeline for completion. A parametric cost model seeks to include all the necessary parameters for network construction such as bandwidth, distance, and geographic terrain. The authors put forth a parametric model that includes these and additional parameters with the goal of optimizing total cost of construction subject to restraints on the parameter values. They bring to light what they term the "Capacity Distance Conflict", which describes the situation where the overall cost for a network can be lower even when the overall link distance is larger than the theoretical minimum total link distance. In other words, due to the nature of traffic and other parameters, provisioning for longer links sometimes can actually save money in the construction of a network even though it raises the total cost of the link media and link construction.

The final article of this issue, entitled "Performance and Complexity Evaluation of OTR-UWB Receiver," deals with an emerging wireless technology that may soon replace infrared and Bluetooth technologies that are common in personal electronics and home appliances. This technology is Ultra Wideband technology (UWB) and it is based on a low power signal with a wide bandwidth footprint for short range communications. Remote controls and wireless mobile phone handsets that currently use infrared or Bluetooth could soon use UWB as the basis for their short range communications. The authors of this article examine a modulation scheme for UWB from an electrical engineering-oriented point of view. UWB represents a form of what is known as "spread spectrum" technology, and as such, presents unique challenges with respect to channel modeling and modulation. The authors give an overview of current approaches with respect to these activities and present their own approach based on orthogonal transmitted reference UWB (OTR-UWB). They use simulation to test this approach against other approaches with promising results.

Francois Cosquer, who is CTO Security and Technology Strategist for the Alcatel-Lucent Enterprise and Verticals Business Group, was a guest speaker at WTS 2009. His article in Issue 4, "Mobility and Enterprise 2.0," which was co-authored with Annie Ohayon-Dekel, strategy director for the group, was based on his conference presentation. The article reviews the emergence of Web 2.0 and its rapid adoption by the corporate world, known as Enterprise 2.0, which has radically modified access to tacit knowledge and significantly reduced business latency. The next step is to combine Enterprise 2.0 with mobility. The authors consider three scenarios selected from different vertical domains (healthcare, education and emergency services) to illustrate the benefits of mobility support for Enterprise 2.0.

In his Issue 4 article, "On the Internationalization of the Wireless Telecommunications Industry: A Market-Based Analysis of Six European Service Providers," Steven Powell analyzes the internationalization efforts from 2002 to 2007 of six European wireless telecommunications service providers which have followed different internationalization paths: Telefonica, Deutsche Telekom, France Telecom, Vodafone, Telenor, and TeliaSonera. Using a proportionate weighting scheme based on the number of subscribers in each market, the author ranks the companies with respect to three aspects of their internationalization efforts: their degree of internationalization; the attractiveness of their foreign market portfolios from a growth, profitability, and risk perspective; and the distance between their foreign market portfolios and their home markets.

In their Issue 4 article, "Revisiting the Gatekeeping Model: Gatekeeping Factors in European Wireless Media Markets," Vassiliki Cossiavelou and Philemon Bantimaroudis examine the impact of wireless technology on gatekeeping practices in the context of the European Union news markets. The authors argue that the role of continuously changing media technology, especially wireless technology, should be incorporated into the established Shoemaker/Reese gatekeeping model. Their argument is supported by a preliminary survey of media content and network related gatekeepers indicating that the digital media landscape is being influenced by ICT technologies.

In the Issue 4 article, "Trust-Based SAODV Protocol with Intrusion Detection, Trust Management and Incentive Cooperation in MANETS," author Floriano De Rango introduces enhancements to the

SAODV routing protocol that will improve its performance and improve its security. Simulation results support the effectiveness of the security features he proposes.

In the final Issue 4 article, "Modeling and Simulation of Self-Similar Traffic in Wireless IP Networks," Dimitar Radev, Izabella Lokshina, and Svetla Radeva examine the self-similar properties of telecommunications traffic data over a wide range of time scales. The authors present new efficient algorithms for modeling fixed-length sequence generators to simulate the self-similar behavior of wireless IP network traffic and analyze simulation results based on the models.

We would like to thank our colleagues on the *IJITN* editorial board – especially Qing-An Zeng and Izabella Lokshina, who edited the special issues 1 and 4, respectively - as well as the many authors and reviewers for their contribution to the journal. We would also like to thank the staff at IGI Global – especially Heather Probst, Director of Journal Publications - for their support for *IJITN* and their help publishing it. We hope that you will enjoy reading this book and find the articles informative.

Michael R. Bartolacci
Pennsylvania State University – Berks, USA

Steven R. Powell
California State Polytechnic University - Pomona, USA

Chapter 1
Interdisciplinarity in Telecommunications and Networking

Steven R. Powell
California State Polytechnic University-Pomona, USA

ABSTRACT

The interdisciplinary approach can be beneficial to researchers, educators, and practitioners for solving complex problems and synthesizing knowledge. While a great deal has been written about interdisciplinary activity in relation to the humanities and the natural and social sciences, it has been important in the telecommunications and networking field as well. This article investigates the subject of interdisciplinarity, especially as it relates to telecommunications and networking research and education. Among the questions examined are: What exactly is meant by "interdisciplinary"? Why is the interdisciplinary approach important in telecommunications and networking research and education and what are some examples of how is it being utilized? How does the International Journal of Interdisciplinary Telecommunications & Networking differ from other telecommunications and networking publications in its treatment of interdisciplinary research and what does the journal hope to accomplish?

INTRODUCTION

Driven by the need to solve complex social, economic, and technological problems, the desire to explore broad issues requiring more than one discipline, and the need to integrate the knowledge generated from these efforts, the late twentieth century witnessed a dramatic increase in interdisciplinary activity among researchers, educators, and practitioners that continues to this day. In its report *Facilitating Interdisciplinary Research* (2005), which reviews the state of interdisciplinary research and education in science and engineering and recommends ways to facilitate them, the Committee on Science, Engineering, and Public Policy of the National Academy of Sciences,

DOI: 10.4018/978-1-60960-505-6.ch001

the National Academy of Engineering, and the Institute of Medicine asserted:

Interdisciplinary research (IDR) can be one of the most productive and inspiring of human pursuits—one that provides a format for conversations and connections that lead to new knowledge. As a mode of discovery and education, it has delivered much already and promises more—a sustainable environment, healthier and more prosperous lives, new discoveries and technologies to inspire young minds, and a deeper understanding of our place in space and time. (p. 1)

Among the great triumphs of interdisciplinary inquiry and collaboration cited in the National Academies' report are the discovery of the structure of DNA, the Manhattan Project, laser eye surgery, and manned space flight. Current "hot" interdisciplinary research topics cited by the Committee include nanotechnology, bioinformatics, neuroscience, conflict, and terrorism. Other examples of vital interdisciplinary research and educational activity mentioned in the report include: the strong attraction of students to interdisciplinary courses; the growth in the number of university departments resulting from a blending of previously distinct fields; the growth in the number of interdisciplinary centers, institutes, and similar units at universities; the excellence of interdisciplinary research and development at industrial laboratories such as IBM's Research Center in Yorktown Heights, New York, Alcatel-Lucent's Bell Laboratories in Murray Hill, New Jersey, and Xerox's Palo Alto Research Center in California; strong interdisciplinary research programs at national laboratories such as the Oak Ridge National Laboratory in Tennessee, Lawrence Livermore National Laboratory in California, the Jet Propulsion Laboratory in California, and the National Institute for Advanced Interdisciplinary Research in Japan; funding for interdisciplinary research projects by the National Science Foundation, National Institute of Health, and the Defense Advanced Projects Research Agency; and the proliferation of government-industry-university collaborations in fields such as the life sciences.

The literature on the subject of interdisciplinarity is extensive (see Klein 1990; Klein 2006). A search of Ulrichsweb.com (2008) for periodicals whose title keyword contained the word "interdisciplinary" yielded 307 results, primarily periodicals pertaining to the humanities and the natural and social sciences. Key issues that researchers into the subject of interdisciplinarity have investigated include: the meaning of the term interdisciplinary; how interdisciplinary efforts relate to disciplinary efforts; the benefits of interdisciplinary thinking; under what conditions interdisciplinary projects should be undertaken; how interdisciplinary activities should be performed and managed; and applications of interdisciplinarity in research and education.

The remainder of this article will focus primarily on interdisciplinary activities in research and education in telecommunications and networking. The article is structured as follows: the next section examines what is meant by the interdisciplinary approach and investigates some characteristics of it; the following section describes why interdisciplinarity in telecommunications and networking research and education is important and provides some examples of interdisciplinary activity in the field; and the last section describes how telecommunications and networking journals have been treating interdisciplinary research and how the *International Journal of Interdisciplinary Telecommunications and Networking* will differ from them. The article concludes with a summary and a statement of the journal's vision.

THE INTERDISCIPLINARY APPROACH

Although the term "interdisciplinary" is frequently used, there is sometimes confusion over what exactly is meant by it. Webster's Third New International Dictionary (1971) defines "interdisciplinary" as: "characterized by participation or cooperation of two or more disciplines or fields of study: drawing on or contributing to two or more disciplines." The term "discipline" is defined as "a subject that is taught: a branch of learning: field of study". Klein (1991) is more precise in her definition of a discipline: a discipline "signifies the tools, methods, procedures, exempla, concepts and theories that account coherently for a set of objects or subjects" (p. 104). Since these change over time, a discipline develops a particular perspective or worldview.

The eminent philosopher Karl Popper (1963) recognized the need to cross disciplinary boundaries when dealing with certain kinds of problems:

Disciplines are distinguished partly for historical reasons and reasons of administrative convenience (such as the organization of teaching and of appointments), and partly because the theories we construct to solve our problems have a tendency to grow into unified systems. But all this classification and distinction is a comparatively unimportant and superficial affair. We are not students of some subject matter, but students of problems. And problems may cut right across the borders of any subject matter or discipline. (p. 88)

By definition interdisciplinary efforts require the participation or cooperation of two or more disciplines. What is not clear from the definition, and a subject for debate, is the nature and degree of this participation or cooperation. At issue is the nature of knowledge. Believers in the instrumental nature of knowledge view interdisciplinarity as a functional, problem-centered activity. Bridges are built between disciplines; the tools, methods, procedures, etc. of one discipline are borrowed by another discipline. Believers in the synoptic nature of knowledge view interdisciplinarity as a synthesizing activity whose objective is new knowledge, methods, and concepts. Disciplines are changed in the process. A variety of terms have been introduced into this debate. At one end of the interdisciplinarity spectrum are "multidisciplinary" activities in which several specialized disciplines are combined for a common purpose. According to Klein (1991) this combination is essentially additive, not integrative. Researchers represent disciplines, bringing their disciplinary perspectives to bear on a problem. There is a lack of interaction and real cooperation, so that the participating disciplines are neither changed nor enriched. At the other end of the spectrum are highly integrative activities that attempt to transcend disciplines, making them irrelevant. Such interdisciplinary activities have been called "transdisciplinary", "metadisciplinary", or "non-disciplinary".

According to Newell (2001) an interdisciplinary study by definition will draw insights from relevant disciplines and integrate those insights into a more comprehensive understanding. The interdisciplinary approach is justified only if the object of study is multifaceted and its facets cohere. If not multifaceted, a single-discipline approach is sufficient. If it is multifaceted but not coherent a multidisciplinary approach is sufficient. He argues that the interdisciplinary approach is justified only if a complex system and its behavior is under investigation.

Interdisciplinary work need not be team-based, although it often is. An activity engaged in by an individual who integrates more than one discipline, applying the tools, techniques, procedures, perspectives, etc. of each discipline, is interdisciplinary. On the other hand, an activity involving multiple researchers from more than one discipline, collaborating on a single problem, is also interdisciplinary. The idea that interdisciplinary work need not be team-based, and can be instrumental or synoptic, is reflected in the definition of interdisciplinary research presented in *Facilitating Interdisciplinary Research* (Committee on Facilitating Interdisciplinary Research 2005):

Interdisciplinary research (IDR) is a mode of research by teams or individuals that integrates information, data, techniques, tools, perspectives, concepts, and/or theories from two or more disciplines or bodies of specialized knowledge to advance fundamental understanding or to solve problems whose solutions are beyond the scope of a single discipline or area of research practice. (p. 26)

INTERDISCIPLINARITY IN TELECOMMUNICATIONS AND NETWORKING

Telecommunications and networking is an example of a subject that is very complex and multifaceted, and as such, certain problems associated with it may require an interdisciplinary approach for their solution. In its report *Renewing Telecommunications Research* (Lucky, R. W., & Eisenberg, J. 2006) the Committee on Telecommunications Research and Development of the Computer Science and Telecommunications Board, representing the National Academies of Science, Engineering, and Medicine, asserted that "major telecommunications advances can require longer-term efforts often continuing for a decade or more and require risk-taking broad-based, interdisciplinary, multifaceted support" (p. 21). Disruptive change in technology "requires an interdisciplinary approach, namely diverse participants that bring different perspectives to the table" (p. 51). The Committee argued for support for interdisciplinary research efforts such as was demonstrated at Bell Laboratories, where "researchers spanned a broad range of disciplines from physics to economics and thus were able to tackle the cross-disciplinary problems that often arise in designing, deploying, and operating telecommunications systems" (p. 45). In recommending the establishment of a new research organization, the Advanced Telecommunications Research Activity, the Committee felt that part of the organization's mission would be to identify, coordinate, and fund telecommunications research for the U.S. focusing on critical under-invested telecommunications research including "interdisciplinary research at the intersection of technology, economics, and policy, including work that studies the technical, economic, and sociological issues underlying regulatory decisions"(p. 54). Furthermore, it recognized that a number of research problems might require research centers which "provide a nucleus for academic-industry collaboration and for interdisciplinary research" (p. 66). The report emphasizes the importance of collaboration across academia and industry so that researchers outside industry can have the "specifically necessary information or access to facilities that are crucial to a quality and relevant outcome" (p. 46).

George Rittenhouse, Vice President of Technology Integration at Alcatel- Lucent Bell Laboratories, concurs with the Committee's findings about the importance of interdisciplinary research to the telecommunications industry, and the impact that it has had at Bell Laboratories. According to Rittenhouse (2008):

Some of the largest challenges facing our industry today are end-to-end communication issues. Focusing on one piece of the communication chain is not enough. Security, architecture, and application performance are examples of this type of problem. It is extremely difficult for a single individual or teams comprised of a single discipline to solve these. In Bell Labs we've found that the most innovative solutions come from interdisciplinary teams.

According to Rittenhouse, one clear example of interdisciplinary teamwork at Bell Laboratories dealt with a problem in laptop security. Keeping a laptop in constant communication with the enterprise is preferable to simply keeping it behind a firewall, where it can be lost or stolen. To fill this market need Bell Laboratories created a novel, award winning solution with an interdisciplinary team of computer scientists, networking experts, and wireless experts. Together they created a computer within a computer. This additional computer, which resides on a PCMCIA card with its own battery, CPU, memory, network connection, GPS, and OS, is always connected to the network regardless of whether the laptop is turned on or not. If necessary, it can remotely power up the laptop, apply security patches, or erase the disk.

What kinds of skills must interdisciplinary telecommunications researchers possess to conduct these and other projects? Crossing disciplinary boundaries to solve complex, multifaceted problems often requires breadth of knowledge, as well as depth of knowledge, and the ability to integrate and synthesize it. Leonard Lewin, an IEEE Fellow and recipient of the 1993 IEEE Microwave Career Award, argued for breadth of knowledge in telecommunications. In the 1984 book *Telecommunications: An Interdisciplinary Survey*, Lewin wrote:

It would be a great mistake to see telecommunications too narrowly as only the technical aspect of the design and operation of the communication network. Features concerned with government regulation at the international, national and local level, the economics, the management, the legal and social impact; these are but a few of the important aspects that determine the shape and growth of the industry. (Lewin 1984a, p. i)

Lewin cited the example of the CATV industry, in which FCC regulations determined in large measure the structure of the industry, copyright laws and federal taxation policy determined programming and finance, and developments in optical fiber technology determined growth.

More recently, a Gartner Research report asserted the need for IT professionals, including those focusing on telecommunications and networking, to have breadth of knowledge. The report stated:

By 2010, IT professionals will need to possess expertise in multiple domains. Technical aptitude alone will no longer be enough. IT professionals must prove they can understand business realities – industry, core processes, customer bases, regulatory environment, culture and constraints. Versatility will be crucial. (Morello, 2005, p. 1)

Ali Tabassi, Vice President of Technology Development at Sprint Nextel, recognizes the need for engineers to see the big picture and not just focus on the specifics. According to Tabassi (2007) the days of having "silo engineering" in a telecommunications environment are becoming fewer. A more corrective understanding is necessary. For example, an RF engineer designing radio access has to understand how that access is going to work with connectivity to the core network and how that connectivity is going to work with the billing system. Since the RF engineer has to have a good understanding of the whole concept, an end-to-end view is needed. "More and more we are looking for students who are well-rounded and have a better understanding of an overall picture."

In response to the need for telecommunications and networking professionals who possess breadth of knowledge and the ability to integrate and synthesize it, some universities offer interdisciplinary telecommunications programs at the undergraduate and graduate levels. As Leonard Lewin (1984b) later wrote in his book *Telecommunications: An Interdisciplinary Text*:

One of the great advantages of an interdisciplinary program is that it gives an opportunity for a holistic approach to otherwise disparate subject matter, and prevents the student from specializing too narrowly. (p. xxi)

Lewin was a professor of Electrical Engineering at the University of Colorado at Boulder and coordinated the MS Interdisciplinary Telecommunications Program. This program, the oldest such program of its kind, began in 1971 and now has over 2000 alumni. The program offers an MS and ME in Telecommunications and graduate-level certificates in Wireless and Security. Students can emphasize their studies in many areas, including wireless, networking, security, business strategy, telecommunications policy and more. About one-third of the degree is technical, one-third is economics and business, and one-third is law and policy. Students may change the weighting within certain guidelines to emphasize any given area (Interdisciplinary Telecommunications Program 2008). Other universities that emphasize the interdisciplinary nature of telecommunications in their educational programs include Oklahoma State University, the University of Pittsburgh, Michigan State University, Syracuse University, New York University, Polytechnic University, the University of Southern California, George Washington University, Southern Methodist University, and Columbia University. Many of these universities have interdisciplinary research centers associated with them. Although interdisciplinary telecommunications programs "infrequently serve as a pathway into a career in telecommunications research" (Lucky & Eisenberg, 2006, p. 44), graduates seem to be well suited for positions where a broad industry view is required, such as policy-making and management.

Some telecommunications and networking conferences also provide a platform for the presentation of interdisciplinary work. The Wireless Telecommunications Symposium (2008), for example, brings together representatives from all levels of industry, government, and academia to discuss the latest developments in global mobile communications and wireless network technology, business, policy, applications, and security. Other conferences offering interdisciplinary telecommunications research include the INFORMS Telecommunications Conference, the International Conference on Telecommunication Systems Modeling and Analysis, the International Conference on Networking and Electronic Commerce, and some of the larger IEEE Communications Society conferences, such as the IEEE International Conference on Communications (ICC) and the IEEE Global Communications Conference (GLOBECOM).

INTERDISCIPLINARITY AND THE INTERNATIONAL JOURNAL OF INTERDISCIPLINARY TELECOMMUNICATIONS AND NETWORKING

The body of publishable telecommunications and networking research that is generated by academic, industrial, and governmental researchers each year is diverse as well as voluminous. Research areas range from highly technical subjects related to hardware, software, networks, and standards to qualitative studies dealing with media, policy, and social impact. Researchers represent a variety of disciplines, including electrical engineering, computer science, operations research, information systems, management, economics, media, and law. As we have seen, a good deal of the important research being performed is interdisciplinary, and its volume is growing. How have the many professional journals publishing telecommunications and networking research responded to this trend? Unfortunately, most of them have ignored it, choosing to approach the subject from the narrow perspective of a single discipline - usually technical. Those journals that have not ignored interdisciplinary research may devote small sections or special issues to it, or if they publish it on an ongoing basis tend to focus on a set of closely related disciplines, such as electrical engineering and computer science, or law and economics. In addition, most telecommunications and networking journals tend to focus on a specific level of analysis – society, industry, organization, or end-user - and rarely attempt to bridge the communications gap between researchers and practitioners at the technical level and high-level executives and policy-makers.

Given the subject's significance the editors of the *International Journal of Interdisciplinary Telecommunications & Networking* feel that the coverage of the interdisciplinary telecommunications and networking landscape has been inadequate. How does *IJITN* address this problem? How does it differ from other telecommunications and networking journals? *IJITN* will differ in at least five ways. First, the scope of topics is broader. Areas of interest include, but are not limited to, telecommunications technology, business, policy, applications, economics, and social impact. Second, the disciplinary perspective is wider. *IJITN* emphasizes the cross-disciplinary viewpoints of many disciplines, including electrical engineering, computer science, information technology, operations research, business administration, economics, sociology, and law. Third, *IJITN* does not restrict the level of analysis. Research dealing with societal, industrial, organizational, and end-user issues are publishable in the journal. Fourth, *IJITN* seeks to bridge the communications gap between researchers and practitioners at the technical level and high-level executives and policy-makers. To accomplish this, thc journal publishes case studies, surveys, articles, interviews, and roundtable discussions by and with leaders and experts in the field, in addition to theoretical and empirical research findings. Original journal submissions originating from the industrial and governmental sectors, as well as the academic sector, are all welcome. Fifth, and finally, the editors intend to fill each issue with high-quality interdisciplinary content, not just provide a small section for it or reserve it for special issues. Regular issues will contain a diversity of timely and interesting topics and viewpoints; special journal issues may be devoted partially or entirely to very timely topics of great interest. In this way the journal will be relevant as well as informative.

CONCLUSION

In order to solve society's complex problems and integrate and synthesize the voluminous amount of knowledge being generated, researchers, educators, and practitioners are finding more and more that they must cross disciplinary boundaries. This is particularly true in the field of telecommunications and networking, a complex subject involving a wide variety of interrelated technological, management, policy, economic, and social factors. The solution of many problems in telecommunications and networking now requires a multidimensional perspective; the traditional approach of dividing up the subject into specialized topic areas and then examining them from the perspective of a single pertinent discipline – for example, electrical engineering for transmission technology or law for regulation – is no longer workable. Employers increasingly seek telecommunications and networking professionals who can see the "big picture".

This article has addressed the subject of interdisciplinarity in telecommunications and networking. The article examined what is meant by the interdisciplinary approach and why it is beneficial, and provided some examples of interdisciplinary research and education in the telecommunications and networking field today. A great deal is happening of an interdisciplinary nature, and this article reviewed some of the key developments. The article also investigated how the *International Journal of Interdisciplinary Telecommunications and Networking* differs significantly from other professional telecommunications and networking journals in its treatment of interdisciplinarity. The journal's vision is to be a focal point for the state-of-the-art in interdisciplinary telecommunications and networking research, practice, and education. By playing this role *IJITN* hopes to stimulate greater interdisciplinary activity and promote an understanding of its benefits. Although this goal is ambitious, the editors feel that it is achievable. A publication like *IJITN* is long overdue. We hope that you agree and will help to bring our goal to fruition.

REFERENCES

Committee on Facilitating Interdisciplinary Research, National Academy of Sciences, National Academy of Engineering, and Institute of Medicine. (2005). *Facilitating Interdisciplinary Research*. Washington, DC: The National Academies Press.

Interdisciplinary Telecommunications Program. (2008). ITP Fast facts. University of Colorado at Boulder. Retrieved July 11, 2008, from http://telecom.colorado.edu/

Klein, J. T. (1990). *Interdisciplinarity: History, Theory, and Practice*. Detroit, MI: Wayne State University Press.

Klein, J. T. (2006). Resources for Interdisciplinary Studies. *Change*, (March/April): 52–58.

Lewin, L. (Ed.). (1984a). *Telecommunications: An Interdisciplinary Survey. Dedham, MA: Artech House Publishers*. Dedham, MA: Artech House Publishers.

Lewin, L. (Ed.). (1984b). *Telecommunications: An Interdisciplinary Text*. Dedham, MA: Artech House Publishers.

Lucky, R. W., & Eisenberg, J. (Eds.). (2006). *Renewing U. S. Telecommunications Research*. Washington, DC: The National Academies Press, Committee on Telecommunications Research and Development, Computer Science and Telecommunications Board, Division on Engineering and Physical Sciences, National Research Council of the National Academies.

Morello, D. (2005). The IT Professional Outlook: Where Will We Go From Here? ID Number: G00130462. September 14, 2005: Gartner Inc.

Newell, W. H. (2001). A Theory of Interdisciplinary Studies. *Issues in Integrative Studies*, *19*, 1–25.

Popper, K. R. (1963). *Conjectures and Refutations: The Growth of Scientific Knowledge*. New York, NY: Routledge and Kegan Paul.

Rittenhouse, G. (2008). Private communication to the author. June 24, 2008.

Symposium, W. T. (2008). Retrieved July 11, 2008, from http://www.csupomona.edu/wtsi

Tabassi, A. (2008). Presentation at Wireless Telecommunications Symposium 2007. April 27, 2007, Pomona, CA.

Ulrichsweb.com. (2008). Proquest. Retrieved July 9, 2008, from http://www.ulrichsweb.com

Webster's Third New International Dictionary (1971). Springfield MA: G. & C. Merriman Company.

This work was previously published in International Journal of Interdisciplinary Telecommunications and Networking, Volume 1, Issue 1,edited by Michael R. Bartolacci, and Steven R. Powell, pp. 1-8, copyright 2009 by IGI Publishing (an imprint of IGI Global).

Chapter 2
Wireless Communications:
Is the Future Playing Out as Predicted?

William Webb
Ofcom, UK

ABSTRACT

Much of the rationale for the launch of 3G networks from 2000 onwards was to promote the expected rise in wireless data usage. Consumers were predicted to make widespread use of mobile internet and mobile TV but it was felt that the data rates offered by 2G networks were insufficiently fast to give them the experience that they expected. But more recently success has been achieved with different types of phone such as the iPhone and with new applications rather than higher data rates. This article takes a look at predictions of the future, considers whether they need to be modified in the light of events and makes its own suggestions as to where things might go from here.

INTRODUCTION

In late 2007, arguably, the mobile internet arrived. A new phone was launched with which users generated 50 times more data traffic than the average. But this dramatic increase in data traffic was not based on higher speed access – indeed, the phone was not even capable of accessing 3G networks. The phone, of course, was Apple's iPhone, and the increased data usage came about because of the superior user interface. The success of the iPhone demonstrated clearly that it was not insufficient data rates that were discouraging the use of wireless data; it was the ease of use and practicality of small devices.

And yet, despite this clear evidence, the industry continues to focus on ever higher data rates – spending much time and effort debating 4G, LTE, WiMax and others and showing charts predicting data rates in the region of 100Mbits/s. We predicted in our book that this was completely unnecessary. In this article we will take a look at

DOI: 10.4018/978-1-60960-505-6.ch002

why this is, how the industry has developed over the last two years and where we predict it is going.

Understanding Technological Progress

One of the key drivers of the future is technological change. Understanding how technology might evolve over the next two decades is therefore one of the key components of making any prediction. Equally, though, it is easy to get carried away. Just because something is technically possible does not necessarily mean that it will happen – video calling has been technically possible for decades and yet is still very rarely used. In essence, a new service cannot happen until it is technically viable but it must also be economic and become socially accepted. For example, controlling the home heating system remotely from a cell phone is technically simple but the cost of installing wireless heating control systems is currently much higher than the utility many would perceive to get from moving from the current systems to one that can be remotely controlled. Video calling is both technically possible and, for users with broadband connections and PCs, inexpensive, but has not yet become socially acceptable and hence is not in widespread use.

There are two key ways to predict technological progress. One is through broad trends, making use of physical and empirical laws such as Moore's law. The other is to look at specific technologies that are currently in the research stage and ask how long it will take for them to become commercial propositions and what impact they might have. Both approaches are discussed below.

The Laws

Shannon's Law (Shannon, 1948, 1949). This is more of a physical limit than an empirical law. It sets out the capacity of a single channel in the presence of noise. While directly applicable to wired connections, it cannot be easily applied to wireless systems where frequencies can be reused in neighbouring areas. Much work is underway to estimate the maximum capacity of a wireless network with a given spectrum allocation and number of cells and results vary depending on the approach taken. However, it seems likely that we are within a factor of 10 of the maximum we can achieve, and may well be much closer than that in practice – perhaps as close as a factor of 3. Hence, there is little room for technological breakthroughs that dramatically increase the capacity per cell. However, as will be explained shortly, this can be circumvented simply by increasing the number of cells.

Moore's Law (www.intel.com). Easily the most well-known of the empirical laws, this predicts that the number of transistors that can be placed on a chip will double every 12-18 months (in practice it has been doubling every 24 months). The implications for wireless of this are mixed. It does imply that handsets will continue to increase in memory and processing power. However, as we approach the limits of what is possible over wireless channels, massive increases in processing are needed for even small gains in capacity.

Cooper's Law (www.arraycomm.com). This is a most intriguing prediction that notes that the number of wireless voice channels in the world doubles around every 30 months, and has done so since 1901. What is most interesting is that of the million-times improvement that has occurred since 1950 roughly 15 times was the result of being able to use more spectrum (3 GHz vs. 150 MHz) and five times was from using frequency division, that is, the ability to divide the radio spectrum into narrower slices (25-kHz channels vs. 120-kHz channels). Modulation techniques (such as FM, single sideband, time-division multiplexing, and various approaches to spread spectrum) can take credit for another five times or so. But the lion's share of the improvement - a factor of about 2,700 – Cooper suggested was the result of effectively confining individual conversations to smaller and smaller areas by spatial division or spectrum re-use. Again, this gives us a pointer that smaller cells may well be more important in the future than any "wonder" technology.

Edholm's Law (Cherry, 2004). Edholm pointed out that data rates over wired and wireless networks increase over time. Wired networks typically support data rates about two orders of magnitude (ie a hundred times) greater than wireless. While wireless improves steadily over time so does wired making convergence between them unlikely. This suggests that wireless will not replace wired networks where high speed is needed, although it may form a short "tail" on the end of a high speed wired connection.

Hard disk size (Webb, 2007). There has been a very interesting trend in hard disk size. Since 1980 the average hard disk size on a mid-range PC has increased by an order of magnitude every six years. Of course, all trends of this sort come to an end eventually, but if this were to continue then hard disks 18 years from now would have 1,000 times the capacity of hard disks today. If this also applies to portable devices then the sort of hard disks currently found within iPods and similar devices, which currently have a capacity of around 20GB could store 20TB by 2025 or thereabouts. This suggests that storage capacity on mobile devices over the next two decades will, for most people, become effectively unlimited.

Technologies

There are many technologies in various stages of development that have yet to become commercial reality. Technologies can take many years to make it out of the lab and into the hands of consumers. For example, OFDM was a research topic in the 1960s but not a product until around 2000. More recently, ultra wideband (UWB) was being aggressively developed by start-ups from around 1998 but is only now just becoming commercially available and will possible take another 3-5 years to become widespread. 3G standardisation started around 1992 but 3G still does not have as many customers as 2G. In considering emerging technologies that were viable for prototypes and demanding users such as the military and tracking their evolution it seems that it takes of the order of 10 years for emerging technologies to make it into consumer devices, and a few more years for these to become widespread.

So there is a long lead-time on many technologies. Just by looking at those currently under development it is possible to be near-certain that anything likely to materialise in the next decade has been considered - and indeed most things likely to turn up in the next 20 years. The key emerging technologies are discussed below.

Software defined radio (Tuttlebee, 2002). Many future visions of wireless communications involve multi-modal devices connecting to a wide range of different networks such as 2G, 3G, WiFi and Bluetooth. They may even involve devices modifying their behaviour as they discover new types of network or as home networks add additional functionality. At present this is achieved by incorporating the chipsets from each of the different standards into the handset, eg 3G and Bluetooth. While such an approach works, it is relatively inflexible. An alternative is for communication devices to be designed like computers with general purpose processing capabilities and different software for different applications. Such devices could then call up, or download, the appropriate software for the particular communications requirement currently in use. The underlying architecture needed to achieve this is termed Software-Defined Radio (SDR).

In practice, the benefits of SDR appear relatively minor compared to the downsides. The current approach of multi-modal devices works well and will likely always be less expensive than a general purpose SDR radio. Further, since new network technologies are generally introduced much less frequently than users replace handsets there is little need for a handset to download a new standard when this can more readily be embedded in the handset during production. Because of this, we do not expect "true" SDRs that can change their radio behaviour to be implemented during the next two decades. We do, however, expect handsets to be able to download a wide range of new applications and software patches.

Smart antennas (Gross, 2005). Smart antenna technology has the potential to significantly increase the efficient use of spectrum in wireless communication applications. Through intelligent control of the transmission and reception of signals, capacity and coverage in wireless networks could be improved. While smart antennas have strong potential, we do not expect to see them make a significant impact over the next 10 – 20 years. This is because most schemes are difficult to implement, do not always bring gains and require large arrays of antennas at base stations at a time when environmental concerns are high. Comparatively, installing smaller cells brings much greater capacity gains for a smaller cost and lower risk.

Mesh networking (Held, 2005). A wireless mesh network utilises other users in the network as nodes to relay information. In this way information can be transmitted from one user to a distant user via multiple hops through the other users. Many advantages are claimed for mesh networks including a limited need for infrastructure, reducing deployment and ongoing operational costs; increased capacity because each node acts as a mini-base station and increased coverage due to the ability to hop around corners.

However, a significant body of research cast doubts on these claims. Unless a node is connected via backhaul to the core network, it cannot effectively generate capacity. Instead, it needs to relay any communications that it receives. There are some gains due to the lower power that can be used for multiple short hops compared to one long hop but these are outweighed by inefficiencies such as signalling protocols and the fact that the short hops are unlikely to align well with the single "long hop" and so power requirements increase.

Our view is that wireless mesh networks will not make a significant difference to high-volume wireless communications. Mesh networks may have niche applications such as working alongside existing networks to fill in areas of poor wireless coverage, in areas where conventional networks are uneconomic such as the provision of broadband services to rural communities, or in deploying sensor networks.

Cognitive radio (Fette, 2006). Although there are many different definitions of what a cognitive radio is, the basic concept is of a device which, on arriving in a new environment, can "understand" the usage of the radio spectrum and adapt its behaviour accordingly. So, for example, a cognitive radio might detect that the emergency service frequencies were currently lightly used – perhaps because there were few emergencies taking place. It might then move to these frequencies and make a series of short transmissions, checking after each one that the frequencies were still free. If not, it might then move to other frequencies, perhaps those used by broadcasting for example, which were essentially unused in the area.

There are many problems with cognitive radio, including an inability to be certain that the spectrum is not being used, and hence their transmissions might cause interference. Even if all the problems could be overcome, there is still a question as to what the cognitive radio transmits to. A network of base stations which is able to scan all frequencies for possible transmissions would be expensive and hugely risky to construct. Locating other terminals to transmit directly to is also difficult and suffers the problems associated with mesh networks.

Summary of Technological Progress

All these observations suggest that there is no wonderful technology, or technical trend, that on its own is going to revolutionise wireless over the next decade, and with all likelihood over the next twenty years. Hence, there seems little rationale

Figure 1. Technology adoption curve

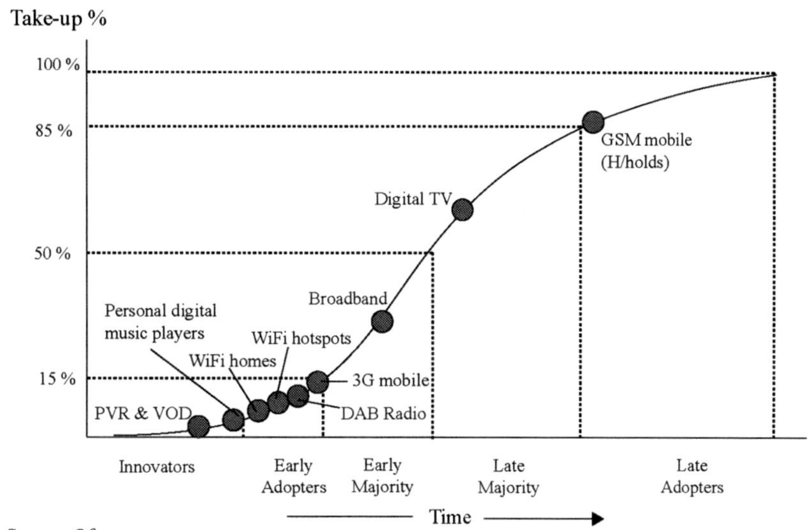

Source: Ofcom

for a completely new "4th generation" of cellular as 3G reaches the limits of what is possible in a radio channel. Fixed wireless access is unlikely to succeed – even with the advent of WiMax technology.

Despite the lack of promise of new technology, we can expect increased capacity and data rates almost entirely as a result of smaller cells. These might be a mix of cellular networks continually reducing cell sizes in urban areas and utilising femtocells in the home and the increasing deployment of W-LANs, in some cases providing coverage across entire cities.

However, different problems emerge. The key cost element for small cells, particularly those offering high data rates, soon becomes the "backhaul"- that is the connection of the cell into the core network. Other costs include site rental and power but for small cells in urban environments these costs are typically low. While backhaul of any required data rate can be provided through the deployment of appropriate copper or fibre cabling, this can be uneconomic for small cells serving only a few customers. Indeed, we can go as far as to say that since small cells are the key route to increased wireless data rates and capacity, and since backhaul is the key constraining factor in the deployment of small cells, that the biggest challenge facing wireless is fixed communications.

Economics and Social Issues

Even if the "perfect" service is introduced, the entire population does not rush out to buy the service immediately. Instead the service is first adopted by a particular type of individual – the early adopter. Depending on their reaction it may then become adopted more widely. This is shown graphically below, with some examples of various levels of device penetration in 2005.

The majority will typically only take up the service if the experiences of the early adopters are good. They will rely on word-of-mouth, reviews and widescale promotion to convince them that the technology is worthwhile and mature. Initially pricing can be high as manufacturers seek to recover development costs and profit from the willingness of early adopters to pay higher prices to be the first to have a new technology. Then price starts to fall as penetration increases. For technology products many now wait for falling prices as a signal to buy, having experienced situations in the past where they have bought products only to find the same or better products available for less shortly afterwards. Lack of an agreed standard can also inhibit uptake – for example many awaited a resolution of the Blu-Ray versus HD-DVD standards battle before buying a high definition DVD player since they did not want to purchase the "losing" standard.

If a new technology gets off to a bad start it may never recover. If the early adopters do not like the product, if the price is not set correctly, if the distribution chains do not champion it, etc, then the launch of the product will fail.

Even if it succeeds the process will take many years. Regardless of how wonderful the product is, the majority will not adopt it until they feel enough time has passed for the product to become proven, for the price to stabilise and for it to be an acceptable thing for them to own.

Structural Issues

The communications industry has developed predominantly in specialised "verticals". For each type of service, such as cellular, or fixed, there is one or more companies, who have built the network, run the network, deliver the customer service and billing, maintain a brand, etc. These companies might include:

- A fixed operator (typically the old "post and telecommunications" entity).
- One or more cable network operators.
- One or more terrestrial TV broadcasters.
- Typically a single satellite broadcaster.
- Multiple cellular network operators.
- Multiple W-LAN hotspot providers.

There have been some attempts to change this. In some countries consolidation has occurred with mobile operators also owning fixed or cable operators. Mobile virtual network operators (MVNOs) act as service providers on top of "network pipes". Some operators are outsourcing the build and maintenance of their networks so they can concentrate on service provision, taking the first steps towards disaggregation. Some cellular operators have also become hotspot providers, and third parties have emerged who build hotspot networks and then sell capacity to service providers such as cellular operators.

Nevertheless, the industry structure is still predominantly one of vertical silos with different communications technologies competing with each other at the edges. Within such a structure there is little incentive to provide the service that the consumer wants – one where their devices roam across multiple different networks, using whatever is most suited at the time. Instead operators try to keep consumers on their network as long as possible, even if it is not the most suitable network, since if they allow the consumer to roam to a different network they will not get any revenue from the call. This is the reason why there are relatively few packages available that provide cellular coverage outside and WiFi coverage in the home (using the fixed network as a link into the network). Instead, cellular operators prefer to deploy femtocells within the home, as discussed in more detail below.

This will eventually change as innovative companies offer consumers what they really want and cause the industry structure to finally change. However, it is likely to be a slow process - companies rarely choose to split themselves apart. This could be a much more significant impediment to the provision of consolidated services in the future than any technological issues or user concerns.

Regulatory Issues

Perhaps surprisingly when looking at the future of wireless, there is little need for regulatory intervention in the area of spectrum. This is partly because we predict that there is little need for additional networks or capacity and partly because many regulators have now moved towards a world of spectrum markets where it is possible to buy any spectrum needed from others and change its use as necessary. The only area where there might be a need for regulation is cognitive access, where new rules are required if it is to be allowed. But since we predict that cognitive access will not play a significant role in the future then whether this regulation is forthcoming is not of great importance.

However, there may need to be much more regulation in the areas related to market failure. We have already seen initial concerns around net neutrality and the ability to use any device or application on particular networks. As MVNOs and service providers seek to use the networks of others to provide service it seems likely that they will seek wholesale access to networks under terms that may need to be regulated. In some cases, regulators may even decide to split apart organisations or at least seek equivalence in the manner that an organisation provides services both to internal and external entities.

The Predictions

The table below provides a summary of my predictions for 2011, 2016, 2021 and 2026 across a number of specific areas. But first, let's consider how the overall user experience might change.

Between now and 2011 the key change for the end user will be the convergence between home and wide area phones. Users will gradually stop using their home phone, reserving it for those times that their cellphone batteries are low. This will lead to a gradual understanding of the problems of convergence, such as the difficultly in separating work and business calls, and the emergence of filtering solutions to deal with this. Wireless homes will offer a range of capabilities to control the home which users will increasingly come to rely on. This will be the period that many will make their first video call and will start to appreciate that there are times when video calling is advantageous. Users will appreciate higher data rates to the home and in the wide area but will not yet have any new applications that take advantages of these. They will also become accustomed to the mobile phone being a multi-purpose device, with camera functionality acceptable for most and music capabilities sufficient to displace iPods and similar devices for many.

Moving on to the period from 2011 to 2016 change will have become more of a continuum than a specific event. Building on the change of the previous five years, users will now become completely comfortable with convergence. The home phone will be relegated to a cupboard "just in case". Users will understand how to structure their communications in such an environment and when to use video calls. Users will perceive a significant revolution in broadcasting – no longer will many use TV guides. Users will also expect wireless control of all electrical devices in the home and office, including the ability to remotely interrogate home appliances from devices such as the cellphone.

By 2021 users will become increasingly familiar with their new converged world. They will perceive increased value in personalised services, allowing their service provider, via their phone, to make all sorts of decisions and provide suggestions for them. This will be the end of a journey – little more will change between 2021 and 2026.

Based on these ideas, it seems likely that the key technologies will be mesh, RFIDs, UWB, low cost sensors and enhanced software. However, equally developments will be held back by slow progress in backhaul and battery technology. Major growth areas will include ever-enhanced handsets, home wireless networks, intelligent software and service provisioning. Manufacturers will need to increasingly focus on handsets, operators will need to prepare for a move from vertical integration to horizontal, service providers need to build competence and credibility and academics needs to switch research from wireless technology to intelligent software.

For manufacturers, this forecast presents a picture of two halves. In the early years there will be considerable deployment of new technology, including 3G networks, mobile TV networks and core IP networks. Sales of handsets, W-LAN routers and BlueTooth devices will grow considerably. However, between 2011 and 2016 networks will become stable and little additional roll-out will occur. Of course, there will continue to be upgrades, replacement of faulty and obsolete equipment and filling in additional cell sites, but the days of complete network builds will be over.

Handset manufacturers are likely to see continued growth, with handsets developing ever increased capabilities and better human interfaces. There will need to be on-going research across a wide range of areas to improve capabilities such as displays, batteries and speech recognition.

For operators too, this is a future with two distinct periods. In the first decade or thereabouts, for most operators this will be a period of stability and profitability. No major change in plans will be needed from any operators. Perhaps operators might use this time to focus on reducing costs, improving networks and building portfolios of new services. Then, at some point, the industry structure will shift. Operators will undergo major and painful splits with a subsequent period of mergers and acquisitions. This will distract the operators for some time, but once over this period there will be a strong growth in service offerings, potentially requiring enhancements to core networks.

Concerning academics and the research community our predictions suggest that the key areas where advances will be needed are:

- Handset technology including batteries, displays, man-machine interface, processing power and storage.
- Software capable of providing contextual information and eventually leading to automated diary management and environment control.
- Systems that can handle the complexity of a large number and wide range of wireless devices, eg in the home, but present a simple plug-and-play interface to the user.
- Backhaul systems that will facilitate the rapid and inexpensive deployment of cells.

How the Future is Unfolding

These predictions were made two years ago. Many would argue that two years is a long time in wireless. What has happened during this time?

In overall terms, in line with our view that changes take a long time in the world of wireless, the answer is that not much has changed. 3G has become a little more established, WiFi hotspots are somewhat more widespread, but otherwise there is little change. For the most-part, people are using wireless in a near-identical manner to that in 2006, or indeed in 2000 for that matter.

As we alluded to at the start of this article, arguably, the biggest change is the iPhone. This was a phone so significant that the carriers were, for the first time, prepared to engage in revenue sharing with Apple. Data usage on the iPhone is around 50 times greater than average, although some care is needed since those who buy iPhones were probably already higher than average data users. The difference with the iPhone was the user interface – a better screen, a new way of interacting via touch and more intelligent software made the device much more usage. We predicted exactly this – that better phones rather than enhanced networks would open up a range of new services and applications for users.

Despite our predictions that another generation of wireless infrastructure was not needed, 4G appears to have substantial momentum behind it. Many carriers are now discussing plans for LTE deployments and vendors are promising equipment around 2009 / 2010. But, more in line with our predictions is the early analysis from some operators who suggest that LTE will not offer any significant enhancement in spectral efficiency over 3G systems with high speed packet enhancements. It will offer higher peak data rates but these can

Table 1.

Area	2011	2016	2021	2026
Fixed networks	IP cores, ADSL2+ in access network	Fibre deployment to the home underway around the world	Fibre deployment mostly complete in developed world	Little further change
Broadcasting	PVRs starting to build individual programmes, high definition used for films and sports events	Majority now use intelligent PVRs to assemble content and distribute to appropriate devices	Little true broadcasting, instead PVRs assemble content from a range of providers and channels	Little further change
Wide area wireless	Little change	Some mesh extension of coverage.	Cell sizes decrease in some areas.	Little further change
Short range wireless	Massive increase in indoor W-LAN deployments leading to a range of converged services	Indoor wireless coverage ubiquitous, outdoor widely available (but not covering entire cities), many appliances now wireless enabled	Little change in technology, but large growth in home control services	Little further change
Handsets	Storage capacity grows, speech recognition improves	Better displays, increased functions integrating other personal devices	Incremental improvement in most areas	Further incremental improvement
Services	Mobile TV emerges	Broadcasting becomes a personal service integrated into wireless devices, contextually aware and personalised services widely used	Growth in personalised services providing "remote control on life"	Ever wider range of personalised and general services built on top of stable wireless platform
Convergence	Huge growth in single phone for home and wide area, and increasingly office.	Broadcasting well on the way to converging with telecoms, telecoms completely converged	Complete – the time when different devices were used for different services will seem old-fashioned	Now fully converged
Industry structure	Little changed	Change from vertical to horizontal integration	Stable in the new horizontally integrated form	Pipe providers remain stable but change in service providers.

only be supported in broad bandwidths which are only available at high frequencies, making LTE more of a competitor to WiFi than 3G. Perhaps more appropriate is the view of operators using systems other than 3G who see that migrating to LTE will provide them with a way out of their existing systems as they become less widely supported.

Also, despite our predictions that it would be of limited value, it appears that MIMO will be widely offered. It is already available on some WiFi routers and is promised across much of the WiMax range of equipment. But here again, all is not quite as it seems. Operators and others accept that MIMO will only likely be deployed in a few specific situations where cell splitting is not possible or greater range is required. Manufacturers appear to be offering it more as a marketing tool to boost their maximum data rate than because of a strong need to deploy it. And simple diversity schemes which have been deployed for many years are sometimes labelled "MIMO" in an effort to market equipment.

Another change from our predictions is the apparent popularity of femtocells. Although few have been deployed yet, many operators are trialling them and many manufacturers are offering them. We predicted that WiFi, rather than femtocells, would be the preferred means of providing wireless coverage within the home. However, again, all is not what it might seem. Operators appear to be considering femtocells as a mechanism to compete with the wireline providers in the home. We argued that a better approach would be for a service provider to offer a package containing both cellular and home wireless (likely using WiFi) but that this would not occur until operators became less vertically aggregated. There are some early signs that operators are becoming less interested in owning their own network, with a growing use of outsourcing arrangements to manufacturers like Ericsson. We predict that these various forces will continue to play out over the coming years leaving it far from certain that femtocells will be successful, at least in the longer term.

A prediction that does appear to have been correct is our view that metropolitan WiFi deployments would not succeed. Almost all of these now appear to have been discontinued and no new schemes have been proposed for some time.

Finally, we suggested that the future would be brighter for handset vendors than infrastructure vendors. Perhaps it is too early to really understand whether this is happening, but the infrastructure vendors have generally performed poorly overly the last two years, with falling share prices and redundancies. Conversely, most of the handset manufacturers, with the exception of Motorola, have generally performed well.

So, two years on from our prediction, much of what we suggested still appears to be correct. Perhaps the biggest change from our predictions is that 4G does appear to be happening, but more for reasons of migration towards a single, widely-supported air interface across the world than because it offers higher data rates, or indeed because these higher data rates are actually needed. Overall, we are still of the view, as evidenced by the iPhone, that the networks are broadly sufficient for most applications that can be envisaged, it is the user interface to the phone where the most important gains can be achieved.

REFERENCES

Cherry, S. (2004, July). Edholm's law of bandwidth. *IEEE Spectrum*,58–60. doi:10.1109/MSPEC.2004.1309810

Fette, B. (2006). *Cognitive radio technology (communications engineering)*. Newnes.

Gross, F. (2005). *Smart antennas for wireless communications*. McGraw Hill.

Held, G. (2005). *Wireless mesh networking*. Taylor & Francis. doi:10.1201/9781420031263

http://www.arraycomm.com/serve.php?page=Cooper

http://www.intel.com/technology/mooreslaw/index.htm

Shannon, C. E. (1948, October). A mathematic theory of communications. *Bell Systems Technical Journal, 27*, 379-423 ; *27*, 623 – 656.

Shannon, C. E. (1948, January). Communications in the presence of noise. *Proc IRE, 37*, 10-21.

Tuttlebee, W. (2002). *Software defined radio – enabling technologies*. Wiley.

Webb, W. (2007). *Wireless Communications: The Future*. Wiley. doi:10.1002/9780470065129

This work was previously published in International Journal of Interdisciplinary Telecommunications and Networking, Volume 1, Issue 1,edited by Michael R. Bartolacci, and Steven R. Powell, pp. 9-18, copyright 2009 by IGI Publishing (an imprint of IGI Global).

Chapter 3
Analysis of the Forces Reshaping the Mobile Internet Business

Hannu Verkasalo
Helsinki University of Technology, Finland

ABSTRACT

This article studies the emergence of the mobile Internet business from six strategic perspectives, highlighting the ongoing transformations of the mobile industry. The article suggests that companies of the mobile industry can reconstruct their business models with six different kinds of choices: 1. Positioning differently in the value network, involving both vertical and horizontal movements. 2. Acquiring and developing critical assets and competencies to build sustaining and dynamic competitive strategies. 3. Inventing new mobile business models by utilizing the lessons from the Internet. 4. Integrating end-users in making and redefining services, optimizing pricing logic, introducing new business models with end-user originated content creation in mind, and making services easy to adopt by end-users. 5. Leveraging and focusing on disruptive, winning technologies that are changing the rules of the business. 6. Adapting to relevant policy schemes, and proactively planning and reconsidering business decisions under a dynamic regulatory environment.

INTRODUCTION

Mobile services constitute a growing market because of two reasons. First, mobile devices in their various forms are growing in number rapidly. In particular, advanced devices (including e.g. "smartphones", "converged devices", "internet tablets" and "multimedia computers") that have built-in multimedia functionalities, Internet connectivity and modular software design (enabling user-installed applications) are increasing in penetration in developed markets. Second, along

DOI: 10.4018/978-1-60960-505-6.ch003

with the diffusion of more capable mobile devices also the variety of mobile services is increasing, ranging today from mobile Internet browsing to music streaming, and from mobile gaming to various location-based services.

Technologies are available to capitalize on the mobile extensions of Internet services. These enabler technologies include e.g. wireless broadband connectivity, advanced operating systems of mobile devices, and architectures such as IMS (IP Multimedia Subsystem) and mobile web servers (Nokia S60 Apache). Commercial challenges of WAP (wireless application protocol) and MMS (multimedia messaging service), however, suggest that the technology push model is not enough, and it is increasingly important to understand how users perceive the new services introduced. The openness provided by the mobile Internet and advanced devices is likely to shift the mobile services industry towards the pull model, in which people choose and customize themselves whatever services they prefer.

Traditionally mobile services have been provided through the "silo" model, meaning that typically one company (or a closely networked "group" of companies) provides the different elements of mobile service delivery. This delivery typically encompasses a network access solution, core network operations, service provisioning and maintenance, content creation, both service and network management functions and customer accounts (including marketing, charging and billing). The silo model has emphasized the vertical integration of the value chain (Vesa 2005; Verkasalo 2007a). Mobile operators as we know them today, including the likes of Vodafone, TeliaSonera and Sprint, have typically represented as focal actors in the delivery of mobile services. Mobile operators have managed this "silo delivery", controlling the different layers on the supply side of the mobile services industry.

The Internet world is different in that it has little vertical integration. In rare occasions "silos" have emerged. Broadband operators have focused on their bit-pipe strategies and provisioned the network access, while the services itself (e.g. Google, Skype, Ebay) are provided by other companies on another conceptual layer, on the "edge of the Internet", meaning the application layer of the OSI model. The Internet model emphasizes the customer pull model. For example, people visit the Internet sites they want, subscribe to blogs and social community sites they prefer the most, and choose their online stores based on price comparison engines. Technologies exist in the back-end, but end-users care only about services and their added value. The Internet makes it easy to choose among a high number of alternative services that are independent of e.g. local broadband operators.

Now the worlds of Internet and mobile services are clashing, as wireless broadband connectivity and modular mobile platforms (the computer like evolution) are spreading rapidly and providing the push for the "horisontalization" of the mobile services business (Verkasalo 2007a). In addition, the rapid emergence of services around e.g. entertainment (digital music, blogging), productivity/office (email services, calendar) and location (Google Maps, Nokia Maps) are touching both the Internet and mobile worlds. Therefore, rather than being just an extension of the traditional "wired Internet", the mobile Internet provides more with truly ubiquitous, location-based, personal and context-sensitive nature of mobile devices. All this spawns major transformations in the conduct of mobile business and the roles of different companies. The transformations of the mobile business have high impact, as some estimates claim that the value of the general mobile communications market will approach $200 billion in 2009 (iSuppli 2005).

This article discusses the transformations taking place in the mobile industry. The presented approach provides an "external angle" at industry dynamics, this meaning that instead of providing tools to understand a given company's internal operations, the article helps in understanding the dynamics in the competitive environment of the transforming mobile services industry. The main research question of the article is:

How to model the transformations taking place in the competitive environment of the mobile services business, and to guide theoretical and empirical studies of the future with a fact-based evaluation of the current market situation?

The article first provides background on modeling competitive business environments and business models. In addition, the specific change drivers of the mobile services industry are identified. The research article then presents the six strategic perspectives of the mobile services industry, providing examples in each of the perspectives. The study claims that all of the defined perspectives are relevant in modeling the ongoing transformations of the industry. The article presents findings from the Finnish smartphone studies (see e.g. Verkasalo 2008b), providing empirical observations on the emergence of the mobile Internet. The smartphone studies include both survey and usage-based data from the market (see Verkasalo & Hämmäinen 2007). In conjunction with the discussion on the strategic perspectives, the article suggests areas for future research.

BACKGROUND

This section provides background on two separate fields, both of which are important from the perspective of this article. First, earlier academic literature on the dynamics of competitive business environments and industry transformations is introduced. Second, the ongoing changes in the mobile Internet business are identified and relevant concepts provided to the reader. The mobile Internet is defined in this article as the provisioning of services to mobile devices via internet technologies (e.g. TCP/IP, WWW). The mobile device is considered as a handheld, pocket-size device that includes at least the functionalities of a phone. Primarily the assumption, however, is that in the future these devices run operating systems that support Internet-based services with either dedicated clients (e.g. mobile Google Maps client) or via standardized Internet interfaces (WWW), like Google's standard mobile search engine[1].

Modeling of Competitive Business Environments

When studying competition between companies, a *business model* is one of the key concepts. Business models have been extensively studied in the past (for a good review, see e.g. Pateli and Giaglis 2004). There are many definitions for a business model. Some of the popular ones include the consideration of a business model as a description of components defining a business idea consisting of products, services, actors, roles, information, revenues and benefits (Timmers 1998 and Weill & Vitale 2001); Linder and Cantrell's (2000) approach of defining a business model as the company's core logic of creating value; Magretta's (2002) description of a business model being a story that explains how a company works; and Osterwalder's and Pigneur's (2002) perception of a business model as the link between practical business processes and a company's strategy plan. The research on business models faces a new fresh wave of articles highlighting the complexity and uniqueness of business models in electronic and Internet based services (see e.g. Afuah & Tucci 2001; Rappa 2003; Petrovic et al. 2001; Timmers 1998; Papakiriakopoulos et al. 2001 and Weill & Vitale 2001). Zero marginal costs of production (Shapiro & Varian 1998), complex relationships between companies (Verkasalo 2007), and constant technical innovation around electronic services (Christensen 1997), among others, raise the concern for new approaches in understanding the business models behind new mobile Internet services.

A business model is the key object of study in this research article. In this article, a business model is defined to mean everything that specifies how companies are to make profits in the industry. In this article the business model is considered as both a strategic and operative issue. In addition to defining the organization of company operations to produce the products, this article sees business models as ways of positioning value offerings in relation to competitors, suppliers and customers of the industry. Because of this definition, two important strategic questions arise. First of all, how to model a company's relationships with external companies in competitive environments? Secondly, how to model the industry evolution, and to internalize this to the analysis of business models?

The best known work on the modeling of competitive business environments is Michael Porter's work on the so called *five forces analysis* (Porter 1979 and 1980). Porter identifies five distinct dimensions in which companies should consider their position in the market: threat of substitute products, threat of the entry of new competitors, intensity of rivalry in the industry, bargaining power of customers, and bargaining power of suppliers. According to Porter, a company's strategic positioning in these dimensions could establish a *competitive advantage* inside the industry. In line with his view on competitive environments, Porter has also highlighted the role of a value chain in explaining how products and services accumulate value-added and how different companies participate in the linear processing of products from raw materials to end-users (the *value chain*). (Porter 1980)

This classical view on the dynamics of value creation has to be reconsidered in the modeling of today's mobile services (or any other electronic services). Mobile services should be understood through *value networks* (see e.g. Feng & Whalley 2002; Allee 2000; Christensen & Rosenbloom 1995 and Stabell et al. 1998) or *value webs* (Tapscott et al. 2000). Instead of a linear view of value-creation, value network concepts typically assume that the value is created in complex networks of companies collaborating with each other. Feng and Whalley (2002) argue, however, that it is difficult to construct a holistic value network of the whole industry, but instead each company of the industry constructs a unique value network around itself.

A transaction cost theory (Williamson, 1975, 1985, 1989) explains the movement from the value chain perspective towards value networks. The transaction cost theory suggests that companies bear costs in organizing their internal operations (transactions inside a company) and external relationships (e.g. relationships with suppliers). Internet technologies force transaction costs to zero, provided that a company utilizes the Internet as a medium of organizing its activities and providing products or services (Butler et al., 1997). In essence, the Internet has contributed to the birth of more complex value networks, in which companies increasingly specialize on something, and integration of separate activities inside one organization becomes economically suboptimal. In the era of the expanding electronic services industry, and particularly in the Internet business, the movement towards value network structures is evident. This is what is happening in the mobile Internet business (see e.g. Müller-Veerse et al. 2001 for mobile data service value webs) currently, too.

The evolution of the mobile industry is the key object of this study, assuming that transformations are taking place in the industry. McGahan (2004) argues that there are four kinds of industry evolution: a progressive change, radical change, intermediating change and creative change. In the *progressive change* the key assets of the industry are easy to understand and they do not face threats. In this kind of industry change companies should set out to only organize their operations optimally. In Christensen's *The Innovator's Dilemma* (1997) the concept of "sustaining technological change" is all about surviving through a progressive, stable, change of the industry. The *radical industry change* presents another end of the spectrum. All the relevant assets and business processes of the past become rapidly worthless during the radical industry change. The mobile operator business is facing this kind of evolution when looking from the correct angle. For example, the dominant way of providing mobile messaging services to subscribers through SIM cards (SMS) is facing a threat of completely Internet-based messaging services (e.g. instant messaging and email). Particularly assets related to services that typically involve older technologies now face a relatively strong threat from a number of Internet services that are easy to "port" into the mobile domain. The so called *intermediate change* occurs when companies of the industry face serious threats on the way they do business – but many of the underlying assets are still relevant. If the mobile business is investigated from another angle, the industry evolution can be considered as intermediate. For example, mobile operators having a wide user base for voice services (having thereby also a high number of phone numbers) face a threat from Skype kind of VoIP services that can be free-of-charge to an end-user. However, the voice service infrastructure (networks and centralized management) and phone number domains are important assets, and regulation of voice services in Finland does not make it possible for Internet voice services to use a similar number space than cellular voice services, neither does the gateway regulation regarding interconnectivity from Internet to PSTN voice services provide any advantage to Internet-based voice. No doubt, however, there is a serious threat from mobile Internet voice to cellular voice – through intermediate change. The last kind of industry evolution – *creative change* – involves relatively mature industries that do not pose the characteristics of intermediate or radical change, but are nevertheless not dying. In other words, in creative change you invent new ways of making money or change your way of doing business, but your relationship with customers stays the same. The mobile industry can be seen to experience creative, intermediate or radical change – depending on the standpoint.

Hardagon (2003) claims that revolutionary breakthroughs in technology and business take place when different paradigms are combined together. When applied to the evolution of the mobile industry, the open Internet innovation is now clashing with the "silo" mode of the mobile business, raising thus a threat of radical change in the ways of doing business (Verkasalo 2007b). However, a static look at industry dynamics is not the only thing that matters. In today's academic research the so called dynamic strategy view is pressed (Rumelt 1991). This strategy discipline claims that a static analysis of the industry does not help. Instead, companies should develop a dynamic and interactive process of strategy making, in which they evolve together with the industry and create business opportunities by themselves. This contrasts with the approach of considering the competitive environment as static. Companies can do well in unattractive industries – and equally well it is possible to do poorly in attractive industries. Therefore in the long-term the company should look for company-specific superior sources of long-term success (i.e. resource-based view, see e.g. Wernerfelt 1984 and Grant 1991). Also Mäkinen and Seppänen (2005) conclude that future business model conceptualizations should better provide answers regarding how to utilize unique resources and competencies as the foundation of a successful business model. The new rules of building sustaining advantages are relevant in emerging business models of the mobile industry that strongly challenge the dominant business models of the past. Not only have the conditions of the industry changed, but also the conditions have expanded for individual companies to evolve interactively, possibly through trial-and-error logic, but anyways with unique, distinct business models.

Figure 1. Correlation of service usage and ABPU (Finnish smartphone study 2007)

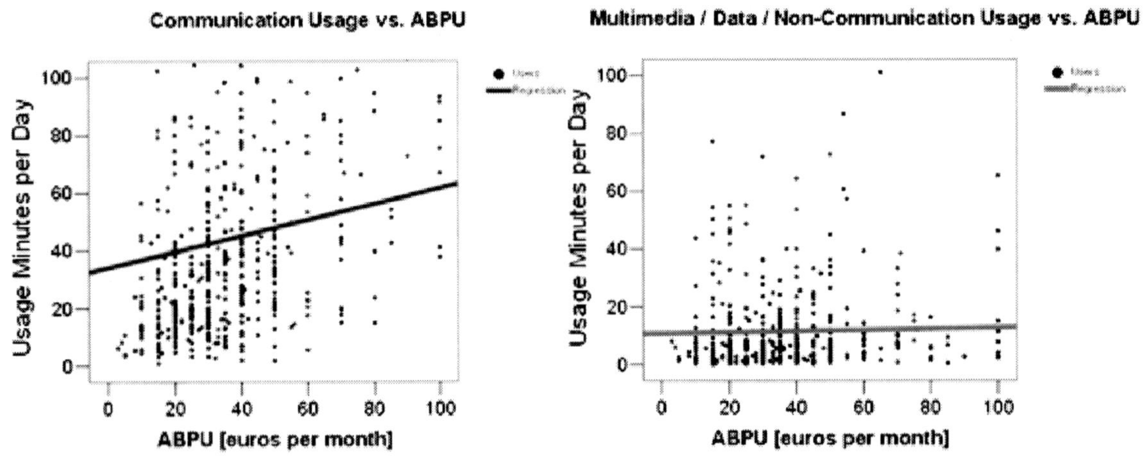

CLASHING MOBILE AND INTERNET BUSINESSES

The mobile industry has been dominated by large operators (e.g. Vodafone, TeliaSonera, Telefonica) earlier. They have typically provided both the network access and services. This has been possible because of two things. First, it has been expensive to establish a wireless access network from scratch. Second, once an access network has been in place, it has been a proprietary one, in most cases a local entity (e.g. national). Typically operators have restricted the utilization of the network without a permission (consider e.g. the SIM card technology of 2G/3G networks). In other words, many fundamental factors have driven the mobile world towards operator dominance, and economies of scale. (Verkasalo 2007a; Vesa 2005; Kiiski 2007)

Even before the era of mobile communications, internet technologies have pushed the evolution of the wired communications industry towards a new direction. Due to the rapid innovation at the edge of the Internet, the providers of wired broadband connectivity (typically legacy operators) saw that it is not viable to keep on doing business on the application/service layer of the industry. Instead, broadband operators pushed for bit-pipe strategies, and quickly deployed flat-rate charging schemes and optimized their wired broadband networks for a variety of Internet traffic. Internet service businesses - the likes of Google, Yahoo, Amazon and Skype – then developed unique business models that relied on advertising revenue instead of consumer charging, for example. This business model utilizes the ideology of two-sided markets (see e.g. Rochet and Tirole 2003; Eisenmann et al. 2006), in which the services can be provided in a subsidized form to consumers, provided that another party (e.g. advertiser) is more willing to pay for the service given that they receive benefits in another form (e.g. the attention of a consumer) from the value network.

Fundamentally the wired Internet is based on a horizontally layered mode of operation, whereas the classical mobile market is more vertically oriented. Jarkko Vesa (2005) discusses the silo model of mobile service delivery and its applications in three separate markets. Some studies exist, which study the ongoing transition of the mobile industry towards more horizontally oriented ecosystems (see e.g. Verkasalo 2007b; Saarikoski 2006). Verkasalo (2007b) studies the topic from both theoretical and empirical standpoints, suggesting that some signs of the mobile Internet diffusion can be seen among early-

Figure 2. Ranking of mobile services (Finnish smartphone study 2007)

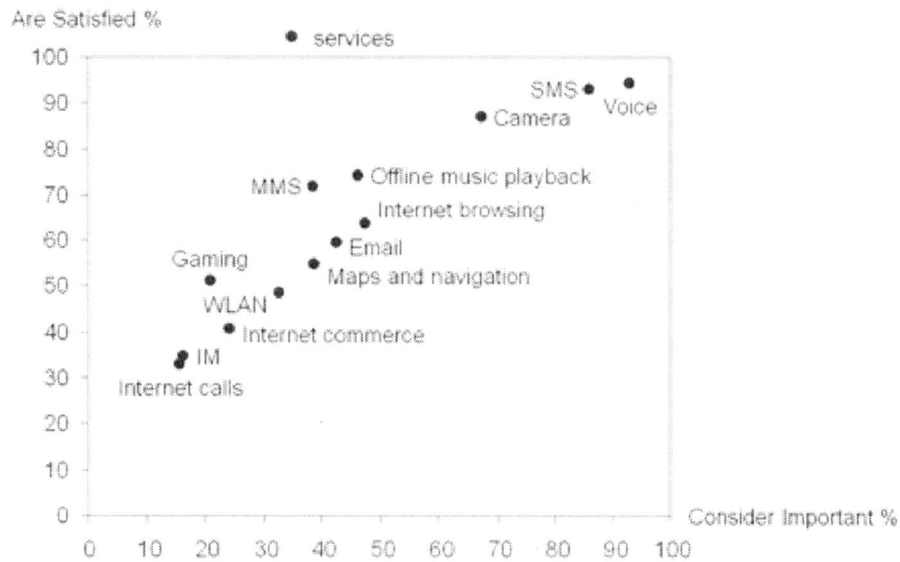

adopter mobile subscribers. Saarikoski (2006) favored the Japanese way of rapidly pushing content and Internet like services to mobile networks, criticizing the western operators' strategy of pushing "non-scale-free" services such as SMS and sticking to usage-based charging in mobile data services.

Essentially the emerging trends in conducting mobile Internet business (e.g. free or flat-rate connectivity charges, advertising-based business logic, end-user driven open innovation etc.) are being pushed by several forces. However, no correlation can yet be found between average revenue per user and data/multimedia usage (see the figure above), and instead most of the money is still coming, even among smartphone users, from voice and messaging. According to Maitland et al. (2002) the evolution of the mobile data industry is shaped towards the Internet world because of economic, regulatory and technical factors. In addition, Maitland et al. suggest that the structural changes of the mobile data market rely on value network kind of inter-organizational structures. Therefore a holistic view is needed in modeling the business dynamics of the mobile market evolution.

From a technical perspective, the increasing complexity of the mobile services market pushes the industry towards open innovation and mobile Internet. Advanced devices that support add-on applications play an important role here, as they force the whole industry towards a computer-like industry evolution. The mobile industry is currently facing disintegration of vertical value structures and a modular design of technologies is emerging. Advanced mobile devices support add-on services and applications that do not rely on any specific kinds of networks or proprietary interfaces.

The evolution of the mobile communications industry can be perceived from Maslow's hierarchy of needs point of view (Maslow 1943). The first mobile services were to fulfill the basic needs of communication. The ongoing merging of industries moves mobile services towards the next steps, in which people fulfill also their needs for social belongingness (e.g. mobile Web 2.0 services) and entertainment content consumption (purely hedonic services, e.g. multimedia streaming or mobile games). The traditional voice and messaging services, however, will remain as necessity services still in the future (see the figure above). Along with increasing complexity of mobile services, however, the strategic positioning of services becomes easier. For example, mobile multimedia portals providing social networking functionalities, and secured email connectivity solutions, can be positioned as premium services targeted at specific niche segments of the market, in contrast to simple text messaging services that appeal to all subscribers.

Accenture divides the mobile handset industry evolution into two phases (Accenture 2006). In the first phase the overall diffusion of mobile handsets will proceed, and also advanced handset capabilities will be pushed to the market gradually. In the second phase mobile phones will emerge as personal computing devices, leveraging new kinds of services. The figure above suggests the first phase has already taken place and the mobile subscriber penetration has for long grown faster than the global Internet penetration. Accenture's view corresponds well with the overall perspective of this article, foreseeing the movement to a computer-like evolution path in which both advanced features and overall modularization of technologies will evolve further. This technical point of view should be understood along with the ongoing merging of industries in the mobile domain. The traditional communications industry is merging with entertainment and media industries. The introductions of e.g. TeliaSonera Surfport, Google Android, Nokia Ovi and Operator Three's X-Series suggest that the key players are looking for new ways of doing business in the mobile domain. Depending on the role of a given company in the value network, different strategies should be applied (Camponovo & Pigneur 2003).

MODELING THE TRANSFORMATIONS OF THE MOBILE INDUSTRY

This section discusses the dynamics of the mobile Internet business. The discussion is structured around an analysis of the opportunities and challenges of the mobile Internet. The analysis sheds light on industry transformations and helps in assessing the key drivers having an impact on the competitive environment. From each perspective the key directions of future research are identified, along with a set of key implications for the business world.

Structure of the Discussion

The discussion is structured on six different themes. These themes reflect the different perspectives at the transformations of the mobile industry. The *ownership of critical assets and competencies* is a critical point of view with regards to understanding how to build competitive advantages. It equally concerns all actors of the value network. *Changes in value network structure and power positions* provide room to either extend or focus business – at the same time this introduces new competitors to incumbent players. A more technical perspective – the *emergence of supporting infrastructure* – deals with new technical innovations emerging in the mobile domain, that possibly have an impact on business dynamics. *Regulation and policy on mobile Internet* is a concept covering both supply and demand sides of the market. The role of policy making, legal systems and regulation is non-trivial in the evolution of the industry. *Business model transformations* and *service demand modeling* are concepts relevant when studying how to capitalize on end-users and the demand side of the market in general.

The figure above illustrates the structure of the discussion in this article. The strategic perspectives are illustrated on top of a generic value network of the mobile industry. The figure illustrates that some of the perspectives are closely tied to a certain interrelationship (e.g. business model transformations are most interesting between service providers and customers), whereas some perspectives should be holistically analyzed from

Figure 3. Global growth of mobile phone and Internet users (adapted from ITU 2006)

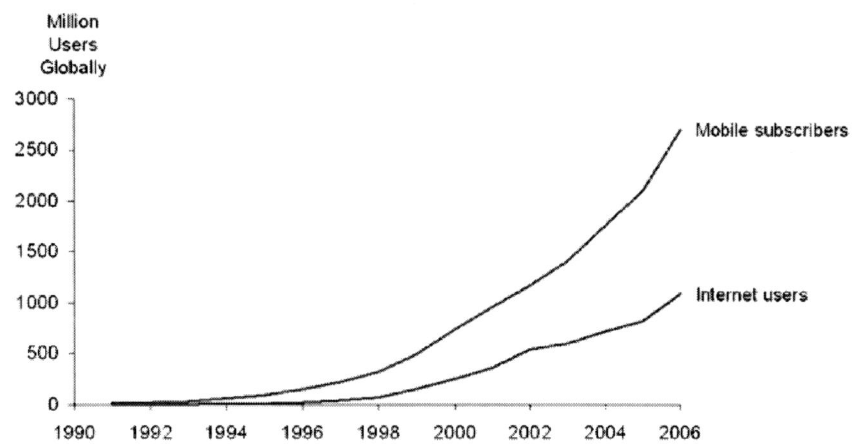

the viewpoint of the whole value network (e.g. the value network structure and positions).

The different perspectives provide a breakdown of the larger strategic shift that is taking place around the mobile Internet business. The perspectives also help in the assessment of challenges arising from emerging business opportunities.

Value Network Structure and Power Positions

The first perspective is ***value network structure and power positions***. The number of players in the mobile industry is increasing. Handset vendors are exploring the possibilities of vertically extending to mobile services and content. Apple is bringing turbulence into handset business with its new IPhone and associated business models with mobile operators. In addition, Apple participates in the platform game by releasing its Apple App Store[2], designed for the creation of a 3rd party application market for IPhone. Nokia, on the other hand, has launched several new mobile Internet services recently (Nokia Music Store, Share on Ovi) and acquired companies such as Navteq and Plazes. Internet players are speculating on the possibility to participate in the mobile business by capitalizing on their services that already exist in the Internet. For example, Google has contributed to the development of more open mobile phone

Figure 4. Analysis of the dynamics arising in the mobile Internet business

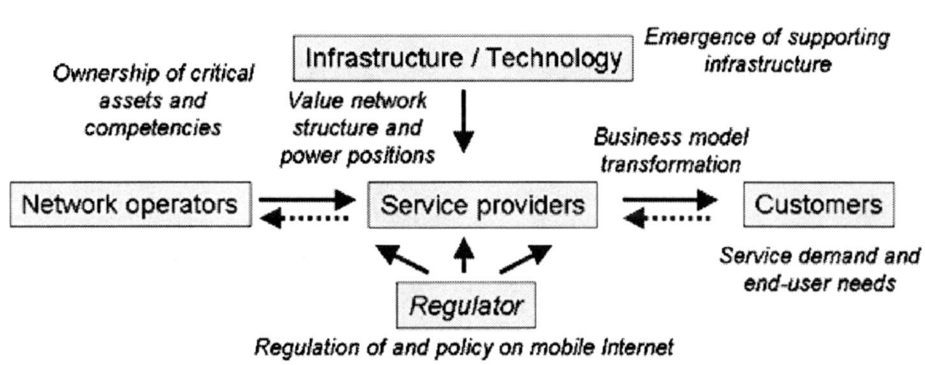

software platforms and publicly supported open wireless spectrum. Google wants mobile devices to openly support Internet-based services. Nokia complies with this public commitment to push mobile Internet services, though the two companies have quite different backgrounds.

Incumbent operators defend from "truly open mobile Internet platforms" by using mobile Internet platforms such as IMS or technologies such as UMA/GAN, and forcing their own interfaces to handset browsers (e.g. TeliaSonera's new mobile SurfPort implementation). Operators leverage their existing "ownership" of the end-user domain and orchestrate service provisioning networks, like NTT DoCoMo in Japan. Orchestrators are players that in a way or another orchestrate companies around themselves in the value network (see e.g. Van Liere et al. 2004). On the other hand, operators also have the possibility of focusing themselves more by implementing for example bit-pipe strategies in which they only sell connectivity and forget the service side.

Strategies of the major players in the mobile industry and consequent future form of the mobile ecosystem determine who is going to own the customers and who is best able to capitalize on new services. Regarding future research, for example the traditional studies on handset and subscription bundling should be extended to handset and service bundling, as some dominant players such as Apple and Nokia can nowadays bundle services and devices together, thus strengthening their own brands and services in the mobile market.

The position of operators in the past can be considered strong (Vesa 2005). However, now the industry trend is to move towards more open environments, in which the power of operators is relatively lower. Figure 5 above illustrates, that in 2007 the use of operator-specific mobile Internet domains experienced a significant fall in comparison to the similar studies of 2005 or 2006. The table above indicates that according to a survey study on 579 smartphone users in 2007, handset vendors and operators are both considered as important actors in the market, as perceived by end-users. These datapoints together suggest that operators are not the only dominating force of the industry any more, and instead also other players, particularly handset vendors, should be seriously taken into account when addressing the future of the industry. The threat to the existing operator-controlled mobile business can be illustrated with three examples.

Figure 5. Analysis of mobile Internet usage (Finnish smartphone studies 2005-2007)

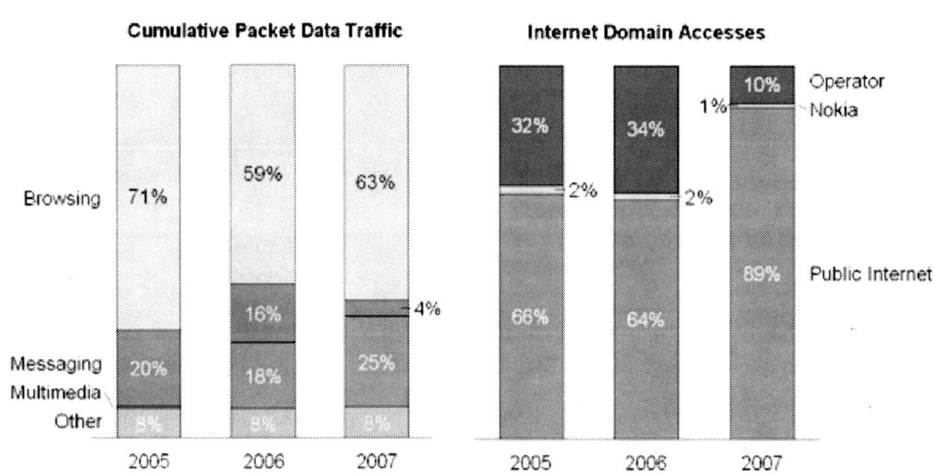

First, Apple has proven that also vendors can push their brands in the eyes of the customers. Apple established a new business with its IPhone, arranging for a significant share of partnering mobile operators' service revenues to themselves. Apple optimally levered its position as an innovative company, pushing cool, new and refreshing devices to the market. Apple leveraged its brand that is has established in other industries. Apple has maneuvered its own devices to the portfolios of numerous operators by 2008. Operators agree to share revenue with Apple based on the assumption of selling more subscriptions and perhaps by boosting ARPU through increasing use of add-on features such as mobile Internet browsing. Nokia, though making significant profits by simply selling devices, has rarely been able to steal a share of direct operator revenues. In general, handset vendors and mobile operators are re-evaluating their relationships today, thanks to Apple.

Second, vendors of technologies have been pushing strongly to the services market. Apple is a serious software company in addition to its role as a mere IPhone manufacturer. Perhaps the best example, however, is Nokia, together with its highly promoted mobile Internet services strategy. Nokia has been aggressively promoting its Ovi Internet service and Music Store, N-gage, Nokia Maps and Gizmo VoIP solutions. With these moves it has positioned itself as a direct competitor of operators and many service companies. Naturally Nokia's communications strategy has stressed close collaboration with operators – though it is clear that not all operators have appraised Nokia's ideas. All in all, technology providers are extending vertically in the value network.

Third, the purely Internet-based companies like Google and MSN (part of Microsoft) have been waiting for the emergence of a truly open mobile Internet. As the penetration of the mobile Internet connectivity has not yet reached the mass market, these companies have taken some moves to help the market. For example, Google has introduced its vast alliance of companies supporting the upcoming Google Android mobile software platform, based on which they aim to sell Internet-based mobile services. Google attempts to break the still existing binding to the past of the mobile business, and push instead the movement towards horizontal ecosystem structures. In general, new technologies, business strategies and opportunities in the mobile market have driven a path towards more horizontally layered ecosystems. In this evolution, also network operators are forced to consider whether they should only focus on vertically narrower bit-pipe strategies, like most broadband operators of the Internet today.

The examples above suggest that there are two main kinds of strategic possibilities to change value network (and power) positions. First, in the vertical dimension companies can either extend (Nokia is moving to services, operators are selling handsets manufactured by subcontractors, Google considered the possibility to participate in radio bandwidth auctions) or narrow down (some mobile operators narrowing down to bit-pipe strategies, handset vendors withdrawing from network hardware manufacturing, Google not extending to devices but narrowing their strategy down to fostering the success of Google Android software platform) vertical value network positions. The other possibility is to consider horizontal movements. In a way the mobile domain can be understood as just a horizontal extension of providing instant messaging services in the wired Internet, for example. The so-called spill-over strategies of existing Internet services belong to this category, as mobile devices are just seen as one new terminal category to which to port services. On the other hand, another kind of example of horizontal extensions comes from the operator business, as operators have for long considered the movements to other than mere voice and messaging services, good examples being content and multimedia services.

The figure above illustrates the possibilities in re-evaluating positions in the value network for two examples (exemplary transformations of a handset vendor and operator). The vertical height of bubbles reflects the extent of vertical reach across the market environment. For example, companies providing both hardware and services are more vertically extended than mere hardware manufacturers. On the other hand, the horizontal width of bubbles reflects the variety of horizontal areas of business. For example operators who start deploying wired broadband in addition to wireless broadband, have higher horizontal reach than mere wireless operators.

All in all, along with the reconsideration of business models in complex value networks, the turbulence in the market raises the question of dominant power positions and changing forms of value network positions. Indeed the optimal positioning in the value network and strategic consideration of partner companies is relevant (Gulati et al. 2000). In Netsize's most recent report (Netsize 2008) almost all of the key mobile market events of 2007 involved some sort of value network changes for the companies in question (e.g. acquisition of companies, moving to new areas of business etc.).

This chapter indicates that technology vendors and Internet companies are causing turmoil in the market, challenging operators. This can be true to some extent. However, also counteracting forces take place. For example, operators have incentives to acquire even more power by migrating to the wired broadband market, and vice versa. Therefore the game for power positions, and competition inside the value network, are far from converging to any particular direction.

Challenges for companies: How to build value-adding relationships with other companies through reconsideration of value network positions? How to best prevent other companies increasing their negotiation power and to best increase the company's own negotiation power by e.g. partnering with orchestrator companies? How to optimally either focus or extend business strategies in both vertical and horizontal dimensions?

Table 1. Importance of the industy actors (Finnish smartphone user survey 2007)

I consider the following actors important in the delivery of mobile services...		
	Consider very important	Consider important
Mobile phone vendors (e.g. Nokia)	31%	79%
Operators	26%	75%
Internet companies (e.g. Google)	10%	45%
Media companies (e.g. MTV3)	6%	33%

This perspective suggests that future research should more deeply study value networks specific to the mobile industry and their consequent success factors (see e.g. Hanbo 2008). In addition, the frameworks introduced in the management research earlier (e.g. Porter's static tools) should be tuned to better support the analysis of the mobile industry.

Ownership of Critical Assets and Competencies

The second perspective is ***ownership of critical assets and competencies***. The fact who controls critical assets and who can best commercialize new services remains ever more relevant. Critical assets of the future mobile industry include e.g. ownership/rights of content (digital maps, music, videos, movies), contracts of key distribution channels (e.g. handset distribution, ownership of retail stores) and brand value/extent (value already embedded in the brand). The emergence of the mobile content business raises the challenge of being able to control the new channels from content providers to end-users. This, among others, is a new challenge to the players of the mobile market.

On the other hand, one should not overlook the skills of bringing new services quickly to the

Figure 6. Examples of strategic moves in horizontal/vertical value network dimensions

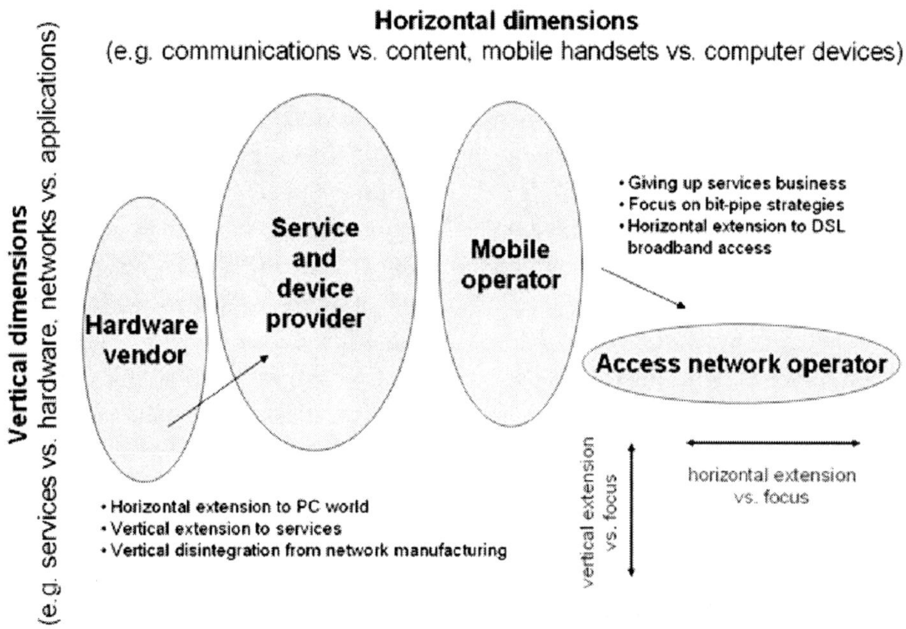

market, insights in doing the right business moves with correct timing, management competencies is acquisition/outsourcing decisions etc. Network edge based innovation has prospered in the Internet due to open interfaces and platforms. As the mobile industry is likely to move away from walled garden kind of business models, the choice of an optimal strategy with regards to service and content deployment is important. The mobile Web 2.0 provides a new angle to look at mobile services and content business. Companies who already understand this new business (like Google and Facebook) have important intellectual, technical and business assets already in place.

New skills are also required among older players such as operators to understand in which assets to invest and what competencies to acquire, because organic growth in most cases is not enough. Handset vendors are pushing for the Internet strategy. For example Nokia, despite having high profits from its device business, faces new challenges. Only the future will show whether they will be able to make profits with Internet services. Needless to say, services and software are totally different from consumer electronics. In the optimal situation handset vendors such as Apple and Nokia could lever efficiently their valuable brands and power positions in bundling services and applications with devices. In addition, for example Nokia's high market share in devices makes it possible for them to establish critical assets, such as the newly introduced global mobile advertising network.

Although many new phenomena (mobile Internet services, mobile content) have been mentioned above, one should not forget the critical question of connectivity. Still today the operators are the ones who own nation-wide wireless broadband networks (e.g. 3G). This is one of the most fundamental assets they have, and they are likely to have it for years to come. A more important question is whether an operator emerges who wants to play a pure bit-pipe strategy. A wide-scale emergence of this kind of operators would provide a window of opportunity for Internet companies to make money in the mobile market. In the other end of the spectrum is the situation in which mobile operators collectively agree to sustain their walled garden business models by not collaborating closely with Internet actors. In its rudest form mobile operators have the possibility to block other IP traffic than their own. The competition from alternative radio access technologies (e.g. WiFi and WiMAX) is likely to push down the relative value of existing wireless broadband networks in the 3G evolution (Accenture 2006).

In addition to network infrastructure, also the ownership of common platforms provides a possibility to rule the market. This is why Microsoft, Google, Nokia and Apple are so interested in the platform game of the mobile industry. Nokia has established a strong position in the converged device market through its Symbian/S60 platform (see the table above). Because of the significant installed base of S60 devices globally, Nokia has already been able to build viable business models in selling applications and software through the Nokia Download client, for example. Because platforms can be pushed through the penetration of devices, there is a direct link between the platform game and manufacturing scale economies (Accenture 2006) that Nokia currently possesses. Nokia can achieve not only high margins with its device business, but it can also push its software platforms in ways that best contribute to its own entry to the services and software business. Nokia has the advantage of controlling the user interface. To take a recent example, a dedicated Google Search client has been now agreed as part of the Nokia Search service concept[3]. Although this is a new kind of entry from Google into the mobile search market, Nokia is the company owning and controlling the customer interface, and thus Nokia has the real power in the delivery of this particular Internet service.

User domains should be considered as valuable assets. Mobile operators have significant customer domains. These domains can be leveraged in pushing new services, content, or mobile advertising. It is easier to promote new services to existing customers than to build new customer domains from scratch. In the future the information and understanding of customer domains becomes even more valuable. For example, mobile advertising is likely to play an important role in the future, and along comes a need to better understand the differences between customer segments. Not only advertising, but also product and service differentiation, price discrimination and bundling strategies ask for more information regarding customers.

Challenges for companies: How to build new dominant designs and platforms? What constitutes of a critical asset/skill in the mobile market of the future? How to convert the ownership of critical assets and competencies into optimal business strategies, market power and actual profits? How to extract more information on current and potential customers?

Based on this perspective, the key academic questions of the future include: How the platform games can be modeled (e.g. game theory can be used here) and how to better acquire insights regarding the success factors of business models? More specifically, for example, the strategic roles of bundling should be discussed more thoroughly (an exemplary analysis from the handset/subscription bundling business can be found in Tallberg et al. 2007).

Business Model Transformation

The third perspective is ***business model transformation***. Business models are likely to be reshaped in the future mobile industry. In this article a business model was defined to mean the logic through which a company aims to position itself with regards to existing or emerging competitors, suppliers and customers. In essence the business model transformation in this context boils down to three things.

First, companies can decide on the operational structure of their value networks (not only strategic positioning in the value network that was explained

Figure 7. Sales of converged mobile devices Q4/2007 (Canalys 2008)

Vendor	Q4 2007 shipments	% share	Q4 2006 shipments	% share	Growth Q4'07/Q4'06
Total	35,522,360	100.0%	20,667,200	100.0%	71.9%
Nokia	18,802,480	52.9%	11,114,630	53.8%	69.2%
RIM	4,046,860	11.4%	1,829,260	8.9%	121.2%
Apple	2,320,840	6.5%	-	0.0%	NA
Motorola	2,301,260	6.5%	1,463,090	7.1%	57.3%
Others	8,050,920	22.7%	6,260,220	30.3%	28.6%

above). The decision of collaborating actively with external companies (e.g. operator Three's X-Series service platform is closely integrated to the services of actors such as Skype) should be balanced with the decision to do things internally. Operators are outsourcing service development, whereas e.g. Nokia is trying to orchestrate more things by itself than earlier. Nokia and Telefonica, for example, have agreed that Nokia deploys its Ovi service in Telefonica's network with a revenue sharing idea. This is a good way of transforming business models through restructuring the operational logic with suppliers and collaborators. Apple IPhone has showed that it is possible for handset vendors to push for a new kind of revenue logic and to better promote their brands – with or without operators.

Secondly, companies can change their relationships with customers (i.e. customer revenue logic). Not only can service and product portfolio be changed, but the delivery logic can be radically changed as well. Advertising provides promising business opportunities in the mobile domain (see e.g. Blyk's[4] idea). The survey study on 579 smartphone users (see the table below) in Finland suggests that already 24% of panelists are interested in trying advertising-based services in the mobile domain, with 76% in total being positively oriented. Mobile devices are personal devices, providing certain authentication functions built-in. In addition, people carry mobile devices everywhere. Therefore the idea of capitalizing on mobile advertising based business models remains trivial – but still it has not really kicked off. Another example of changing customer relationship logic comes from bundling of software and content with devices. With bundling some power can be transferred from the customer to the service provider. In economics this phenomenon is called as one form of price discrimination. On the other hand, bundling in general may facilitate service adoption. In addition to Apple and IPhone, Nokia is an example of a company who has been looking for new business models with customers. It not only sells handsets, but bundles various kinds of content (new Nokia N95 8GB includes the new Spiderman movie) and applications (e.g. Gizmo and Nokia's application portfolio available at the Nokia Download client), and looks for more service-oriented (continual) revenue instead of mere product-specific (transaction-based) revenue flow. The new Nokia advertising network (Nokia 2008) is an example of Nokia's search for new kinds of relationships with both customers and other players of the market.

Third, user-originated content and social networking communities are emerging as new forces driving the business evolution of the traditional Internet. In general, end-users will increasingly participate in the interaction with services, and create actively own content. These forces are likely to emerge also in the mobile industry. Operators have had the exclusivity of owning charging and billing systems, the key interface with the customer. Now the mobile industry is moving away from transaction-based charging and the idea of mobile

Table 2. Propensity to mobile advertising (Finnish smartphone user survey 2007)

Would you be willing to receive advertising, if you got free applications and content in exchange?	
Yes	24%
Maybe	52%
No	24%

broadband with flat-rate pricing is emerging. According to the Finnish smartphone user survey in 2007, 84% of respondents claim that flat-rate pricing is the better than usage-based charging in data services. End-users will have more freedom of choice with regards to what they want and can do with mobile devices and networks.

In the future of the mobile Internet it will become increasingly difficult to have centralized user domains. People will subscribe to services based on their individual preferences. People will also create new services and content by themselves. The real business model of the future will efficiently leverage and involve end-users in "making the service experience". Good examples can be found from the widely used Web 2.0 services of the Internet, including e.g. Facebook and Youtube. The mobile industry has compelling elements for the application of the Web 2.0 business logic: a truly ubiquitous context, location-based adaptation and personal nature of the device. It would be stupid to ignore possibly the strongest force of innovation and new business creation – the end-users themselves.

Challenges for companies: How to transfer the business logic of the Internet to the mobile domain with or without other companies, possibly with an optimal modification of business logic to mobile specific opportunities (e.g. personal and ubiquitous nature of devices)? How to apply the rules of the Web 2.0 game in the mobile industry?

Future research should study the specific elements of mobile services, and how they should be linked to business models. For example, how the logic of the Web 2.0 (e.g. user created content) can be leveraged in mobile services? Or, what is the impact and role of service pricing?

Service Demand and End-User Needs

The fourth perspective is ***service demand and end-user needs***. Both WAP and MMS were solid and working technologies. In addition, operators and vendors aggressively hyped these technologies. So why did not these technologies emerge as serious cash cows? Perhaps the technology push model was wrong, and companies should listen to customers more than they did earlier?

Even though many transformations are likely to take place on the supply side of the mobile industry, the demand side of the market should not be overlooked. It is still end-users whose needs determine which services are adopted and which are ignored. Empirical observations from early-adopter smartphone users in Finland reveal that only few of these biased users explore the high-end features of mobile devices such as Internet browsing, email or streaming multimedia, and even less use these services actively (Verkasalo 2007c). Modeling of end-user behavior in light of new issues such as flat-rate charging of mobile services and emerging advertisement-based services remains important. End-users eventually decide whether location-based and content services, for example, are going to fly or not. Also, strategically mobile services face the demand of various user segments from early adopters to late adopters sequentially. This provides a possibility to position new mobile Internet services strategically. Even Richard Rumelt, the developer of the theory of dynamic strategy making, is suggesting that one of the key essentials in forming sustainable dynamic strategies is to be able to understand customer needs (see Lovallo & Mendonce 2007).

In assessing end-users, two main challenges emerge. First, the adoption challenge is concerned of the ways to get people to use certain mobile services. As implied in the results of the Finnish smartphone study in 2006 (see the figure below), not all people who intended to use services actually ended up using them. Therefore a gap exists between end-user needs and actual use, and this gap is all about the adoption challenge. Second, the question how to push the extent of mobile service usage is relevant. Appendix A from the same Finnish smartphone study shows that only two services (voice and SMS) are used frequently, and most new services do not experience active usage. The challenge therefore is to study how to make the new Internet-based mobile services into people's everyday use. This is perhaps not possible for other than basic communication services (voice and SMS), and consequently a more relevant challenge is to better identify the relevant segments in which to promote the new services, and achieve success in those particular segments. This leads to an ever increasing need to better understand customer preferences and segmentation approaches. Companies providing mobile services in the future, in order to be profitable, should know how to capitalize on the needs of niche user segments.

Preferences of end-user are likely to become more heterogenic in the future. Positioning, differentiation, price and timing of new services determine how services are going to succeed. The long tail effect (Anderson 2006 and Kilkki 2007) suggests that there is value even in being able to fulfill needs of niche segments with low volume content, for example. With Internet services (e.g. on-demand music services, online book sales) it is now possible to fulfill the needs of the "long tail" customers. The long tail is just an exemplary concept highlighting the heterogeneity of customer preferences and how it links to business.

One conclusion of this discussion on end-user needs and demand-side understanding is that new, more comprehensive empirical research methods should be developed. If companies cannot make more money any more by simply developing new services and technologies and pushing them to the market, they have to do it the other way round. In other words, they have to establish relationships with end-users and try to understand what they need, and why they are not using some of the already existing services. Some of the ongoing research on mobile end-user research (see e.g. Verkasalo 2008a; 2008b; and Kivi 2007) is revealing new observations in the dynamics of service adoption. Models that explain the demand-side dynamics are likely to improve the possibilities to optimally capitalize on emerging business opportunities.

Companies of the mobile domain should in some cases let end-users decide what services they use, how they use the service, and how they want to customize and recreate their services. These are some of the core ideas of Web 2.0. This is the pull model of service deployment, rather than the push model. In the future end-users should be considered as part of the value-creation process. They create content, customize services, innovate and even implement new features, not to talk about their role as marketers of the services to others (e.g. their role as recommenders). The earlier concept of early-adopters (Rogers 1962) is being today replaced by the concept of alpha users – powerful influencers of other people and individuals that have a significant impact in promoting services and brands (see e.g. Ahonen & Moore 2005).

Challenges for companies: How to better segment end-users into more actionable groups? How to move end-users from mere intentions to actual use of services? How to push actual service usage? How to optimally involve end-users in "making" the service experience?

One of the key areas of the future research is to address the end-user part of the industry, by conducting rigorous end-user research. This not only means the use of surveys and interviews, but also the deployment of various new technologies (see e.g. Verkasalo & Hämmäinen 2007 and Kivi 2006) in understanding the behavior and motivation of end-users.

Figure 8. Adoption of mobile services (Finnish smartphone study 2006)

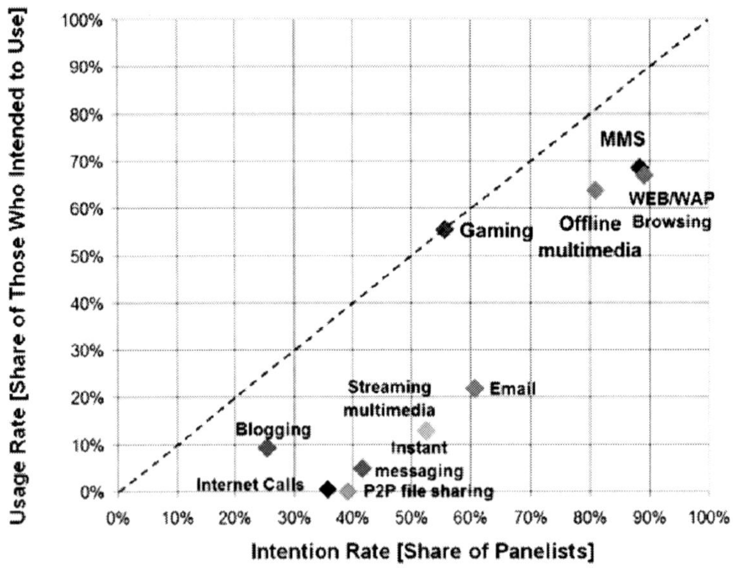

Emergence of Supporting Infrastructure

The fifth perspective is ***emergence of supporting infrastructure***. New mobile Internet services are dependent on the evolution of certain technologies. Therefore the uncertainties of technical development should be understood. To take two examples, the diffusion of radio access technologies and mobile software platforms determine who controls services and what kind of services can be deployed in forthcoming mobile handsets. The battles between different technologies and platforms are an important topic (see e.g. Suarez 2004; Smura 2006) of academic research. Standardization battles have taken place for long (see e.g. Shapiro & Varian 1998), but today the different service platforms (Symbian vs. Windows Mobile vs. Google Android) and network technologies (mobile WiMAX vs. WiFi vs. the 3G evolution) represent new battles, radically affecting the outlook of the mobile industry.

The ownership and technical development of WiFi hotspots, UMA/GAN technologies, LTE, HSDPA and mobile WiMAX are all relevant topics on the level of radio access. Earlier in this article the ownership of critical assets and technologies was found important. However, these assets can quickly become obsolete or less important; given that another technical trajectory will take over. Christensen's theory on disruptive technologies (Christensen 1997) is one example of a theoretical framework in which a poor, but radically cheaper, technology takes over dominant technologies. Google, for example, even considered participating in a radio frequency auction in the U.S., as this could have contributed to their possibilities in pushing alternative technologies and more open platforms in the market of wireless broadband.

On the other hand, open software platforms such as Symbian and Windows Mobile face competition from originally more closed platforms such as Apple IPhone that do not support to a similar extent for example installation of native add-on applications. However, even Symbian devices can be said to have elements of control (e.g. the difficult application signing procedure in the development process). Linux has been said to invade to the mobile domain, but the truth is that Nokia (Symbian/S60), RIM (Blackberry) and Windows Mobile have maneuvered strong positions in converged handset solutions. In fact Nokia's situation is similar to that of WinTel (Microsoft Windows + Intel processor combination) in computers, other than in Nokia Symbian/S60 devices Nokia provides both the hardware and software. Serious challenger to Nokia will emerge, most notably Windows Mobile, various proprietary Linux platforms, RIM Blackberry, Apple Iphone and Google Android. Different kinds of technology battles will reshape the competitive environment of the future mobile market. Companies need to develop long-term technology strategies. Dominance of only one technology provides benefits in terms of interoperability and positive network externalities, but the shortcomings include weaker competition and possible tipping of technically immature technologies.

Challenges for companies: How to leverage the emerging technologies in capitalizing on new business models? How to strategically determine the optimal technology policy e.g. among competing radio access technologies? How to correctly predict the technology change and technology drivers of the transformation of mobile business models?

Studies on the diffusion of innovations in the mobile industry are important in the future research. Both technical and economic factors should be analyzed, for example with regards to the deployment of next generation wireless broadband technologies. The techno-economic analysis of mobile technologies facilitates a deeper discussion on the technological turmoil taking place currently.

Regulation of and Policy on Mobile Internet

The sixth perspective is ***regulation of and policy on mobile Internet***. Regulation and policy making in the transforming mobile industry is likely to have an impact on the emergence of new services. Licensing of wireless spectrum remains critical, as spectrum is a physically limited resource. On the other hand, the emergence of mobile Internet calls by new Internet actors (Skype) is currently facing discriminating policies, as terminal fees to cellular networks are in many countries different for companies operating cellular networks from companies owning mere PSTN access points (VoIP providers). In addition to regulation of issues originating from the traditional mobile industry, new issues such as digital rights management (DRM) of content and privacy of mobile subscribers in the case of mobile advertising should be addressed.

Regulators of the industry have already taken steps towards the Internet-like world, in which customers should have more freedom to choose among service components that are provided by different vendors (see e.g. Melody 2000 and 2001; Seaberg et al. 1997). Local differences, however, still exist in the regulatory field. Even in the EU the national regulatory agencies have independent power in applying EU-level directives. For example in Finland the national regulator decided to allow handset and subscription bundling only in 3G handsets, and only for an experimental period between 2006 and 2008.

According to Maitland et al. (2002) there are two important regulatory factors that are likely to play a greater role in the future of the mobile business. The first is competition policy, which takes care that no one company can achieve monopoly rents through a dominant market position. This should be considered with regards to e.g. inflated prices of mobile roaming (particularly e.g. current European data roaming; Kumar & Hämmäinen 2007) in which network operators collectively have received exceptionally high profits. Needless to say, high roaming and interconnectivity charges do not favor the truly open mobile Internet diffusion. Competition policy should look into difficult topics of the future mobile industry. For example, regulators should not only look at operators, but instead to focus on potentially monopolistic positions of more complex networks of companies involving operators, technology vendors, content providers and Internet players. These companies are likely to do also cross-industry acquisitions, which raise new questions for competition policy. For example Nokia recently acquired Navteq in a multibillion transaction, thereby merging content to its devices. Naturally an extension of the competition policy in mobile markets should look upon the redefinition of the industry itself – as some companies are making an entry to the mobile market from other industries, such as media and entertainment. And this takes place also vice versa.

The second important piece of mobile industry regulation according to Maitland et al. (2002) is access regulation (Maitland et al. 2002), in which for example the discrepancies between different types of access networks might affect competition dynamics. A good example comes from the mobile VoIP business, in which currently the high termination fees to PSTN networks make the business of voice services discouraging for operators deploying solely mobile VoIP (de Bijl & Peitz 2006; Lindqvist 2007).

Another relevant regulatory topic is handset bundling with subscriptions (Tallberg et al. 2007), that has been common in the developed mobile markets for long. The Finnish regulator, however, for long prohibited the bundling of devices with subscriptions. According to Tallberg et al. (2007), this might have had adverse effects on mobile data diffusion, as adequate devices did not reach the market (without bundling the turnover rates of devices are lower). In addition, the prohibition of handset bundling potentially reduced operators' willingness in Finland to invest in new 3G technologies, as ARPU levels were lower. The whole Finnish market focused on intense price competition instead of service innovations (Kiiski 2007; Verkasalo 2007a). On the other hand, handset bundling has several adverse effects, most importantly the increasing power of operators and decrease in end-user freedom of choice. Currently the Finnish regulator is experimenting by allowing handset and subscription bundling with 3G devices. The issue of device and subscription bundling is complex. For example, some operators have already turned off e.g. WiFi and Bluetooth chips in their lineup of mobile devices in order to prevent customers from using other radio networks than their own. This is against the general ideas of fostering end-user choice and freedom (Intven 2000) with regards to services they intend to use. In the US at least one law suit[5] has been filed due to the restriction of consumer choice in a case where an operator disabled Bluetooth connectivity of the device.

The future mobile industry is going to face other kinds of bundling issues that pose new challenges to regulators. One of the emerging issues is the handset vendors' possibility to prebundle services into devices that favor their own solutions or ones of their partners. For example Nokia is pushing its own Music Store, N-gage and Gizmo services into their Symbian-based devices. In addition to services and applications, the bundling of content becomes also relevant. Operators and handset vendors can bundle music and movies into handsets, and potentially achieve additional economic rents through price discrimination.

Last but not least, intellectual property rights are an issue belonging to this regulatory perspective of the mobile industry. IPRs will play a role indirectly in technology competition, in which players will wage war in court rooms with patents. Recent examples include for example the battle between Nokia and Qualcomm in patents related to 3G technologies, though the parties recently settled the case out-of-court. The IPR system can even in the short-term reshape competitive advantages of companies. There are trade-offs. Companies developing future technologies should have the incentive of putting billions of dollars into R&D by knowing that they hold the IPRs to the technology as a result of the development process. Otherwise it would be possible to free-ride easily by leveraging technologies developed by other companies for free. This would in the extreme case lead to a system in which nobody wants to do innovation and new technology development.

On the other hand, strict patent laws also weaken competition, if the use of an important technology is forbidden from other companies. The terms of licensing should be fair.

However, a totally new area of an application of IPRs in the mobile industry is born in the emerging content business, in which the property rights of movies and music should be reconsidered amidst the rapid penetration of different kinds of legal (Apple ITunes) and illegal (mobile P2P file sharing networks) media of mobile media consumption. Entertainment, games and multimedia are the key cornerstones of the potential upside in the value of the mobile services market. Therefore also the rights and policy with regards to them has direct impacts on the market. Let's take the example of the mobile music business. It is clear that music will transform to purely digital format and form of distribution in the future. The policy makers should fight for assuring working conditions for a legal expansion of the mobile industry in this area. Content copyright infringements have been already seen in the P2P space of the wired Internet, including law suits against Napster[6] and operators[7]. No copyright infringements by mobile subscribers have been publicly discovered yet. However, there are many things happening in the mobile space today. For example, Warner withdrew from Nokia's new Music Store service because of the fear that Nokia's MOSH (Mobile Sharing) service could contribute to illegal distribution of music under copyright (Engadget 2007).

Challenges for companies: How to optimally leverage emerging opportunities in the rapidly changing industry, in which the regulator is typically a step behind? How to step away from areas in which the regulatory context imposes temporal or sustainable difficulties? How to prevent the company's own operations from harmful regulatory decisions, and how to optimally participate in making an economically viable market of the future mobile business together with other companies and policy makers?

The future research regarding this perspective should look at key regulatory points, and to apply specific academic tools in modeling the relevant regulatory aspects. To take an example, the handset bundling business experiences interesting interrelationships, as it can be used both for boosting the penetration of handsets in the market, and also for pushing add-on services to customers. A system dynamic modeling framework could be used in an academic study of these interrelationships, possibly indicating the key areas and arguments of decision making from the regulatory point of view.

CONCLUSION

The article presents six strategic perspectives on the business dynamics of the mobile industry. The study provides examples in each of the perspectives, thereby claiming that they are relevant standpoints in the ongoing mobile industry transformation. In general, the article suggests that there are various perspectives in analyzing the transformations of the turbulent mobile industry. In order to reduce complexity in the analysis, these perspectives can be discussed as separate topics in the future research.

The article finds that companies of the mobile industry have six main ways of changing their business models and adapting to the changing business environment:

1. Positioning of the company differently in the value network, involving both vertical (e.g. from hardware to services) and horizontal (e.g. provisioning of new kinds of services) strategic choices.
2. Acquiring and developing critical sustainable assets (content, networks, IPRs, and brands) and competencies (Internet business, advertising, and software technologies).
3. Inventing new business models by utilizing the lessons from the Internet, e.g. subsidizing and bundling products, using two-sided

markets and advertising revenues, deploying innovative revenue sharing schemes, dropping of old pricing models (e.g. usage-based pricing) and radically restructuring business models (e.g. Apple's IPhone device business with operators).

4. Integrating end-users in making and redefining services, better optimizing pricing with end-users in mind, and making the services easy to adopt (e.g. bundling and preconfiguration of services).
5. Leveraging and focusing on new, disruptive, and winning technologies that may change the rules of the business.
6. Adapting to new regulatory schemes and policy initiatives, and proactively conducting business in a dynamic regulatory environment.

The future research should focus on two kinds of things. First of all, the future research should discuss whether all of the business critical transformations can indeed be discussed under the presented analysis – and if not, to redefine the approach. Secondly, the future research should conduct follow-up studies in interpreting actual market events and industry transformations in a longitudinal research setting. This study suggests that a holistic approach should be taken in analyzing emerging business models. No single business model will dominate others, but instead a variety of successful strategies can be found from the mobile market of today and the future (Hanbo 2008).

The developed discussion framework helps in elaborating some of the empirical observations and measurements of the mobile market dynamics. This article provides a structured approach of analysis for a deeper discussion around specific (empirical or theoretical) topics of the mobile industry transformation.

REFERENCES

Accenture. (2006). *Handset industry evolution: navigating a path to high performance.* Accenture's reports on Supply Chain Management.

Afuah, A., & Tucci, C. (2001). *Internet Business Models and Strategies* (International Editions). New York: McGraw-Hill.

Ahonen, T. T., & Moore, A. (2005, March). *Communities Dominate Brands*. Business and marketing challenges for the 21st century. Futuretext Ltd.

Allee, V. (2003). *The Future of Knowledge: Increasing Prosperity through Value Networks.* Butterworth-Heinemann.

Anderson, C. (2006). *The Long Tail.* London: Random House.

Butler, P., Hall, T. W., Hanna, A. M., Mendonca, L., Auguste, B., Manyika, J., & Sahay, A. (1997). A revolution in interaction. *The McKinsey Quarterly, 1/1997.*

Camponovo, G., & Pigneur, Y. (2003). Business Model Analysis Applied to Mobile Business. *Presented at 5th International Conference on Enterprise Information Systems* (ICEIS 2003, Angers, 23-26 April).

Christensen, C., & Rosenbloom, R., S. (1995). Explaining the attacker's advantage: Technological paradigms, organizational dynamics, and the value network. *Research Policy, 24*, 233–257. doi:10.1016/0048-7333(93)00764-K

Christensen, C. M. (1997). *The Innovator's Dilemma.* Harvard Business School Press.

de Bijl, P. W. J., & Peitz, M. (2006). *Access Regulation and the Adoption of VoIP*. International University in Germany; Working Paper No. 43/2006.

Eisenmann, T., Parker, G., & Van Alstyne, M. (2006, August). Strategies for two-sided markets. *Harvard Business Review.*

Engadget. (2007). Nokia Music Store launches without Warner Music due to file sharing fears. http://www.engadget.com/2007/11/02/nokia-music-store-launches-without-warner-music-due-to-file-shar/. Referred 20.2.2008.

Feng, L., & Whalley, J. (2002). Deconstruction of the telecommunications industry: from value chains to value networks. *Telecommunications Policy*, *26*, 451–472. doi:10.1016/S0308-5961(02)00056-3

Grant, R. (1991). The resource-based theory of competitive advantage: implications for strategy formulation. *California Management Review*, (Spring): 119–135.

Gulati, R., Nohria, N., & Zaheer, A. (2000). Strategic Networks. *Strategic Management Journal*, *21*, 203–215. doi:10.1002/(SICI)1097-0266(200003)21:3<203::AID-SMJ102>3.0.CO;2-K

Hanbo, Z. (2008). *Emerging Business Models of the Mobile Market*. Master's thesis. Department of Communications and Networking, Helsinki University of Technology, Finland.

Hargadon, A. (2003). *How Breakthroughs Happen - The Surprising Truth About How Companies Innovate*. Boston, MA: Harvard Business School Press.

Intven, H., Oliver, J., & Sepulveda, E. (2000, November). *Telecommunications Regulation Handbook.* InfoDev program, World Bank. iSuppli. (2005). Mobile communications market will approach $200 Billion in 2009. *Market Tracker. Wireless Systems*, *Q3*, 2005.

ITU. (2006). *Key Global Indicators for the World of Telecommunication Service Sector.* http://www.itu.int/ITU-D/ict/statistics/at_glance/KeyTelecom99.html. Referred 20.2.2007.

Kiiski, A. (2007). *Impacts of Virtual Operators on Mobile Market*. Licentiate thesis, Helsinki University of Technology, 2007.

Kilkki, K. (2007, May). A practical model for analyzing long tails. *First Monday, 12(5)*. http://firstmonday.org/issues/issue12_5/kilkki/index.html. Referred 2.7.2007.

Kivi, A. (2006). *Mobile Internet Usage Measurements: Case Finland. Master's Thesis Series.* Networking Laboratory. Department of Electrical and Telecommunications Engineering. Helsinki University of Technology, Espoo, Finland, May 2006.

Kivi, A. (2007). Measuring mobile user behavior and service usage. *Presented at 6th Global Mobility Roundtable*, 1-2 June 2007, Los Angeles, USA.

Kumar, R., & Hämmäinen, H. (2007, October). Competition in International Mobile Roaming: Alternative Clusters. *Communications and Strategies*, *3*(67), 151–172.

Linder, J. C., & Cantrell, S. (2000). *Changing business models: surveying the Landscape.* Working Paper, Accenture Institute for Strategic Change.

Lindqvist, K. (2007). *Scenario Analysis of Mobile Voice Market.* Master's Thesis Series. Networking Laboratory. Department of Electrical and Telecommunications Engineering. Helsinki University of Technology.

Lovallo, D., & Mendonca, L. (2007). Strategy's Strategist: Interview with Richard Rumelt. *The McKinsey Quarterly*, (August): 2007.

Magretta, J. (2002, May). Why business models matter? *Harvard Business Review*, 86–92.

Maitland, C. F., Bauer, J. M. B., & Westerveld, R. (2002). The European market for mobile data:evolving value chains and industry structures. *Telecommunications Policy*, *26*, 485–504. doi:10.1016/S0308-5961(02)00028-9

Maslow, A. H. (1943). A Theory of Human Motivation. *Psychological Review*, *50*, 370–396. doi:10.1037/h0054346

McGahan, A. (2004). *How Industries Evolve - Principles for Achieving and Sustaining Superior Performance*. Boston: Harvard Business School Press.

Melody, W. H. (2000). Shaping liberalized telecom markets. *Telecommunications Policy*, *24*(10–11), 803–806. doi:10.1016/S0308-5961(00)00073-2

Melody, W. H. (2001). Assessing highly imperfect mobile markets. *Telecommunications Policy*, *25*(1–2), 1–3. doi:10.1016/S0308-5961(00)00083-5

Müller-Veerse, F., Kohlenbach, B., Bout, D., Singh, S., Golub, G., & Häyrynen, J. (2001). *UMTS Report - An Investment Perspective*. Durlacher Research Ltd, Eqvitec Partners Oy.

Netsize. (2008). *The Netsize Guide 2008. Mobile 2.0, you are in control*. Mobile Industry Report. http://www.netsize.com/downloads/intro.aspx. Referred 20.2.2008.

Nokia. (2008). *Nokia launches premier mobile advertising network*. Nokia press release. February 11, 2008. http://www.nokia.com/A4136001?newsid=1190110. Referred 20.2.2008.

Osterwalder, A., & Pigneur, Y. (2002). An ebusiness model ontology for modeling eBusiness. In *Proceedings of 15th Bled Electronic Commerce Conference – eReality: Constructing the eEconomy*, Bled, Slovenia, June 17–19, 2002.

Papakiriakopoulos, D., Poulymenakou, A., & Doukidis, G. (2001). Building E-Business models: An analytical framework and development guidelines. In *Proceedings of 14th Bled Electronic Commerce Conference,* Bled, June 25-26, 2001.

Pateli, G. A., & Giagli, M. G. (2004). A research framework for analysing eBusiness models. *European Journal of Information Systems*, *13*, 302–314. doi:10.1057/palgrave.ejis.3000513

Petrovic, O., Kittl, C., & Teksten, R. D. (2001). Developing business models for eBusiness. In *Proceedings of International Conference on Electronic Commerce 2001,* Vienna: October 31 – November 4, 2001.

Porter, M. E. (1979). How competitive forces shape strategy. *Harvard Business Review*, (March/April): 1979.

Porter, M. E. (1980). *Competitive Strategy*. New York: The Free Press.

Porter, M. E. (1985). *Competitive Advantage*. New York: Free Press.

Rappa, M. (2003). *Managing the digital enterprise – Business models on the Web*. http://ecommerce|.ncsu.edu/business_models.html. Referred 1.11.2007.

Rochet, J. C., & Tirole, J. (2003). Platform Competition in Two-Sided Markets. *Journal of the European Economic Association*, *1*(4), 990–1029. doi:10.1162/154247603322493212

Rogers, E. (1962). *Diffusion of innovations*. New York: The Free Press.

Rumelt, R. (1991). How Much Does Industry Matter? *Strategic Management Journal*, *12*, 167–185. doi:10.1002/smj.4250120302

Saarikoski, V. (2006). *The Odyssey of the Mobile Inter*net. Doctoral Dissertation. University of Oulu, Finland.

Seaberg, J. G., Hawn, J., Dincerler, G. E., Eugster, C. C., & Rao, N. (1997). Attackers versus incumbents: The battle for value in an IP-networked world. *The McKinsey Quarterly*, *4*, 138–153.

Seppänen, M., & Mäkinen, S. (2005). Business model concepts: A review with case illustration. *Proceedings of the Engineering Management Conference*, *2005*, 376–380.

Shapiro, C., & Varian, H. R. (1998). *Information Rules: A Strategic Guide to the Network Economy*. Boston, MA: Harvard Business School Press.

Smura, T. (2006). Competition between Emerging Wireless Network Technologies: Case HSPA vs. WiMAX in Europe. In *17th European Regional ITS Conference*, August 22-24, 2006, Amsterdam, The Netherlands.

Stabell, C. B., & Øystein, D. F. (1998). Configuring value for competitive advantage: On chains, shops, and networks. *Strategic Management Journal*, 19.

Suarez, F. F. (2004). Battles for technological dominance: an integrative framework. *Research Policy*, *33*, 271–286. doi:10.1016/j.respol.2003.07.001

Tallberg, M., Hämmäinen, H., Töyli, J., Kamppari, S., & Kivi, A. (2007). Impacts of handset bundling on mobile data usage: The case of Finland. *Telecommunications Policy*, *31*, 648–659. doi:10.1016/j.telpol.2007.08.007

Tapscott, D., Lowy, A., & Ticoll, D. (2000). *Digital Capital - Harnessing the power of business webs*. Boston, MA: Harvard Business School Press.

Timmmers, P. (1998). Business models for electronic markets. *Journal on Electronic Markets*, *8*(2), 3–8. doi:10.1080/10196789800000016

Van Liere, D. W., Hagdorn, L., Hoogeweegen, M. R., & Vervest, P. H. M. (2004). *Organizational Performance of a Firm in a Modular Business Network*. ECIS 2004.

Verkasalo, H. (2007a). *A Cross-Country Comparison of Mobile Service and Handset Usage*. Licentiate's thesis. Helsinki University of Technology, Networking Laboratory, Finland.

Verkasalo, H. (2007b). Empirical Findings on the Mobile Internet and E-Commerce. *Presented at 20th Bled e Conference: E-Mergence: Merging and Emerging Technologies, Processes, and Institutions*. Bled, Slovenia, June 4 - 6, 2007.

Verkasalo, H. (2007c). Measurement of Smartphone Service Evolution in Finland. Forthcoming in *Journal of Targeting, Measurement and Analysis in Marketing; Special Issue; December 2007*.

Verkasalo, H. (2008a). Dynamics of Mobile Service Adoption. Forthcoming in *Journal of E-Business Research, 4(3) July 2008*.

Verkasalo, H. (2008b). Handset-Based Measurement of Mobile Service Demand and Value. Forthcoming in *INFO: The Journal of Policy, Regulation and Strategy, 10*(3); May 2008.

Verkasalo, H., & Hämmäinen, H. (2007). A Handset-Based Platform for Measuring Mobile Service Usage. *INFO: The Journal of Policy, Regulation and Strategy, 9*(1).

Vesa, J. (2005). *Mobile Services in the Networked Economy*. Hershey, PA, USA: IRM Press.

Weill, P., & Vitale, M. R. (2001). *Place to Space: Migrating to E-Business Models*. Boston: Harvard Business School Press.

Wernerfelt, B. (1984). A Resource-Based View of the Firm. *Strategic Management Journal*, *5*, 171–180. doi:10.1002/smj.4250050207

Williamson, O. E. (1975). *Markets and hierarchies. New York*. Free Press.

Williamson, O. E. (1985). *The economic institution of capitalism. New York*. Free Press.

Williamson, O. E. (1989). Transaction cost economics. In Shemlenese, R., & Willig, R. D. (Eds.), *Handbook of industrial organisation, 1*. Amsterdam: Elsevier Science Publishers BV.

ENDNOTES

1 http://www.google.com/mobile/

2 http://www.apple.com/uk/iphone/appstore/

3 http://www.nokia.com/A4136001?newsid=1190585

4 http://about.blyk.com/

5 *Opperman, et al. v. Cellco Partnership d/b/a Verizon Wireless*, Case No. BC 326764, Superior Court of the State of California for the County of Los Angeles, 2005.

6 2001 US Dist. LEXIS 2186 (N.D. Cal. Mar. 5, 2001), aff'd, 284 F. 3d 1091 (9th Cir. 2002).

7 Recording Industry Association of America v. Verizon Internet Services, 2003 U.S. Dist. LEXIS 681 (D.D.C. 2003).

APPENDIX

Figure 9. Smartphone service positioning

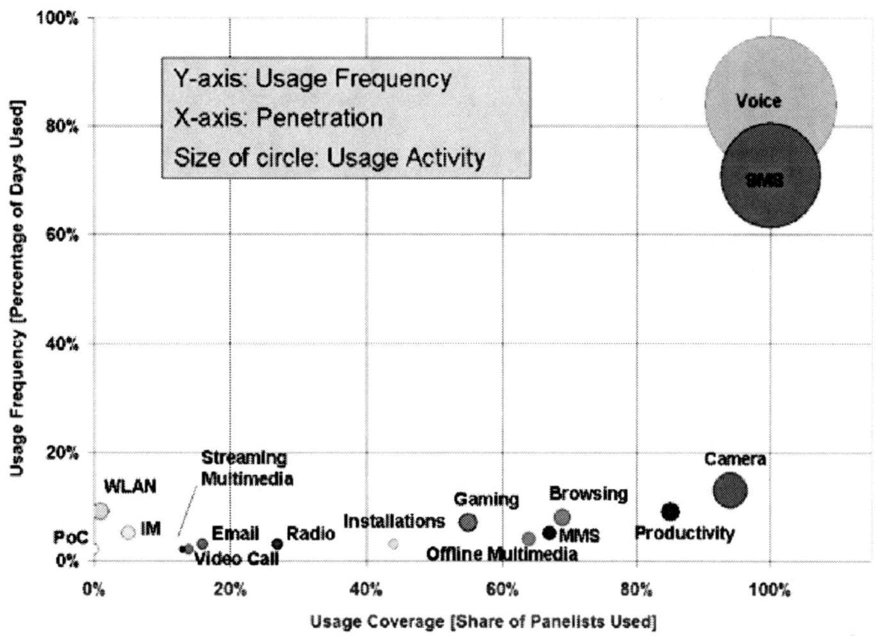

Dataset acquired in fall 2006, analysis in 2007. Base: 695 Finnish consumers using Nokia S60 devices

This work was previously published in International Journal of Interdisciplinary Telecommunications and Networking, Volume 1, Issue 1, edited by Michael R. Bartolacci, and Steven R. Powell, pp. 19-44, copyright 2009 by IGI Publishing (an imprint of IGI Global).

Chapter 4
Strategic Technology Options in the Wireless Industry:
A Case Study for U.S. Wireless Carriers

Hak J. Kim
University of Houston-Clear Lake, USA

Martin B. H. Weiss
University of Pittsburgh, USA

Benoit Morel
Carnegie Mellon University, USA

ABSTRACT

The major U.S. wireless operators already have announced their plans for the evolution of their networks towards 3G, but some uncertainties remain, such as emergence of new technologies (WiMAX and WLAN) and the consolidation among operators (AT&T Mobile and Sprint Nextel). The article discusses a real option based model for technology decisionsand applies it to the U.S. wireless industry as a case study. We also discuss what decisions must be made, what the outcomes are, and how the options model is validated. The preliminary results show that the evolution of wireless network technologies between generations (inter-generations migration scenario) is desirable (a positive net option value), but not desirable (a negative net option value) within generations (intra-generation migration scenario), in the U.S.

INTRODUCTION

In the late 1990s, AT&T Wireless and Cingular Wireless each faced a critical decision. The TDMA technology that they chose earlier that decade was approaching obsolescence; the data communications features that were beginning to be demanded by the marketplace had not been developed for this technology platform, nor would they be by equipment manufacturers. Thus, to remain competitive in the marketplace, these mobile service providers had to switch to a CDMA or GSM-based platform. This transition meant replacing all of their base station and switching hardware as well as customer handsets in an orderly fashion. Although both carriers ultimately chose GSM

DOI: 10.4018/978-1-60960-505-6.ch004

(and ultimately merged to become today's AT&T Mobile), the decision process has received little attention in the literature despite the enormous business and financial risks involved. This article uses this decision and the information available at the time, to propose a real option based model for technology decisions.

Recently the real options approach (ROA) has emerged in the strategic management field (Dixit & Pindyck, 1994 & 1995; Benaroch & Kauffman, 1998; Amram & Kulatilaka, 1999; Kim & Sanders, 2002). The ROA provides a structure linking of strategic planning and financial analysis tools to evaluate potential opportunities and uncertainty (Dixit & Pindyck, 1994). The ROA is appealing to firms because of its distinctive ability to capture managers' flexibility (Trigeorgis, 1996) in adapting their future actions in response to evolving markets or technological conditions. For example, when managers evaluate new projects, they may face several choices beyond simply accepting or rejecting the investment. Other choices include delaying decisions until the market conditions are more favorable, or deciding to start small and expanding later if the results are good.

The network industry is perhaps particularly sensitive to these issues, because firms in this industry require large, up-front capital investment before revenues can be earned, the ROA works best under high uncertainty with large investment costs, so the network industry fits the ROA methodology well (Alleman & Noam, 1999). With the demands for various services in their markets, network operators are challenged by the rapid development of technologies (Lawless & Anderson, 1996; Balachandra & Friar, 2004) as well as the existing network migration dilemma (Podhradsky, 2004; Akhtar et al., 2005). Since the complete replacement of existing networks is not practical, a firm's preferred approach is to create more flexibility (Langlois & Robertson, 1992; Sanchez, 1999; McDysan et al., 2000; Schilling, 2000) in networks that allow for the customization of services for users and the easy upgrading of their networks when better components, with competitive advantages, come along.

The U.S. wireless industry is a good laboratory for a study like this, since multiple technologies exist and the industry was undergoing a major transition from second generation (*2G*) to the third generation (*3G*) during the study period and is currently moving toward fourth generation (4G) technologies. So, each service provider had to choose a particular network transition strategy, indicating when, how, and at what pace to introduce new technologies. The chosen strategy determined the firm's focus and sometimes required the coexistence of new and existing network technologies.

The goal of this study is to develop a theoretical framework for wireless operators to support their strategic decisions when considering technology choices as they move to the next generation network technology. Our study does not give an absolute value for the choice of technology, but provides some inferences by attempting to quantify the value of the technology migration strategy as a basic element of strategic decision-making. As a result, this study intends to raise core issues concerning the transition towards *3G*.

U.S. Wireless Market

By early 2007, the U.S. wireless market (Budde, 2007; CTIA, 2006; FCC, 2006) was dominated by AT&T Mobility, Verizon Wireless and Sprint Nextel, together accounting for more than 70% market share. T-Mobile, Alltel and U.S. Cellular make up the next tier, accounting for about 20%. In 3G networks, Verizon Wireless and Sprint Nextel used cdma2000 technology, while AT&T Mobility and T-Mobile offered WCDMA technology. Recently, Sprint Nextel began deploying a WiMAX network, preparing for the next round of competition, in 4G technologies (Polivka, 2007).

Figure 1 (world) and Figure 2 (U.S.) plot the number of subscribers in each wireless technology from 1992 to 2006. Based on the number of subscribers in generation, the chart clearly shows

Figure 1. World wireless market

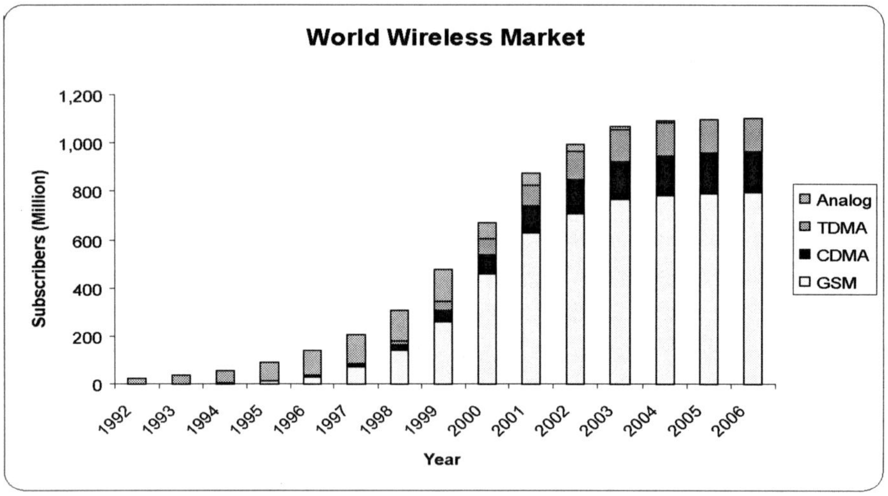

Figure 2. U.S. wireless market

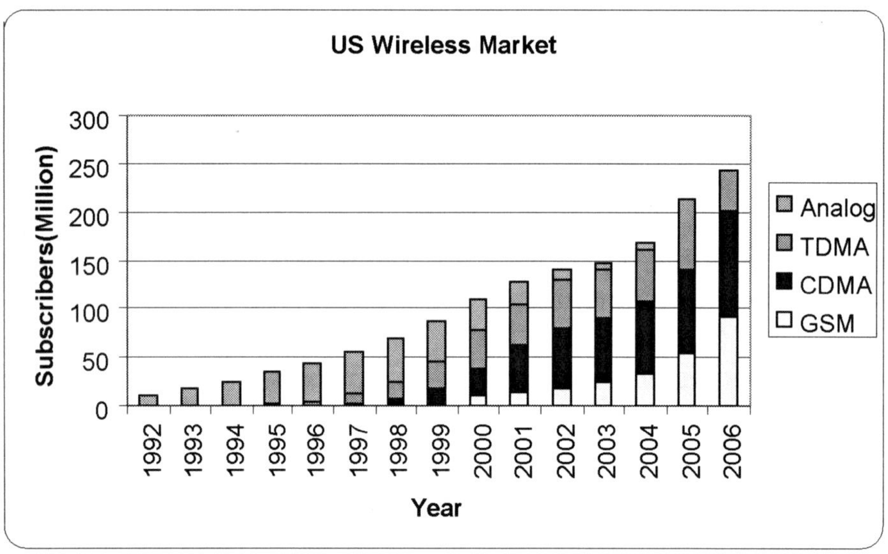

the dramatic growth of GSM in the world wireless industry (Figure 1), while CDMA dominates the U.S. wireless industry. Unlike GSM's dominant position in world wireless market, CDMA has experienced high growth and is the lead technology in the U.S. wireless market. TDMA also has a significant market share, but will eventually be obsolete as providers upgrade to more advanced technologies, such as GSM, GPRS, EDGE, and WCDMA. Analog was completely phased out by 2004 in the world and U.S. wireless market.

As seen in Figure 3, the current U.S. wireless market simply can be broken down into two dimensions; three types of customer groups and three types of network operator groups. In the customer groups, 1) the first group are new cus-

Figure 3. Market structure and operator groups

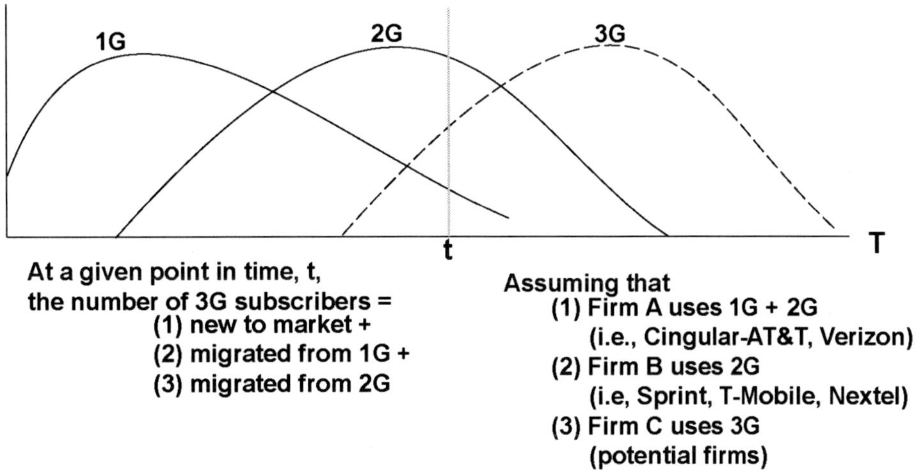

tomers who have never used wireless services; 2) the second group is made up of customers who moved from their current *1G* services (actually 1G service was stopped in 2004 in the U.S.); and 3) the third group consists of customers who moved from their current *2G* services. In the network operator groups, 1) the Firm A-type group are the existing hybrid service providers offering *1G* and *2G* technologies, like Verizon and AT&T Wireless; 2) the Firm B-type group consists of the existing service providers only offering *2G* services, like Sprint PCS and T-Mobile, and Nextel; and 3) the Firm C-type group is made up of potential new service providers only offering *3G* services, i.e., *WCDMA* and *cdma2000*. Based on this simplified market structure, the issues are; 1) How can we assess the value of technology migration as a basis of technology migration strategy? 2) What is a firm's migration strategy for 3G including what technology, when, and how to migrate?

This study begins by tracing the evolution of technologies in wireless networks, and continues by developing the real options approach to assess technology migration from old to new technology. Next, we apply it to the U.S. wireless industry as an empirical study. Finally, we attempt to get some inference from the results of this study.

Technology Options for U.S. Wireless Carriers

Historical Evolution of Wireless Technologies

Over the past decade, wireless networks have moved rapidly from first-generation (1G), to second generation (2G), and further to third generation (3G) wireless networks (Rapport, 2002; Garg, 2001; Prasad, 1998). The 1G wireless networks are voice-only communications with analog technology. The main 1G technology standards are *AMPS*, *TACS* and *NMT* (Rapport, 2002). The 2G wireless networks replaced analog networks (1G) with digital technology which is the most fundamental change in telecommunications. There are several 2G wireless technologies, such as *TDMA*, *GSM*, *cdmaOne* and *PDC* (Garg, 2001). One stage before 3G wireless systems comes 2.5G which is a technology that allowed second generation users to get a taste of what 3G would eventually present. 2.5G systems, such as *GPRS*, *EDGE* and *HSCSD* (Rapport, 2002), can be seen as straightforward upgrades of 2G networks, since in most cases, the 2G infrastructures underwent simple software/hardware developments. Nowadays the wireless network architecture is moving

Figure 4. Wireless technology options

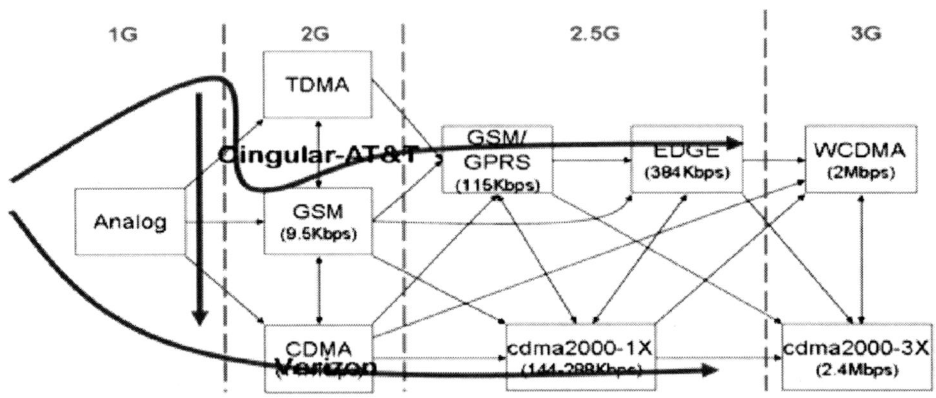

to the *3G* wireless technologies (Carsllo et al., 1997; Dahlman et al., 1998) which provides high-rate voice and data services (i.e., 2 Mbps). The *3G* wireless system is currently split into two groups (Rapport, 2002): the *UMTS* group (3GPP) and the cdma2000 group (3GPP2). The Third generation Partnership Project (3GPP) is collaboration between organizational partners (OPs) which study the W-*CDMA/TD-SCDMA/EDGE* standards and the Third Generation Partnership Project 2 (3GPP2) is collaboration between OPs which examine the cdma2000 standards.

Options for Technology Migration

Figure 4 shows the possible technology transition scenarios for U.S. Wireless carriers. A carrier with 1G can implement one of three 2G technologies: TDMA, GSM, and CDMA. TDMA and CDMA are more popular in the U.S., while GSM is prevalent in Europe as a single standard. For more high-speed data services, *2.5G* technologies, *GPRS, EDGE,* and *cdma2000-1XRTT*, have been developed. *2.5G* is always on, provides simultaneous voice and data, and delivers more speed than today's 2G circuit-switched data connections. Wireless network operators can implement *2.5G* much more cheaply than *3G* because the former uses existing *2G* spectrum and doesn't require a new network infrastructure, although some system upgrades are necessary. So, *2.5G* is often considered a stepping-stone to *3G*. As the wireless industry moves toward *3G* technologies, the current coexistence of three major technologies will most likely evolve into two competing technologies within the *3G* market: *WCDMA* and *cdma2000*.

GSM-Based Network Migration Path

Figure 5 shows the migration path from *GSM* (2G) to *GPRS* (2.5G) and to *WCDMA* (3G). When *GPRS* service is provided in the *GSM* network, several components are added, like *SGSN* and *GGSN* (yellow shaded boxes). In the transition from *GSM/GPRS* to *UMTS* (*3G*), the access network section (blue shaded boxes) must be totally replaced (or added) in the carriers' networks. In addition it is not possible to operate in a *GSM* mode within the same 5 MHz band. Since *UMTS* does not reuse the *GSM* base station hardware, operators must install new hardware cabinets adjacent to existing systems.

CDMA-Based Network Migration Path

Figure 6 shows the evolutionary path from a *cdmaOne*(2G) to a *cdma2000-3X*(3G) network

Figure 5. GSM-based network architecture

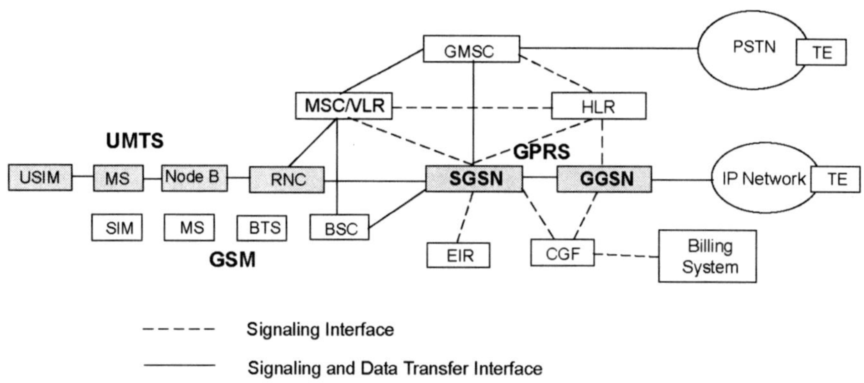

Figure 6. CDMA-based network architecture

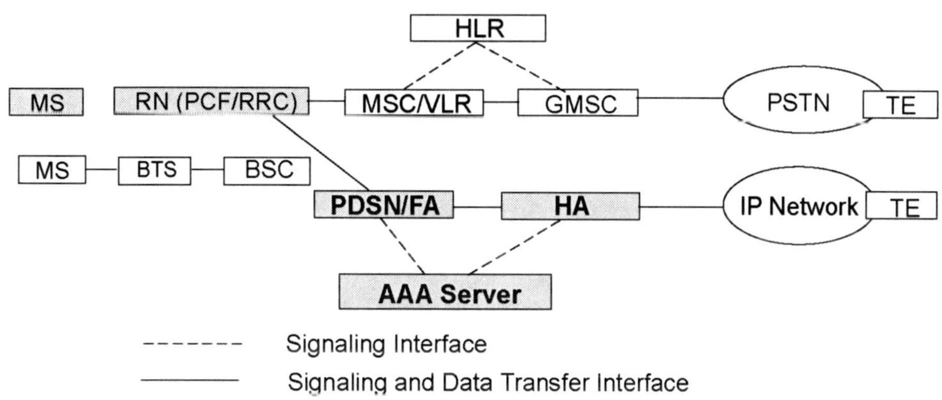

architecture. Since *cdma2000* is the evolution of *IS95*-based systems, it is the natural *3G* evolution of *CDMA* technology which requires only minor upgrades to the network and small capital investment. Because of this, the transition from *cdmaOne* to *cdma2000-1X* is relatively easy for operators and transparent for consumers. Wireless service operators can gradually migrate from '*cdmaOne*' to *cdma2000* at the *cdma2000-1X* (1.2288 Mcps) rate. As users migrate to the new standard, network operators can swap out cdma2000 1X radios and insert a *cdma2000-3X* radio to increase cell capacity. They also have the choice of using three *cdma2000-1X* radios or converting to a single *cdma2000-3X* radio. *Cdma2000* reuses the same 9.6 kbps *vocoder* from *cdmaOne*.

THEORETICAL FRAMEWORK AND THE MODEL

Real Options Theory

Concepts

Real options are simply defined as *the right, but not the obligation* to take some actions (i.e., projects and business) in the future (Trigeorgis, 1996). That is, real options involve discretionary decisions or rights without any obligation to acquire or exchange an asset for a specified alternative price (Amram & Kulatilaka, 1998). The basic idea of real options, namely the ability to provide access to significant upside potentialwhile containing downside losses, makes options more valuable with greater volatility (Dixit & Pindyck, 1994).

Real option is a tool to assess potential opportunities and uncertainty as a positive value through managerial flexibility. Managerial flexibility is a set of real options, for example, the options to defer, abandon, contract, or expand investment, or switch investment to an alternative (Trigeorgis, 1996). It is often not explicitly taken into account when comparing a project's tangible costs and benefits by the traditional, discounted cash flow approach (e.g., NPV and IRR), even though it provides great benefits. The ROA may be a useful tool for a firm's technology management because firms are often faced with higher degrees of uncertainty when making strategic investment decisions (Amram & Kulatilaka, 1998).

History and Applications

The options paradigm in investment decision-making began with Black-Scholes and Merton (1973)'s option theory, which offered some valuation tools under the underlying sources of uncertainty with continuous time, assuming 'Brownian motions (random walk)'. Brennan & Schwartz (1985) and McDonald & Siegel (1986) were the first to actually employ options thinking in the valuation of real assets (i.e., projects), which has become known as 'real options'. After following them, Dixit & Pindyck (1995), Smith & Triantis (1995), and Trigeorgis (1996) deal with the issue of the timing of investment when there is competition in the product market. After that, real options have been popular in several industries, such as oil (Paddock et a;., 1988; Pickles & Smith, 1993; Dias & Rocha, 1999), mining (Brennan & Schwartz, 1985), pharmaceutical research (Micalizzi, 1999), information technology (IT) (Benaroch & Kauffman, 1998; Amram & Kulatilaka, 1999; Benaroch, 2002; Kim & Sanders, 2002; Jeffery & Leliveld, 2004; Pateli & Giaglis, 2007), telecommunications & networking (Alleman & Noam, 1999; Economides, 1999; d'Halluin et al., 2002; Basil & Fontini, 2003; Kim, 2006), and other investment activities (Car, 1988; Pindyck, 1989; Trigeorgis, 1993; Kemma, 1993; Sanchez, 1995; Flatto, 1996; Stonier, 1999; Deng et al., 2001; Chen et al., 2005).

Real Options in ICT

Research studies in the ICT sector (e.g., Clemons, 1991; Dos Santos, 1991; Kambil *et al.*, 1993; Kumar, 1996; Chalasani, Jha & Sullivan, 1997; Alleman & Noam, 1998; Benaroch, 2002; Alleman & Rapport, 2006) have recognized the importance of utilizing the theory of real options to justify the option-like characteristics of ICT investments.

For evaluating investment in information systems and technologies, the theory of real options has been applied to assess the potential business value of investments (Panayi & Trigeorgis, 1998; Taudes, 1998; Benaroch and Kauffman, 2000; Benaroch, 2002; Clemons & Gu, 2003; Fichman, 2004). Similarly, Schwartz & Zozaya-Gorostiza (2003) develop options models which take into consideration the effect of uncertainty in the costs and benefits associated with ICT investment opportunities, using data on the deployment of point-of-sale debit services by the Yankee 24 network as reported in Benaroch and Kauffman (2000). Clemons (1991) has shown that the ICT related benefits include flexibility and increased responsiveness, both of which can be evaluated with real options. Kumar (2002) develops a framework that is intended to aid in understanding decision support system value.

The application of real options to the information systems and technology field has mostly been used to justify strategic investments in information technologies that provide a long-term view of business value, for example: Integrated Services Digital Network (ISDN) services (Dos Santos, 1991), Credit Card Services (Clemons & Gu, 2003), electronic banking (Benaroch and Kauffman, 2000), SAP R/3 software (Taudes *et al.*, 2002).

Software platforms may not generate value directly, but they enable value-added applications to be implemented (Taudes, 1998; Fichman, 2004). Their value lies in the options they create around building applications or serving as platform investments (Taudes *et al.*, 2000; Dai *et al.*, 2000). Software design (Sullivan *et al.*, 1999) is analyzed as an investment activity and introduces options thinking to improve design decision-making. For instance, Taudes (1998) evaluates sequential exchange options in order to obtain estimates for the value of software growth options. Taudes *et al.* (2000) used an options model to quantify the benefits of switching from an earlier-generation software, SAP R/2 to the next version, SAP R/3, for a real-world manufacturing company.

A Model for Assessing Technology Options

In this section, we attempt to develop a theoretical framework to assess strategic technology options in wireless networks. The model is equivalent to using options to exchange one risky asset for another in financial market (Margrabe, 1978) and is applied to assess technology options for migrating into the next generation networks in the wireless industry; that is, we show the transition value when moving from generation-to-generation (inter-generational transition) and within the same generations (intra-generational transition).

Definitions and Assumptions

Let the value of a technology investment in the revolutionary technology (i.e. *CDMA*-based) compared with the evolutionary (i.e. *GSM*-based) be '***H***'. Also let ***P*** and ***B*** be the net value of two alternatives of network migration by the choice of strategy at time *t*.: One (***P***) is a revolutionary technology change with a larger risk and investment ('aggressive') and the other (***B***) is a stepping-stone technology change with a smaller risk and investment ('conservative').

Assuming that the level of investment for improving network performance is directly related to their revenues, the key issue in the choice of strategic options is how to quantify a trade-off between the level of performance improvement and the value of premium in a risk neutral situation. Risk neutrality means comparing one portfolio where an investment is in *stepping-stone architecture* with a premium to the other portfolio where an investment is in the revolutionary architecture with potentially higher value.

Technology Transition Option Value

We treat the choice between the two scenarios as a comparison between two wireless network technology migration portfolios. Again, let ***P*** correspond to a high level of uncertainty (potentially high value) with a much larger investment cost, and ***B*** correspond to a lower level of uncertainty with a much smaller investment cost. The two scenarios are defined as:

- Revolutionary portfolio (W_{REV}) = $v_P P$ (i.e. *CDMA*-based architecture)
- Evolutionary portfolio (W_{EVO}) = $v_B B$ (i.e. *GSM*-based architecture) where v_P and v_B are amounts invested in each scenario.

To compare the two "portfolios", we introduce a quantity $H(P, B)$ which is defined as:

$$v_H H + W_{EVO} = W_H + W_{EVOt} = W_{REV}$$

Using the derivative, it can be described as:

$$v_H dH\,(P, B) = v_P dP - v_B dB$$

By combining the above two formulae, we can write:

$$W_H \frac{dH}{H} = W_{REV} \frac{dP}{P} - W_{EVO} \frac{dB}{B} \tag{1}$$

One way to interpret Equation (1) is to interpret *H*(*P*, *B*) as the value of the option of investing in the revolutionary technology instead of the evolutionary one and to treat $W_H = W_{REV} - W_{EVO}$ as the value of the premium that should be paid to accomplish higher network performance, under the assumption of risk neutrality. *H*(*P*, *B*) quantifies the maximum premium that should be paid to reduce the uncertainty associated with the evolutionary approach to technology migration. In other words, as long as the actual value of the premium paid for the higher network performance is smaller than *H*(*P*, *B*), it is more advantageous to go for the revolutionary technology.

Now let us consider the time horizon τ to deal with a continuous option, like European-type option which can be exercised at τ. This option is simultaneously a call option on asset one with. Clearly *H*(*P*, *B*, τ) depends also on the time horizon τ. Remembering that:

$$W_H = W_{REV} - W_{EVO},$$

$$W_H + W_{REV} = W_{EVO},$$

Thus, Equation 1 can be rewritten as:

$$\begin{aligned} W_H \frac{dH}{H} &= (W_H + W_{EVO})\frac{dP}{P} - W_{EVO}\frac{dB}{B} \\ W_H\left(\frac{dH}{H} - \frac{dP}{P}\right) &= W_{EVO}\left(\frac{dP}{P} - \frac{dB}{B}\right) \\ \frac{W_{EVO}}{W_H}\left(\frac{dP}{P} - \frac{dB}{B}\right) &= \left(\frac{dH}{H} - \frac{dP}{P}\right) \end{aligned} \tag{2}$$

H(*P*, *B*, τ) depends on the two stochastic variables ***P*** and ***B*** (i.e. it is a *derivative*) and on the time horizon τ. Using Ito's lemma, the instantaneous rate of change of that derivative $\frac{dH}{H}$ can be written as:

$$\frac{dH}{H} = \beta\, dt + \gamma\, dz + \eta\, dq \tag{3}$$

where:

$$\beta = \frac{1}{H}\left\{\frac{\partial H}{\partial t} + \mu B\frac{\partial H}{\partial B} + \alpha P\frac{\partial H}{\partial P} + \frac{1}{2}\left\{\delta^2 B^2\frac{\partial^2 H}{\partial B^2} + 2\rho\sigma\delta BP\frac{\partial^2 H}{\partial B\partial P} + \sigma^2 P^2\frac{\partial^2 H}{\partial P^2}\right\}\right\} \tag{3a}$$

$$\gamma = \frac{\sigma P}{H}\frac{\partial H}{\partial P} \tag{3b}$$

$$\eta = \frac{\delta B}{H}\frac{\partial H}{\partial B} \tag{3c}$$

After some manipulation (Kim, 2004), we derived the expression of *H*(*P*, *B*, τ) in terms of the value of the evolutionary technology ***P*** and the value of the higher cost technology ***B***, can be deduced as follows:

$$H(B,P,T) = B\cdot\Phi\left(d_1\left(\frac{B}{P},T\right)\right) - P\cdot\Phi\left(d_2\left(\frac{B}{P},T\right)\right) \tag{4}$$

with:

$$d_1(x,T) = \frac{1}{\sqrt{2T}}\left[Log(x) + \frac{T}{2}\right]$$

$$d_2(x,T) = \frac{1}{\sqrt{2T}}\left[Log(x) - \frac{T}{2}\right]$$

$$\Phi(d) = \frac{1}{\sqrt{\pi}}\int_{-\infty}^{d} e^{-\eta^2}\, d\eta$$

In Equation 4, $T = (\sigma^2 - 2\rho\sigma\delta + \delta^2)\tau$ is the cumulative uncertainty over the time horizon "τ". When $\sigma > \delta$, $T \approx \sigma^2\tau$. When the variability is zero, Equation 2 becomes:

$H(P, B, 0) = Max[0, B-P]$.

Equation 2 provides an expression for the equivalent of an option $H(P, B, T)$. $H(P, B, T)$ is the extra value of using high technology in risk neutral condition.

Since our real option model can assess the transition value from old technology to new technology, we attempt to test the value of technology transition, from generation to generation (inter-generation transition), and within the same generations (intra-generation transition). For example, the parameters for our model consist of ***H*** (option value), ***B*** (the value new technology), ***P*** (the value of old technology) and ***T*** (accumulated uncertainty). Market shares in wireless market are used as the value of ***B*** and ***P*** because we consider that the value of technology is proportional to market power (i.e. market share). Modern, complex technologies often display increasing returns to adoption in that the more it is adopted, the more experience is gained with it and the more it is improved (which is directly connected to the value of technology).

Decision Criteria Rule

As seen in Equation 5, the value of the option (H) should at least offset the premium (P-B). The premium is an opportunity cost to give up some benefits from the existing technology.

$$H\left(B,P,T\right) \geq P - B \Leftrightarrow \frac{H\left(B,P,T\right)}{B} \geq \frac{P}{B} - 1 \qquad (5)$$

SIMULATION AND RESULTS ANALYSIS

Research Design

The scenarios demonstrate the possible transitions of wireless technology and how they might change the value of networks. These scenarios are based on assumptions that had to be made about the future. That the future follows these suggestions is extremely unlikely, but still the scenarios may provide a firm's manager with some new views and foster creativity in thinking about the influence of new technology.

Several assumptions are applied when we construct these scenarios as follows:

- First, it is impossible to move backward technologically. That is, a firm always prefers new technologies over old technologies.
- Second, a firm can only use one technology when it decides to migrate.
- Third, there is no limitation to technological choice. At present, *GSM* is standardized in Europe, but we allow that any technology can be chosen, as is the case in the U.S..

Based on these assumptions, the following alternative technology migration paths are developed based on the wireless technology options diagram (Figure 2), addressed in the previous section. Two scenarios of technology migration paths are identified: *GSM-based Migration Path* and *CDMA-based Migration Path.*

- GSM-based Migration Path
  Analog => TDMA => GSM => WCDMA
- CDMA-based Migration Path
  Analog => CDMA => cdma2000

Data Collection

Since the aim of this study is to understand our real option model can be used as a model for technology choice, we simplify matters where possible. For example, taking into account all the problems of reaching relevant data on technological development, we assume that the only available data are on current market shares of competing

Figure 7. The US market share

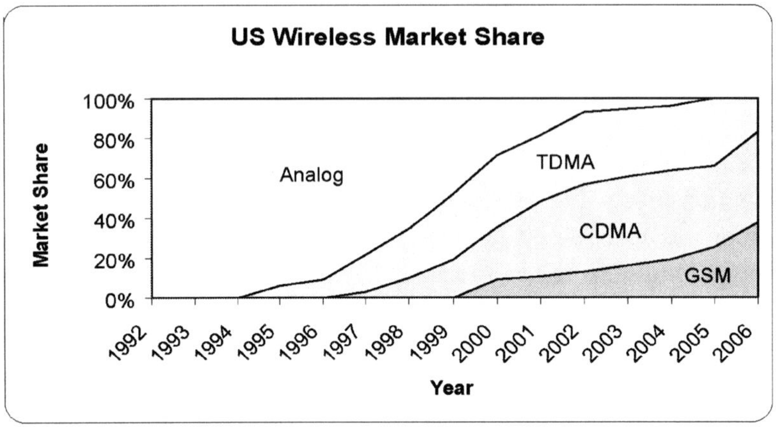

technologies in generation. Because market share data represent actual consumer behavior, the U.S. wireless market share can be used as the proxy of the market value of each technology (Figure 7). They are thus backward looking rather than forward looking, which would be based on *expected* market share.

These historical market share statistics are used to estimate the market value of each technology for two reasons. First, a network's market value depends on their use, so the more the subscribers, the greater the market value, and vice versa. Another reason is that actual market data, like the number of subscribers, is only available on an historical basis. So, this historical data is used for projection purposes.

Although the historical market data do not have strong predictive power, they can illustrate how real options can be applied, in line with the objectives of the article.The future data (until 2010) are constructed using Shuster's estimation model (1998).

RESULTS AND ANALYSIS

GSM-Based Migration Path

Figure 8 is the summary of the option value of the GSM-based Migration path. The shaded box (yellow) shows when the option value is positive, which indicates that it is desirable to migrate during those periods. The detailed analysis for each stage is explained in the next section.

1. **Analog to TDMA** (Figure 8, Column 2). The first scenario is to move from Analog to the *TDMA* network architecture. The premium value begins as positive and gradually decreases, becoming negative after 2000. While the option value is negative in the initial stage, it gradually increases and becomes positive in 2000. The net option value is negative for a long time, but becomes positive after 2000. Analog technology in the U.S. has been popular for a long time, partly because it served as the basic technology in an environment with incompatible 2G standards. The only difference between the two markets relates to timing. Compared to the rest of the world, analog technology in the U.S. has maintained a dominant position for about two years longer, so the transition period to *TDMA* will be longer.
2. **TDMA to GSM** (Figure 8, Column 3). The next scenario displays the value curve when moving from *TDMA* to *GSM* network technology. This analysis shows that the transition is undesirable because the premium value is positive continuously and the

Figure 8. Option value of the GSM-based migration path

Year	Analog=>TDMA	TDMA=>GSM	GSM=>WCDMA	Value
1998	-0.5215			-0.5215
1999	-0.2865			-0.2865
2000	0.0402			0.0402
2001	0.2212	-0.2469		0.2212
2002	0.3611	-0.2663		0.3611
2003	0.4741	-0.2113		0.4741
2004	0.4640	-0.2074	-0.1482	0.4640
2005		-0.2106	-0.1478	-0.1478
2006		-0.1897	-0.1108	-0.1108
2007		-0.1722	-0.0794	-0.0794
2008		-0.1543	-0.0192	-0.0192
2009		-0.1309	0.0782	0.0782
2010		-0.1069	0.1928	0.1928
Total ROA	**0.7526**	**-1.8963**	**-0.2343**	**0.6664**
Net ROA	**1.5606**	**0.0000**	**0.2710**	**1.8316**

option value is always negative. Since the net option value fluctuates in the negative over time, the transition should be delayed or permanently deferred. Since *TDMA* and *GSM* is similar technology there is no need need to invest in this transition. However, in reality, operators prefers to switch from *TDMA* to *GSM* as a stepping stone evolution, as AT&T Wireless did.

3. **GSM to WCDMA** (Figure 8, Column 4). For the transition from *GSM* to *WCDMA* (*3G*) network technology, the premium value decreases continuously, and finally is negative after 2008. The option value is steadily negative, but positive after 2009. NOV is initially negative, but increases rapidly and becomes positive after 2009. So, the transition is desirable starting in 2009 or later.

CDMA-Based Migration Path

Figure 9 shows the summary of option value of the GSM-based Migration path. The shaded box (yellow) shows that the option value is positive, which is indicates that it is desirable to migrate during those periods. Detail analysis for each stage is explained next section.

1. **Analog to CDMA** (Figure 9, Column 2). Moving from Analog to *CDMA* network technology is at odds with the world market, which largely migrated to GSM. Unlike the world market, the transition is desirable starting in 2000 or later. *CDMA* is rapidly growing in the U.S. market, so the transition should take place as soon as possible. However, *CDMA* in the world market is not strong compared to *GSM* which is why the results differ.
2. **CDMA to cdma2000** (Figure 9, Column 3). The next scenario displays the value curve when moving from *CDMA* to *cdma2000* network technology. These results show that the transition is undesirable because the premium value is continuously positive until 2010 (saturation point) and the option value is always negative. Since the net option value ebecomes more negative over time, the transition should be delayed or permanently deferred.

STRATEGIC DECISIONS FOR TECHNOLOGY EVOLUTION

The major U.S. wireless operators already have announced their plans for the evolution of their

Figure 9. Option value of the CDMA-based migration path

Year	Analog=>CDMA	CDMA=>cdma2000	Value
1998	-0.5989		-0.5989
1999	-0.3442		-0.3442
2000	-0.0410		-0.0410
2001	0.2280		0.2280
2002	0.5418		0.5418
2003	0.6531		0.6531
2004	0.6978	-0.4645	0.6978
2005		-0.4587	-0.4587
2006		-0.4196	-0.4196
2007		-0.3693	-0.3693
2008		-0.3010	-0.3010
2009		-0.1909	-0.1909
2010		-0.0564	-0.0564
Total ROA	1.1366	-2.2606	-0.6595
Net ROA	2.1207	0	2.1207

networks, but some uncertainties remain, such as emergence of new technologies (WiMAX, WLAN and LTE) and consolidation among operators (AT&T Mobile and Sprint Nextel). At the time of the 2G transition decsion, the TDMA-based operators were not clear about which technology path should be taken.

Now, we discuss retrospectively what decisions were made, what the outcomes were, and how the options model was validated. Figure 10 summarized the assessment results of all technology transition scenarios in the U.S. wireless industry.

Stage I: Analog to 2G

Moving from analog to any *2G* technology is desirable (all positive); however, the best choice for analog carriers is to move to *CDMA* in 2004 because it showed the highest option value (0.6978) of the three possibilities (TDMA, GSM, and CDMA). From 2000 to 2004, TDMA was a leader in market, but CDMA had a rapidly growing market share with high capacity and call quality, although it was not a mature technology (high uncertainty). As a result, the U.S. wireless market was dominated by CDMA-based operators (i.e., Verizon Wireless and Sprint, together accounting for more than 50% in the U.S. wireless market) in 2004. So, our result using the model showed the right direction in the technology evolution path.

Stage II: 2G to 2G (Intra-Generation Migration)

Three popular 2G technologies (i.e., TDMA, GSM, and CDMA) have competed in the U.S. wireless market. First, let's look at a scenario moving TDMA to GSM. The result shows that it is not desirable because all the transition values during experimental periods (from 2001 to 2010) are negative. This result conflicts with the current U.S. wireless industry. Many TDMA-based operators including AT&T Mobile changed into GSM network. Next, moving from TDMA to *CDMA* is desirable in our model because of the positive net option value of 0.1972 in 2003. However, there was no company to transit from TDMA into CDMA in the U.S. wireless market. The last scenario is to move from GSM to CDMA. In the case of *GSM*-based operators, moving to *2G CDMA* is recommended because of the positive transition option value (0.4654) in 2003.

Why are the results by option model and the operators' decisions not consistent in the intra-generation scenarios? Does it fail to validate our option model? If not, how can we explain this

Figure 10. Summary of technology evolution assessment

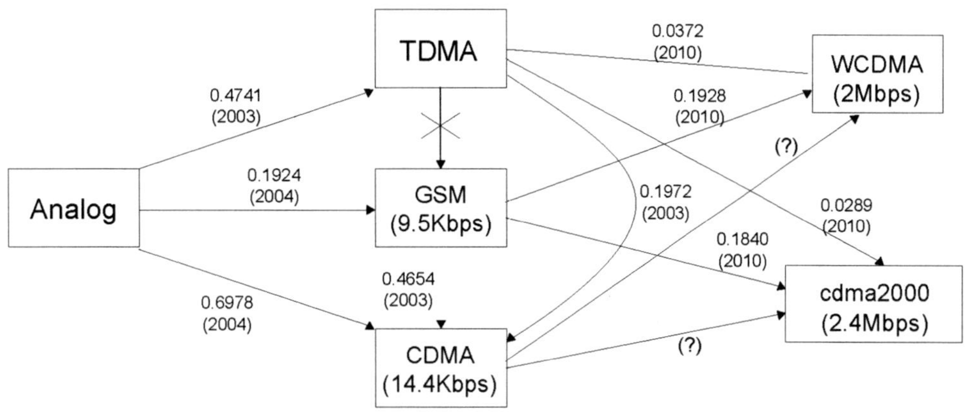

gap? Although our model was focussed only on revenues, not costs; intuitively the transition costs between incompatible technologies (i.e., TDMA and CDMA, GSM and CDMA) may exceed potential revenues (i.e., option value). Another reason is that only market (macro) data is available for technology assessment in our study. If we had access to individual firm data, we may draw more different results.

Stage III: 2G to 3G (Inter-Generation Migration)

The transition from *GSM* to *3G* has a similar positive option value for *WCDMA* (0.1928) and *cdma2000* (0.1840) in 2010. In reality, WCDMA deployment is already the de facto evolution. Many GSM-based mobile operators have chosen WCDMA as the preferred evolution path because many *GSM-based* operators are from Europe, and the only consider *WCDMA* for technical and political reasons. Since a new radio access network (UTRAN) needs to be deployed, WCDMA evolution requires substantial new investment for GSM-based operators. So, GSM-based operator may consider step-by-step evolution (i.e., GSM-GPRS) towards 3G, by minimizing initial infrastructure investment.

If GSM-based operators choose *cdma2000* (however unlikely), there are two possible strategies: one is to maintain a parallel cdma2000 network and the other is the integration of the cdma2000 network within the GSM network. The former is more costly but supports more favorable market requirement and the latter is less costly with a single overlaid network, but higher coverage (i.e., lower population density areas).

CDMA operators should not consider *3G* until 2010 because of the continuing negative transition values, but the transition will occur some time after arriving at the saturation point of the current *2G CDMA* market. The evolution path from 2G CDMA (*cdmaOne*) to *cdma2000* appears technically and economically straightforward, minimizing initial infrastructure investment. The evolution of CDMA-based (i.e., cdmaOne) network to cdma2000 is already taking place on significant scale. For CDMA-based operators, Cdma2000 evolution would require minimum investment and smooth transition without any technical difficulty. cdma2000 evolution refers to the subsequent evolution steps for current cdmaOne operators; from DO (data only) to DV (data and video). Sprint PCS moved directly from cdma 1X to DV.

The WCDMA evolution of CDMA-based operators would be a completely new network being overlaid on the existing CDMA installed base, so it seems unlikely. However, it may happen for regulatory or political factors. For example, partly owned (40%) by the Vodafone Group in UK, Verizon wireless has been under pressure to integrate Verizon's network (CDMA-based network) into Vodafone's network (GSM-based network), with centralized service development and common technology platforms. This suggests that Vodafone may aim to eventually force Verizon onto WCDMA evolution.

Concerning the transition from *TDMA* to 3G technologies, there is not much difference in the transition option value between *WCDMA* (0.0372) and *cdma2000* (0.0289) in 2010. However, the TDMA evolution is controversial. Actually TDMA-based operators cannot evolve directly to 3G, but must choose between a CDMA-based path and a GSM-based path.

CONCLUDING REMARKS

We proposed a real option based model for technology decisions and applied it to the U.S. wireless industry as a case study. We also discussed what decisions were made, what the outcomes were, and how the options model was validated. The preliminary results show that the evolution of wireless network technologies between generations (*inter-generations migration scenario*) is desirable (a positive net option value), but not desirable (a negative net option value) within generations (*intra-generation migration scenario*).

Managerial Implications and Contribution

The findings of our study imply that strategic technology choice is extremely important determinant of firm's competitiveness. Exploring the dimensions of strategic decisions proved to be valuable, as the study found that it is important for a firm to have strategic flexibility is extremely high for improving a firm's value. The study also found that strategic technology choice is important regardless of the level of environmental uncertainty faced by the firm. Since the next generation wireless network technologies and architectures are still a subject of debate with no substantial implementation results, there is much work to do. With the further research, the scope of study can be expanded.

This article intends to raise core issues concerning the transition to *3G*. This study also helps wireless operators make strategic decisions when upgrading or migrating towards the next generation network architecture, by showing which network migration path leads to the most optimum results. That is, wireless operators may find it worthwhile to evaluate new technologies using the ROA as strategic technology options. Through this study, network designers can begin to think in terms of the available network design options and to maximize overall gain in network design. Since the areas of the next generation wireless network architecture and technologies remain the subject of debate with no substantial implementation taking place, there is much work to do. With further research, this study can be expanded and further developed.

Limitation and Future Research

Since one of the implicit aims of this study is to understand how the real options approach can be used as a model for technology choice, we simplify matters where possible. For example, taking into account all the problems of reaching relevant data on technological development, we assume that the only available data are on current market shares (i.e., macro data) of competing technologies in generation. We hope to use more refined and enriched data in future research.

The possibilities for future research on topics related to strategic technology management using the real options approach are extensive. Of them, a few of the possible extensions of the ideas covered in this article:

- First, the U.S. market with a suite of different technologies in use offers an interesting laboratory to test the real options approach as a strategic decision tool. Based on this preliminary practice of our real options model, we would like to develop a theory for a firm's behavior analysis to solve strategic issues in the company level: for example, why do not all of the firms in wireless network industry to migrate or upgrade for the *3G* services at the same time? Or why did some companies choose *WCDMA* instead of *cdma2000* (i.e. AT&T wireless and Cingular), or else?
- Second, real option research is still very much a growing area. Thus there is much more that needs to be done. Although the conceptual foundation for real options is well established, there is scope for further research extensions to some of the basic theories, especially relating to valuation techniques. Options involving real technology choices and strategies are generally much more complex than simple financial options in stock market. First, the uncertainty may be due to several variables instead of simply one variable such as the price in financial options. Further, it may not always be easy to measure the value of underlying assets because of its dynamics and never traded in the market. These complexities may not allow one to find exact valuation model.
- Third, the other future research to come from this study will be the application of our real option theory and techniques to a variety of other industry to solve technology management problems, such as high-tech industry and medical industry. Conceptually any technology choice decision where significant uncertainties are present can be considered our strategic technology transition model using the real options approach.

REFERENCES

Akhtar, N., Callejo, M. A., & Colas, J. A. (2005). Making migration easy: a key requirement for systems beyond 3G. Vehicular Technology Conference, 5, 2974-2978.

Alleman, J., & Noam, E. (1999). *Real Options: The New Investment Theory and its Implications for Telecommunications Economics*. Kluwer Academic Publishers.

Alleman, J., & Rapport, P. (2006). Optimal Pricing with Sunk Cost and Uncertainty. In Cooper, R., Lloyd, A., Madden, G., & Schipp, M. (Eds.), *The Economics of Online Markets and Information Communications and Technology Networks: Contribution to Economics Series*. Heidelberg: Physica-Verlag.

Amram, M., & Kulatilaka, N. (1998). *Real Options: Managing Strategic Investment in an Uncertain World*. Harvard Business School Press.

Balachandra, R., Goldschmitt, M., & Friar, J. H. (2004). The evolution of technology generations and associated markets: a double helix model. *IEEE Transactions on Engineering Management*, *51*(1), 3–12. doi:10.1109/TEM.2003.822463

Basili, M., & Fontini, F. (2003). The Option Value of the UK 3G Telecom licenses. The journal of Policy. *Regulation and Strategy for Telecommunications*, *5*(3), 48–52. doi:10.1108/14636690310487246

Benaroch, M. (2002). Managing Investments in Information Technology Based on Real Options Theory. *Journal of MIS*, *19*(2), 43–84.

Benaroch, M., & Kauffman, R. J. (1998). A Case for using Real Options Pricing Analysis to Evaluate Information Technology Project Investments. *Information Systems Research*, *10*(1), 70–86. doi:10.1287/isre.10.1.70

Benaroch, M., & Kauffman, R. J. (2000). Justifying Electronic Banking Network Expansion Using Real Options Analysis. *Management Information Systems Quarterly*, *24*(2), 197–225. doi:10.2307/3250936

Black, F., & Sholes, M. (1973). The Pricing of Options and Corporate Liabilities. *The Journal of Political Economy*, *81*, 637–654. doi:10.1086/260062

Brennan, M. J., & Schwartz, E. S. (1985). Evaluating Natural Resources Investments. *The Journal of Business*, *58*, 135–157. doi:10.1086/296288

Budde (2007). U.S.A-Mobile Market – Analysis, Statistics and Forecasts. Reports. (http://www.budde.com).

Carr, P. (1988). The Valuation of sequential exchange opportunities. *The Journal of Finance*, *23*(5), 1235–1256. doi:10.2307/2328217

Carsello, R. D., Meidan, R., Allpress, S., O'Brien, F., Tarallo, J. A., & Ziesse, N. (1997). IMT-2000 standards: radio aspects. *IEEE Personal Communications*, *4*(4), 30–40. doi:10.1109/98.612275

CDG. (2007). Quick Market Statistics. CDMA Development Group. (http://www.cdg.org).

Chen, J., Reilly, R. R., & Lynn, G. S. (2005). The impacts of speed-to-market on new product success: the moderating effects of uncertainty. *IEEE Transactions on Engineering Management*, *52*(2), 199–212. doi:10.1109/TEM.2005.844926

Clemons, E. K. (1991). Evaluating Strategic Investments in Information Systems. *Communications of the ACM*, *34*(1), 22–36. doi:10.1145/99977.99985

Clemons, E. K., and GU, B. (2003). Justifying Contingent IT Investments: Balancing the Need for Speed of Action with Certainty before Action. *Journal of Management Information Systems*, *20*(2), 36–47.

CTIA. (2006). CTIA's Semi-Annual Wireless Industry Survey.

d'Halluin, Y., Forsyth, A.P. & Vetzal, R.K. (2002). Managing Capacity for Telecommunications Network Under Uncertainty. IEEE/ACM Transactions onNetworking, 10(4), 579-588.

Dahlman, E., Gudmundson, B., Nilsson, M., & Skold, J. (1998). UMTS/IMT-2000 Based on Wideband CDMA. IEEE Communications Magazine, 70-80. Deng, S., Johnson, B. & Sogomonian, A. Exotic Electricity Options and the Valuation of Electricity Generation and Transmission Assets. *Decision Support Systems*, *3*(30), 383–392.

Dias, M. A. G., & Rocha, K. M. C. (1999). Petroleum Concessions with Extendible Options using Mean Reversion with Jumps to Model oil Prices. The 3rd International Conference on Real Options, the Netherlands.

Dixit, A. K., & Pindyck, R. S. (1994). *Investment under Uncertainty*. Princeton University Press.

Dixit, A.K. & Pindyck, R.S. (1995). The Options Approach to Capital Investment. Harvard Business Review.

Dos Santos, B. (1991). Justifying Investment in New Information Technologies. *Journal of Management Information Systems*, *7*(4), 71–89.

Economides, N. (1999). Real Options and the Cost of the Local Telecommunications Network. In Alleman, J., & Noam, E. (Eds.), *The New Investment Theory of real options and its Implications for the Cost Models in Telecommunications*. Kluwer Academic Publishers.

FCC. (2006). FCC Annual Report on State of Competition in the Wireless Industry.

Fichman, R. G. (2004). Real Options and IT Platform Adoption: Implications for Theory and Practice. *Information Systems Research*, *15*(2), 132–154. doi:10.1287/isre.1040.0021

Flatto, J. (1996) Using Real Options in Project Evaluation. working paper, Life Office Management Association (LOMA).

Garg, V. (2001). *Wireless Network Evolution: 2G to 3G*. Prentice Hall Communications Engineering and Emerging Technology Series.

GSMA. (2007). GSM World: Subscriber Statistics. GSM Association. Jeffery M. & Leliveld, I. Best Practices in IT Portfolio Management. *MIT Sloan Management Review*, *45*(3), 41–49.

Kemma, A. G. Z. (1993). Case Studies on Real Options. *Financial Management*, *19*(3), 259–270.

Kim, H. J. (2004). The Real Options to Technology Management: Strategic Options for 3G Wireless Network Architecture and Technologies. Ph.D. Thesis, University of Pittsburgh.

Kim, H. J. (2006). The Economic Value of Modularity in the IPQoS Network: the Real Options Approach (ROA). [IJBDCN]. *International Journal of Business Data Communications and Networking*, *2*(1), 15–30. doi:10.4018/jbdcn.2006010102

Kim, Y. J., & Sanders, L. G. (2002). Strategic Actions in Information Technology Investment Based on Real Option Theory. *Decision Support Systems*, *33*, 1–11.

Kumar, R. L. (2002). Managing risks in IT projects: an options perspective. *Information & Management*, *40*, 63–74. doi:10.1016/S0378-7206(01)00133-1

Langlois, R. N., & Robertson, P. L. (1992). Networks and innovation in a modular system: Lessons from the microcomputer and stereo component industries. *Research Policy*, *21*(4), 297–313. doi:10.1016/0048-7333(92)90030-8

Lawless, M. W., & Anderson, P. C. (1996). Generational technological change: Effects of innovation and local rivalry on performance. *Academy of Management Journal*, *39*(5), 1185–1217. doi:10.2307/256996

Margrabe, W. (1978). The Value of an Option to Exchange one Asset for Another. *The Journal of Finance*, *33*, 177–186. doi:10.2307/2326358

McDonald, R., & Siegel, D. (1986). The value of waiting to invest. *The Quarterly Journal of Economics*, *101*, 707–727. doi:10.2307/1884175

McDysan, D., Lundberg, T., Bjorkman, N., & Latour-Henner, A. (2000). Multiservice Networking using a Component-based Switch & Router Architecture. Proceedings of the IEEE Conference on ATM 2000.

Micalizzi, A. (1999). Plough Case: Valuing Pharmaceutical Development and Expansion Options. The 3rd Annual Real Options Conference, Netherlands.

Paddock, J. L., Siegel, D. R., & Smith, J. L. (1988). Option Valuation of Claims on Real Assets: The Case of Offshore Petroleum Leases. *The Quarterly Journal of Economics*, 479–508. doi:10.2307/1885541

Panayi, S., & Trigeorgis, L. (1998). Multi-stage real options: The cases of information technology infrastructure and international bank expansion. *The Quarterly Review of Economics and Finance*, *38*(3), 675–692. doi:10.1016/S1062-9769(99)80096-9

Pateli, A. G., & Giaglis, G. M. (2007). The Impact of Value on Governance Decisions for IT-Based Alliances: Evidence from a Joint Venture in the Wireless Networks Industry. Proceedings on the 40th Annual Hawaii International Conference on System Sciences, HICSS 2007.

Pickles, E., & Smith, J. (1993). Petroleum Property Valuation: A Binomial Lattice Implementation of Option Pricing Theory. *The Energy Journal (Cambridge, Mass.)*, *14*(2), 1–26.

Pindyck, R. S. (1989). Irreversible Investment, Capacity Choice, and the Value of the Firm. *The American Economic Review*, *2*, 969–985.

Podhradsky, P. (2004). Migration scenarios and convergence processes towards NGN (present state and future trends): Electronics in Marine. Proceedings Elmar2004, 46th International Symposium, 39–46.

Polivka, J. (2007). Sprint Nextel Cites WiMAX Network Progress For 2007. News Releases on Sprint Nextel, January 8.

Prasad, R. (1998). An Overview of CDMA Evolution toward Wideband CDMA. *IEEE Communication Surveys*, *1*(1), 1–29.

Rapport, T. (2002). Wireless Communications Principles and Practice, Prentice-Hall, Inc., 2002.

Sanchez, R. (1995). Strategic Flexibility in Product Competition. Strategic Management Journal, 16 (summer special edition), 135-159.

Sanchez, R. (1999). Modular Architectures in the Marketing Process. *Journal of Marketing*, 92–111. doi:10.2307/1252104

Schilling, M. (2000). Toward a general modular systems theory and its application. *Academy of Management Review*, *25*(2), 312–334. doi:10.2307/259016

Schwartz, E. S., & Zozaya-Gorostiza, C. (2003). Investment Under Uncertainty in Information Technology: Acquisition and Development Projects. *Management Science*, *49*(1), 57–70. doi:10.1287/mnsc.49.1.57.12753

Shuster, M. S. (1998). Diffusion of Network Innovation: Implications for Adoption of Internet Services, MIT Internet Telephony Consortium Semiannual Meeting, Helsinki, Finland, 20-22.

Smith, K. W., & Triantis, A. J. (1995). The Value of Options in Strategic Acquisitions (book chapter). Real Options in Capital Investment, Trigeorgis (ed.), Greenwood Publishing Group, 135-149.

Stonier, J. (1999). Airbus: Valuing Options in the Airline Industry. The 3rd Annual International Conference on Real Options.

Sullivan, K., Chalasani, P., Jha, S., & Sazawal, V. (1999). Software design as an investment activity: A real options perspective. In Trigeorgis, L. (Ed.), *Real options and business strategy: Applications to decision making*. London, UK: Risk Books.

Taudes, A. (1998). Software growth options. *Journal of Management Information Systems*, *15*(1), 165–185.

Taudes, A., Feurstein, M., & Mild, A. (2000). Options Analysis of Software Platform Decisions: A Case Study. *Management Information Systems Quarterly*, *24*(2), 227–243. doi:10.2307/3250937

TDG. (2007). *GSM continues to have Greatest Gain in U.* S. Cellular Market. TDG Press Releases.

Trigeorgis, L. (1993). The Nature of Option Interactions and the Valuation of Investments with Multiple Real Options. *Journal of Financial and Quantitative Analysis*, *28*(1), 1–20. doi:10.2307/2331148

Trigeorgis, L. (1996). *Real Options: Managerial Flexibility and Strategy in Resource Allocation.* Cambridge, Mass.: The MIT Press.

This work was previously published in International Journal of Interdisciplinary Telecommunications and Networking, Volume 1, Issue 1,edited by Michael R. Bartolacci, and Steven R. Powell, pp. 45-63, copyright 2009 by IGI Publishing (an imprint of IGI Global).

Chapter 5
Time-Based Confidentiality Enhancement Scheme for Mobile Wireless Networks

Qunwei Zheng
Cerqa, USA

Xiaoyan Hong
Rice University, USA

Jun Liu
University of Alabama, USA

Lei Tang
University of Alabama, USA

ABSTRACT

A multi-hop wireless network with highly dynamic members and mobility is vulnerable to many attacks. To address this problem, we propose a novel time-based approach that exploits mobility. In our scheme, the source sends shares at different times. Due to node mobility, these shares will be routed through different intermediate nodes. It is highly unlikely that a particular intermediate node is able to be on many of these routes and to collect enough shares to reconstruct the original message. The scheme is particularly suitable for applications that can tolerate long message delays, as studied in Delay Tolerant Networks. The article focuses on analyzing the feasibility of this scheme. We describe a general approach to calculate the probability of intercepting enough shares by arbitrary nodes, together with simulations. The results show that the probability is small. The scheme provides a valuable alternative for delay tolerant applications to enhance message confidentiality.

DOI: 10.4018/978-1-60960-505-6.ch005

INTRODUCTION

Multi-hop wireless networks, e.g., mobile ad hoc networks, mesh networks, vehicular ad hoc networks, and moveable wireless sensor networks, are peer-to-peer networks where users act not only as hosts but also as routers to forward packets for others. Such networks are self-organizing and highly dynamic due to node mobility, frequent node join and leave, and possible long distances between nodes. In this work, we focus on the challenging mobile ad hoc networks (MANETs). MANETs are vulnerable to many attacks. Possible attackers may want to eavesdrop other nodes' communication, to disrupt communication, or to deplete network resources. In the scenarios where ad hoc networks are deployed in hostile environments, a legitimate node could be captured and turned into malicious. Moreover, the open nature of wireless media allows the attacks to be launched with great ease. Any nodes within the reception range of a transmission can overhear, intercept and alter transmitted messages; or, a malicious node can position itself to be within a network field and emit bogus messages. All these situations require a secure network to protect communications. One important secure aspect is the confidentiality of the messages. In this work, we study the confidentiality issue targeting at defending against the eavesdropping attack that is interested in learning the contents of the messages.

Message confidentiality (or secrecy) can be achieved using encryption or an approach that spreads message shares. Encrypting messages before sending is a common practice. Yet for encryption to work, the keys for encryption and decryption must be available. However, both symmetric key cryptography and public key cryptography face challenges due to the dynamic nature of network members. Several early works have proposed to address the problem (Balfanz, Smetters, Stewart, & Wong, 2002; Zhang & Fang, 2008; Capkun, Hubaux, & Buttyan, 2003; Hubaux, Buttyan, & Capkun, 2001; Luo, Kong, Zerfos, Lu, & Zhang, 2004; Stajano & Anderson, 1999; Zhou, Schneider, & Renesse, 2002). Still, there is no one-fit-all strict security mechanism. In addition, cryptographic approaches usually require additional computation time and bandwidth (for exchanging secure credentials needed in handshake), which could be crucial for nodes that are resource constrained. Also, nodes could be compromised and a compromised node gives away all the keys stored in its memory. Moreover, cryptography cannot defend against adversaries that simply drop messages. After all, with these considerations, sending messages in MANET with needed secrecy remains a challenging issue.

Spreading a message through multiple paths is another approach to achieve secrecy. The basic idea is to split a message into multiple shares and send them to different paths. Usually, the threshold secret share scheme can be used to generate the shares. Several related secure data transmission schemes have been proposed following the idea, e.g., using node-disjoint paths (Lou, Liu, & Fang, 2004; Papadimitratos & Haas, 2003). Multi-path routing in MANET is used to select these paths (Tsirigos, 2001; Wu, 2001; Radunovic, Gkantsidis, Key, & Rodriguez, 2008; Lee, 2001). These schemes fit well for scenarios where the space is large enough for these paths to spread apart. But they are not appropriate when a network is sparse or deployed in a restricted geographic area, where no enough node-disjoint multiple paths can be found.

Our scheme works differently. It explores node mobility by sending shares at different times. Thus it is not limited by geographical features. Notice that nodes in mobile ad hoc networks move all the time. If the time interval is large enough, two shares will be routed through different intermediate nodes. Thus it is very difficult for a node (including the eavesdropping attacker) other than the source and the destination to hear enough shares - unless it physically follows the source and the destination or it has enough collaborators in the network to help collecting the shares. A node having enough shares is able to reconstruct the original message by combining these shares. Otherwise, no information about the original message will be revealed.

Such a time-based multi-path message dissemination scheme has a natural fit for applications and scenarios where delay can be tolerated, similar to applications studied in Delay Tolerant Networks (DTN) (Jain, Fall, & Patra, 2004). It is also suitable for scenarios where multiple paths do not exist or are hard to find, such as a sparse network, or a network deployed in a restricted geographical area. For these scenarios, our scheme can provide message secrecy, yet does not rely on the presence of in-network key distribution and cryptography approaches.

For this scheme to work, the key issue is the guarantee (with probability) that nodes other than the source and the destination will not intercept enough shares. In this article, we introduce the scheme and analyze the feasibility of this scheme. We describe a general approach to calculate the probability, and use the isotropic random walk model to illustrate the calculation. The scheme is further evaluated using simulation. The results show that for the random walk type of mobility, the probability is small. These pieces of work validate the soundness of our scheme and suggest that this time-based message delivery scheme provides a valuable alternative for delay tolerant applications to enhance message confidentiality.

The article is organized as follows. The next section introduces the security and network models and summarizes the related work. The third section describes our scheme with probability analysis given in the fouth section. The fifth section shows a case study with numerical results. The sixth section reports our simulation evaluation. And finally the seventh section concludes the article.

SYSTEM MODEL AND RELATED WORK

System Model

We assume that a mobile ad hoc network runs a routing protocol to find a route for multi-hop data transmissions. We do not specifically require cryptographic operations in the network for message transmissions. We only assume that the two communicating parties are aware of the use of the scheme. When a mobile network has adopted a security mechanism, our scheme can co-exist with it and strengthen the security. For example, if the network uses message authentication code (MAC) to ensure integrity, we can secure the shares with MAC. In addition, our scheme still provides secrecy for the message content.

We assume a reasonable weak threat model, where no global monitor is deployed. It is possible that more than one attackers appear in the network and they could collude to integrate information. But they are sparsely mingled with legitimate nodes. The attackers can eavesdrop wireless transmissions, collect data and perform traffic analysis to obtain information. In our case, they could try to correlate data packets of the same source-destination pairs and perform traffic analysis (for privacy purpose, pseudonyms can be used, but we do not assume they are necessary). The attackers can be passive or appear to be normal network members, e.g, acting correctly according to the routing protocol. However, they lack the ability to follow an active transmitting node continuously due to one (or more) of the following reasons: no knowledge on any specific target identities (or pseudonyms), not able to locate the emitting locations, not able to stay close to an active node for a long time due to own mobility, nor to predict a target node's motion pattern. We assume no available visual information.

Related Work and Our Contribution

Both symmetric key cryptography and public key cryptography face challenges due to the dynamic nature of network members. Some early works have proposed to address the problem. First, establishing symmetric keys between two arbitrary nodes need an infrastructure of Key Distribution Center (KDC). However, in a mobile ad hoc network, the security of KDC cannot be guaranteed. Also, when all nodes are moving, the accessibility of the KDC (to all the nodes at any time) is difficult to guarantee as well. In addition, KDC is a single point of failure. For these reasons, some symmetric key distribution schemes require physical contacts (Balfanz, et al., 2002; Stajano, et al. 1999). Still, there are scenarios where physical contacts are not practical. Another possible solution is to use key pre-distribution as studied in sensor networks (Chan, Perrig, & Song, 2003; Eschenauer & Gligor, 2002). But in an ad hoc network where nodes may join and leave at random - without a possible connection to a trusted authority - these schemes encounter difficulty. Second, in public key cryptography, though public keys can be distributed by the node itself, it usually requires the existence of distributed and online Certificate Authorities (CAs) to verify the public keys. Based on public key cryptography, various schemes have been proposed (Zhang, et al., 2008; Luo, et al., 2004; Zhou, et al., 2002). These schemes assume that a set of nodes are initialized by a trusted party before deployment. A self-organized public key management scheme was proposed in (Capkun, et al., 2003; Hubaux, et al., 2001). In this scheme, nodes issue certificates to each other based on their personal acquaintances. Each node maintains a local certificate repository. Verification of public keys is made possible by finding a certificate chain between two communicating nodes. The scheme is fully self-organized. However it provides only probabilistic guarantees.

A few early schemes have also used the multipath routing to achieve secrecy. A secure data transmission scheme based on Rabin's Information Dispersal Algorithm (IDA) (Rabin, 1989) was proposed in (Papadimitratos, et al., 2003). In this scheme, dispersed message pieces are routed simultaneously through a set of node-disjoint paths. The destination node reconstructs the message after receiving sufficient pieces. The authors discussed how the source can adjust the path set based on feedback from the destination so that the protocol can remain effective in an ever changing network. A similar scheme that is based on Shamir's secret sharing technique (Shamir, 1979) was proposed in (Lou, et al., 2004). Message is divided into multiple shares and sent simultaneously through multiple independent paths. The authors discussed how to split message so that maximum security can be obtained and how to design an effective multi-path routing algorithm.

Our scheme is different from the above schemes in that those schemes utilize spatial redundancy. They can be used in scenarios where multiple paths exist. On the other hand, our scheme utilizes time redundancy. It can be used in scenarios where delay can be tolerated, or where multiple paths do not exist or are hard to find, as in a restricted geographical area. In addition, the scheme of (Lou, et al., 2004) requires link encryption between neighboring nodes, the cost of which is high. Our scheme does not take such an assumption.

TRANSMIT CONFIDENTIAL DATA BY EXPLOITING NODE MOBILITY

We use the threshold secret sharing scheme to divide our messages into multiple shares. A (k,n) threshold secret sharing scheme (Shamir, 1979) divides a secret into n shares ($k<n$). The secret can be reconstructed from k or more shares, but with less than k shares, absolutely no information about the secret is disclosed. In this section, we first give a brief overview of the secret sharing scheme and then we describe our time-based multi-path message dissemination scheme. The

security analysis of our scheme will be given in the next section.

Secret Sharing Scheme

To safeguard a cryptographic key from loss, one way is to make multiple copies. By doing so the reliability increases, but the risk of leaking the key increases too. The larger the number of the copies, the higher the reliability but also the risk. A secret sharing scheme on the other hand manages to enhance the reliability without increasing the risk. Among the various secret sharing schemes, Shamir's (1979) (k, n) threshold secret sharing is the most widely cited algorithm. It starts with a secret S and divides it into n pieces in such a way that knowledge of any k or more pieces allows reconstruction of this secret, but knowledge of $k\text{-}1$ or less leaves the secret completely undetermined.

To divide the secret into n shares, we choose $k\text{-}1$ random coefficients $a_1, a_2, \ldots, a_{k-1}$ and a specific a_0 that equals to S, and define a $k\text{-}1$ degree polynomial

$$f(x) = \sum_{j=0}^{k-1} a_j x_j$$

We then evaluate $f(i)$ for each i=1,2,...,n. Each pair $(i,f(i))$ is a secret share and we have n of them. To reconstruct the secret, with the knowledge of any k pairs, we are able to use Lagrange interpolation (Knuth, 1969) to find all the coefficients of the polynomial and the secret $S=a_0$.

Lagrange interpolation works as follows. Given k points (x_i, y_i), $i = 1, 2, \ldots, k$ of a polynomial of degree $k\text{-}1$, the polynomial can be reconstructed as:

$$f(x) = \sum_{i=1}^{k} \prod_{1 \le j \le k, j \ne i} \frac{x - x_j}{x_i - x_j}$$

Since $f(0) = a_0 = S$, the secret can be expressed as:

$$S = \sum_{i=1}^{k} c_i y_i$$

where

$$c_i = \prod_{1 \le j \le k, j \ne i} \frac{x_j}{x_j - x_i}$$

Some nice properties of this scheme include: it is secure; the size of each share does not exceed the size of the secret; new shares can be calculated and distributed to new users without affecting existing shares; providing a single user with more shares gives him more control over the secret.

Transmit Data Over Time

At the time that the source wants to communicate with the destination securely, the source will split the message into multiple shares using the aforementioned threshold secret sharing scheme. It buffers them and sends them at different times. Figure 1 depicts the idea. At certain locations on the source's trajectory $(s_1, s_2 s_3, s_4)$, it sends the shares. Correspondingly the destination receives the shares at the locations (d_1, d_2, d_3, d_4). The success of the scheme relies on the fact that intermediate nodes are different. Suppose the source sends a share to the destination at position s_i first. The next time that it sends a share to the destination is at position s_{i+1}. Due to mobility, the nodes that relay the current share are different (with high probability) from the nodes that relay the next share. According to our adversary model, attacking nodes are rather weak in terms of forming a highly informative global monitoring network through collusion. Thus, it is safe to say that the only nodes that can hear the shares are those that happen to be on the routing paths where the shares are relayed, or close to nodes on the paths. The destination can recover the message by the (k,n) threshold secret sharing scheme. For

Figure 1. Time-based data transmission scheme. The source sends four shares at four different places s_1, s_2, s_3, and s_4. The destination receives these shares at d_1, d_2, d_3, and d_4 respectively. Due to mobility, shares are relayed through different nodes.

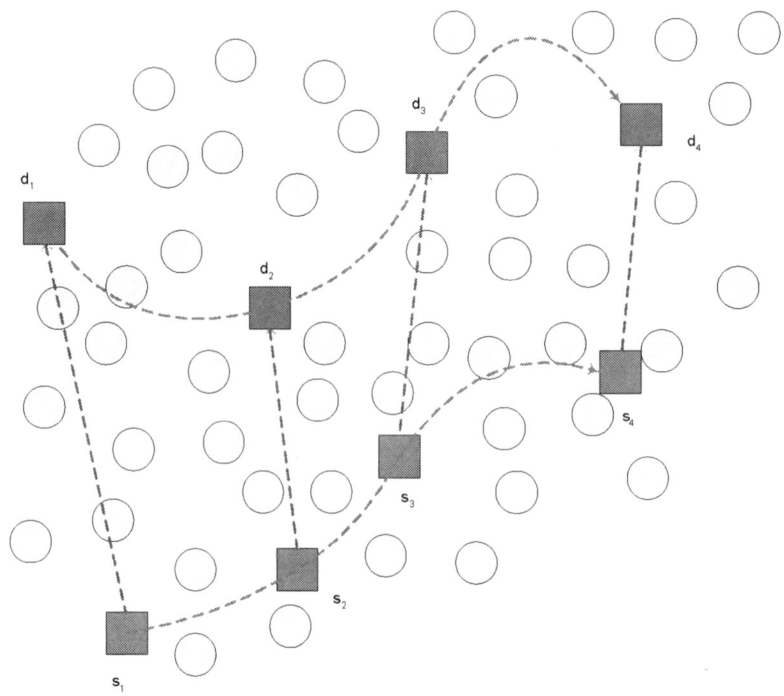

one of any other nodes to obtain enough shares, it must *happen* to be on enough (i.e, k) relaying paths. Since an attacking node can not deliberately follow a target node, its chance of hearing enough shares is very limited. When an application allows long message delay, the time dispersion that the scheme explores brings advantages that are evident through the following analysis.

PROBABILITY ANALYSIS

In this section, we analyze the probability of an arbitrary node other than the source and the destination recovering the original message. We use the isotropic random walk model as the mobility model. Isotropic means that a node's individual steps are isotropically distributed (Hughes, 1995). Without loss of generality, we assume that a node needs to collect all shares (which is M) in order to recover the original message.

Isotropic Random Walk Model

A node's random walk (Hughes, 1995) consists of many steps. A *step* is a movement from one point to another along a straight line. The node re-selects the direction for its next step. The length of each step varies (say, following a distribution), but the time for taking each step is the same - we assume it is 1 time unit. We also assume that there is no pause between two consecutive steps.

The isotropic random walk is a random walk in which the lengths of the individual steps are isotropically distributed. The probability density function of the displacement of each step is

$$g(\vec{r}) = \frac{1}{2\pi\left|\vec{r}\right|}\omega\left(\left|\vec{r}\right|\right)$$

where $\vec{r}$ is a two dimensional vector representing displacement and $\omega(\psi)$ is the probability density function of the length ψ of each step. There are very few cases that the probability density function for the position of a node after n steps can be evaluated explicitly. In our analysis, we will use one of those cases. In this case, the probability density function for the length ψ of each step can be expressed below (Hughes, 1995).

$$\omega(\psi) = \frac{2\psi}{a^2} e^{-\psi^2/a^2} \tag{1}$$

where a is the sole parameter that determines the shape of the distribution. So we have

$$g(\vec{r}) = \frac{1}{\pi a^2} e^{-|\vec{r}|^2/a^2}$$

This result will be used in the following analysis.

Probability Analysis

Suppose that the source sends one share to the destination every n steps. In total, it sends M shares. Notice that because both the source and the destination are moving, they send and receive shares at different locations and most likely through different paths. For a particular path, the vulnerable area that an attacking node can eavesdrop the message consists of many overlapping transmission range based circles along the irregular path. Since a shortest path tends to be geographically close to the straight line linking the source and the destination, we simplify the calculation of the vulnerable area. We approximate the vulnerable area to be a rectangle that surrounds this straight line. The rectangle has a width of $2R$ (R is a node's transmission range) and a length of the distance between the source and the destination. By this approximation, any node outside of the rectangle will not hear the share. To be able to recover the original message, a node must *happen* to be in all M rectangles at the times when the source sends the shares. There are two notes on the size of the rectangle. First, the width $2R$ is more accurate for a dense network because the path can be close to the geographical straight line linking the source and destination; for a sparse network, the width could be larger since the path may diverge away from the line. Second, the rectangle could be extended by including two half-circles of radius R to include the case that the attacker being at the outer sides of the source or the destination but within the transmission range. These considerations will make the calculation more accurate. However they do not change the essence of the analysis and the approximation. So in our analysis, we keep the size of the rectangle as $2R \times$ (the distance between the source and the destination).

We start the analysis using a simple case of $M = 3$ (see an illustration in Figure 2). According to the isotropic random walk model, the source sends each of the three shares every n steps, at locations S_1, S_2, and S_3. The destination receives them at the corresponding locations of D_1, D_2, and D_3. The rectangles are A, B and C. Suppose that an arbitrary node N starts from a location $H(\alpha, \beta)$. It is possible that the node moves into these three areas. For this node's possible locations within the three areas, we denote A_i as an infinitesimal area centered on (x, y) within the area A; B_j and C_k are for areas B and C with similar meanings.

Let $f_{(x_0,y_0),n}(x,y)$ be the probability density function of the position of a node after n steps have been made, starting from (x_0, y_0). Then,

$$\Pr\{\textit{node N is in } A_i \textit{ after n steps}\} = f_{H,n}(x, y)dxdy$$

$$\Pr\{\textit{node N)} \int_A f_{H,n}(x,y)dxdy$$

$$\Pr\{\textit{node N is in } A_i \textit{ after n steps AND is in } B_j \textit{ after another n steps}\} = f_{(x,y),n}(s,t)dsdt \times f_{H,n}(x, y)dxdy$$

Figure 2. Illustration of probability analysis. There are 3 rectangular areas for the paths relating to the positions of the source and the destination at different times. An arbitrary node starts from H

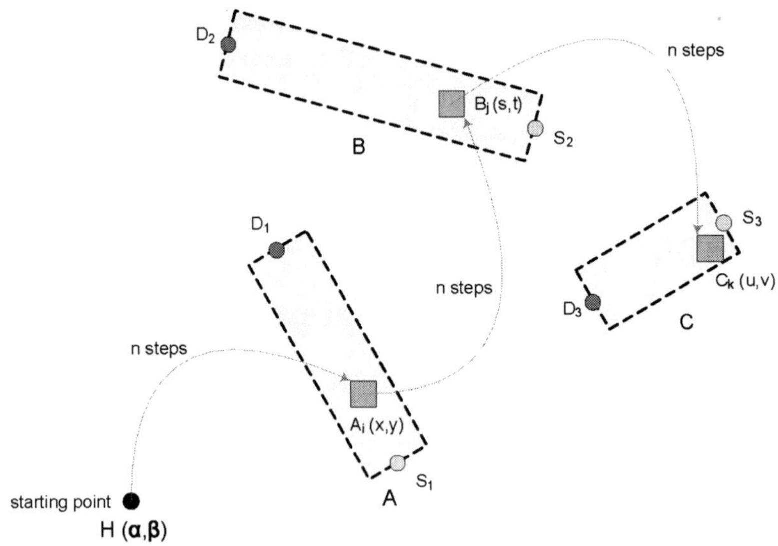

So,

$\Pr\{$*node N is in A after n steps AND is in B after another n steps AND is in C after another n steps*$\} =$

$$\int_C \int_B \int_A f_{(s,t),n}(u,v)dudv \times f_{(x,y)n}(s,t)dsdt \times f_{H,n}(x,y)dxdy$$

A similar result can be acquired if there are M shares (and thus M rectangle areas), by having M iterated integrals.

According to the isotropic random walk model introduced previously, we have

$$f_{(0,0)n}(x,y) = \frac{1}{\pi n a^2} e^{-(x^2+y^2)/na^2} \tag{2}$$

This result can be extended to the case when the node starts from an arbitrary point (x_0, y_0)

$$f_{(x_0,y_0)n}(x,y) = \frac{1}{\pi n a^2} e^{-((x-x_0)^2+(y-y_0)^2)/na^2} \tag{3}$$

So we have

$$\Pr\{node\ N\ is\ in\ A\ after\ n\ steps\ AND\ is\ in\ B\ after\ another\ n\ steps\ AND\ is\ in\ C\ after\ another\ n\ steps\}$$

$$= \int_C \int_B \int_A \frac{1}{\pi n a^2} e^{-((u-s)^2+(v-t)^2/na^2} dudv \; \frac{1}{\pi n a^2} e^{-((s-x)^2+(t-y)^2/na^2} dsdt \; \frac{1}{\pi n a^2} e^{-((x-\alpha)^2+(y-\beta)^2/na^2} dxdy \tag{4}$$

In fact, the above analysis represents a general approach. It can be justified as follows. First, the areas of A, B and C can be any shapes or any reasonable rectangles. For a potential area we should be able to calculate the probability with suitable integrals. Second, the locations of A, B and C are decided by the moving trajectories of the source and the destination. As long as A, B and C are given, we can calculate the probability. Here we have the random walk mobility model, and A, B and C could appear anywhere (with certain probability) in the field. In addition, when the source sends M shares, a similar result can be acquired

Figure 3. The source and the destination move in parallel. The distance apart is l. Both move a distance of ε every n steps. Rectangles A, B and C are of size 2R×l.

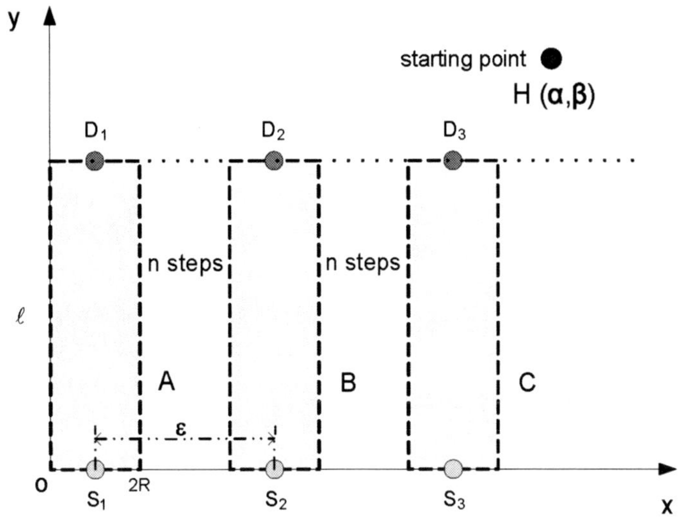

by having M iterated integrals. Finally, it is easy to see that in a general case where there are X adversarial nodes, the probability of at least one adversarial node hearing all the shares is

$$1 - \prod_{i=1}^{X}(1 - p_i)$$

where p_i is the probability of the i th adversarial node entering all those areas at the right time.

A CASE STUDY WITH NUMERICAL RESULTS

In this section, we provide numerical analysis for a case where the source and the destination move along two parallel lines with equal velocity. The rest of the nodes move following the isotropic mobility model. Such a choice does not lose generality because this is still an example of an isotropic random network. We use this case to illustrate how the above analytical approach can be applied. More general situations of the source and destination movements will be simulated in the next section. The network is given in Figure 3. Here, the distance between the source and the destination is l. Thus all the rectangles are of width $2R$ and length l, which are shown in the figure as shaded parts. The source moves along the x axis to the right. The destination moves along the line $y = l$ to the right. We assume that the source and the destination have the same length η for each step. So they move $\varepsilon = n\eta$ every n steps. The step lengths of the rest of the nodes vary according to the distribution.

We denote the function to be integrated in Formula (4) as $h(x, y, s, t, u, v)$, it is

$$h(x, y, s, t, u, v) = \left(\frac{1}{\pi n a^2}\right)^3 \times e^{-\frac{1}{n a^2}[(x-\alpha)^2 + (y-\beta)^2 + (s-x)^2 + (t-y)^2 + (u-s)^2 + (v-t)^2]}$$

Thus, rewriting (4), we have

$$\begin{aligned}&\Pr\{node\ N\ is\ in\ A\ after\ n\ steps\\&AND\ is\ in\ B\ after\ another\ n\ steps\\&AND\ is\ in\ C\ after\ another\ n\ steps\}\\&=\int_C\int_B\int_A h(x,y,s,t,u,v)\,du\,dv\,ds\,dt\,dx\,dy\\&=\int_0^l dv\int_{2\varepsilon}^{2\varepsilon+2R}du\int_0^l dt\int_{\varepsilon}^{\varepsilon+2R}ds\int_0^l dy\int_0^{2R}h(x,y,s,t,u,v)dx\end{aligned} \quad (5)$$

In the following, we investigate how the different parameters, such as the distance between the source and the destination, influence the probability of an arbitrary node hearing all the shares. We consider an arbitrary node starting from position $H(\alpha,\beta) = H(550,450)$. The transmission range $R = 250$ meters and the distance $l = 1000$ meters; shares are sent every $n = 2$ steps; the step length of the source/destination is $\eta = 300$ meters. During this period, the source and the destination move $\varepsilon = n\eta = 600$ meters. The expected value of the length Ψ of each step of an arbitrary node can be calculated from (1) as

$$E[\Psi] = \int_0^{+\infty} \psi\omega(\psi) = \frac{1}{2}a\sqrt{\pi} \quad (6)$$

Let $a = 400$. This gives the expected step length of an arbitrary node $E[\psi] \approx 354$ meters, which is comparable to the step length of the source and the destination. With these values, we calculate Formula (5) by changing one parameter at a time and keep all the other parameters default.

Figure 4 shows that as a (the parameter in Formula (1)) increases, the probability of hearing all 3 shares decreases. Here, a controls expected step length of an arbitrary node (Formula (6)). Since the time taken for each step is a constant (1 time unit), increasing a means increasing the node's speed. Please notice that in order to keep the system consistent, we also increase the source/destination's speed at the same rate. This means that if we double a (an arbitrary node's average speed), we double η too (the source's and the destination's speed). Thus Figure 4 indicates that as nodes' speeds increase, the probability of an arbitrary node hearing all the shares decreases. This will be attested by our simulation. Figure 4 also tells us that as the distance between the source and the destination (l) increases, the probability increases. This is because the rectangles become larger and it is easier for the adversarial node to enter these areas.

Figure 5 shows that as n, the number of steps moved before a share is sent, increases, the prob-

Figure 4. Probability decreases as a increases

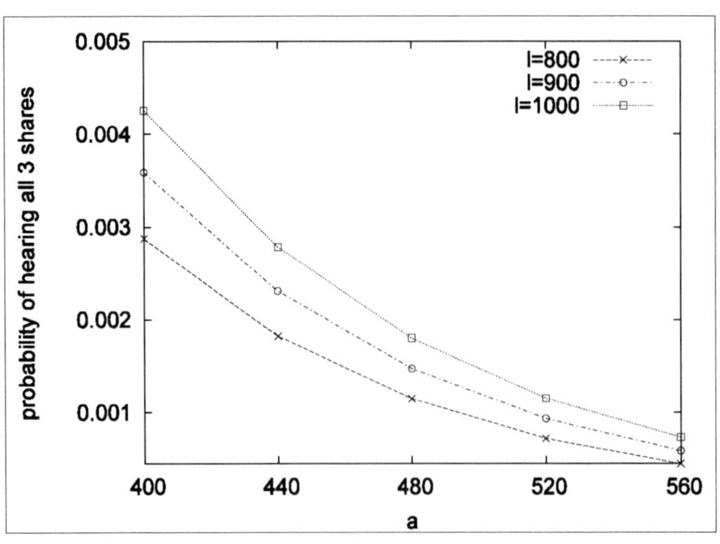

ability of hearing all 3 shares decreases. Since the time taken for each step is constant, increasing n means that the time interval between sending two consecutive shares increases. This reduces the chance for an arbitrary node to be in the areas to collect the shares. The influence from l suggests the same trend as seen from the previous figure.

SIMULATION

We run simulation in QualNet^{TM} to evaluate our scheme. QualNet^{TM} is a packet level simulator for wireless and wired networks, developed by Scalable Network Technologies Inc (Qualnet). The simulated network has 300 nodes moving in an area of $5000m \times 5000m$. *Random Waypoint* model is used to simulate node mobility (Johnson & Maltz, 1996). According to the model, a node travels to a randomly chosen location with a speed picked from the range of [minSpeed, maxSpeed] and stays for a while before moving to another random location. In our simulation, the speed is controlled in such a way that mobility decaying problem is minimized (Yoon, Liu, & Noble, 2003) (by setting the minSpeed larger than zero), and nodes will not pause at the locations that they choose. Initially nodes are randomly deployed in the area. At the network layer, AODV routing protocol is used to select routing paths for data communications (Perkins & Royer, 1999). The distributed coordination function (DCF) of IEEE 802.11 is used as the MAC layer with virtual carrier sensing for unicast packets. The transmission range is $250m$.

We run CBR (Constant Bit Rate) sessions to send shares of a message. A *session* is a lasting connection between a source-destination pair. In a CBR session, the source sends several 32-byte packets to the destination at the constant rate of one packet each time interval (e.g., 10 seconds). The packets are the shares in our scheme: one packet is one share. In the simulated mobile ad hoc network, a packet will be forwarded by several intermediate nodes before it is delivered to the destination.

We collect statistics from all the nodes except the source and the destination for a session in question. This presents the worst case for our scheme, because we treat all the nodes as attackers. We investigate the chance of a particular node to hear a certain number of shares that belong to

Figure 5. Probability decreases as n increases

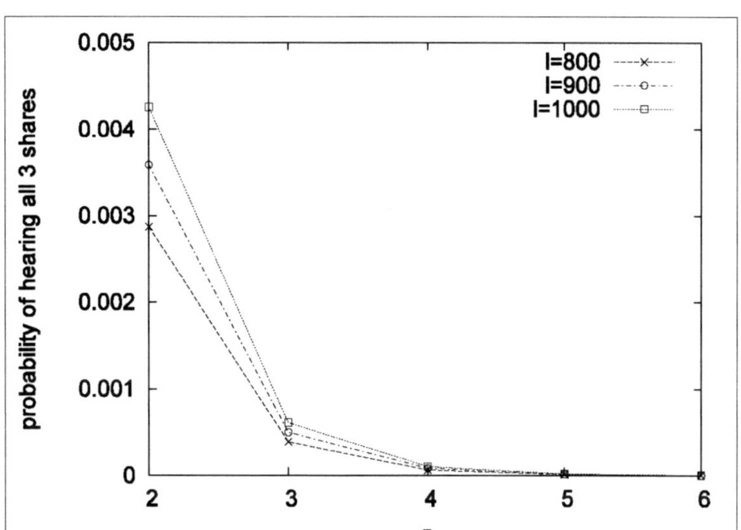

Figure 6. Distribution with different intervals

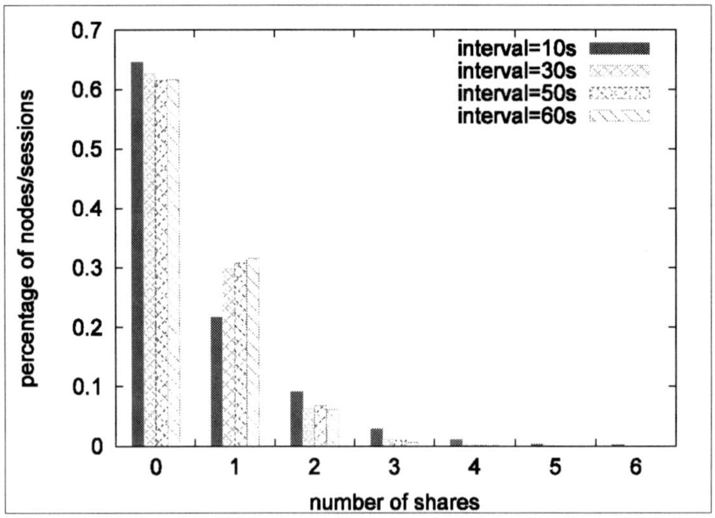

Figure 7. Distribution with different speed ranges

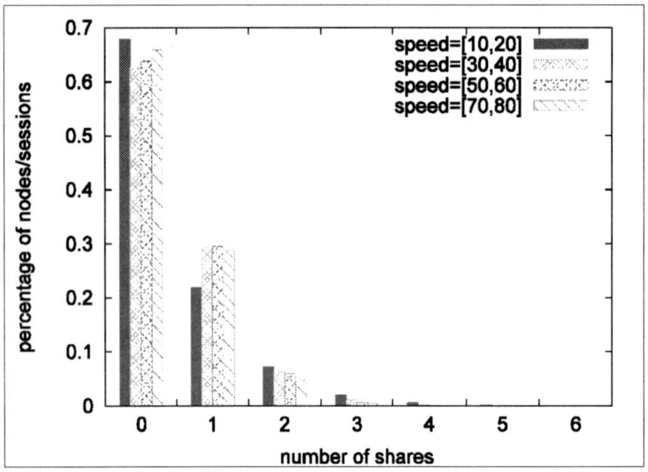

the same session. The results are averaged over all the nodes and sessions. The simulation time is 900 seconds.

In the first set of simulation, we show the probability distribution of shares being heard by an arbitrary node. Here, we set the default CBR sending interval as 30 seconds, the number of shares as 10, the number of sessions as 50, and the speed range [30,40] m/s. In the simulation, we will change CBR sending interval (Figure 6), speed range (Figure 7), and the number of shares per session (Figure 8). These figures show distributions of the percentage of nodes hearing a specific number of shares averaging over the 50 sessions. They also show the percentage of sessions during which a node hears a specific number of shares averaging over the 300 nodes. Notice that these two metrics are in fact the same. From these figures, it is clear that a large portion of nodes hear no share at all. Moreover, the percentage of nodes that hear more than 2 shares is very low, no matter what interval lengths or motion speeds are

Figure 8. Distribution with different numbers of shares

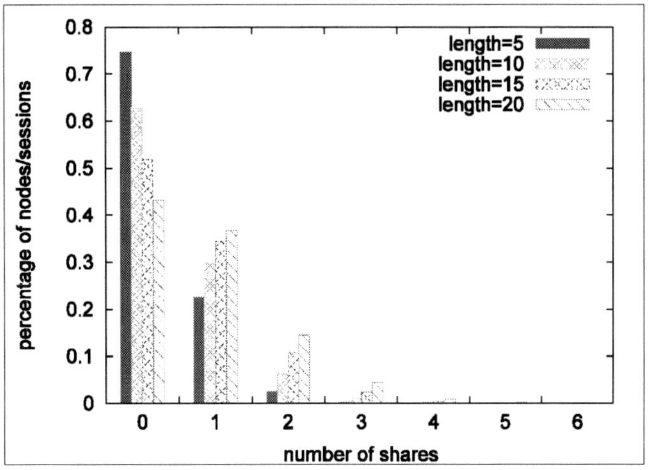

Figure 9. Percentage of nodes hearing enough (3 or more) shares

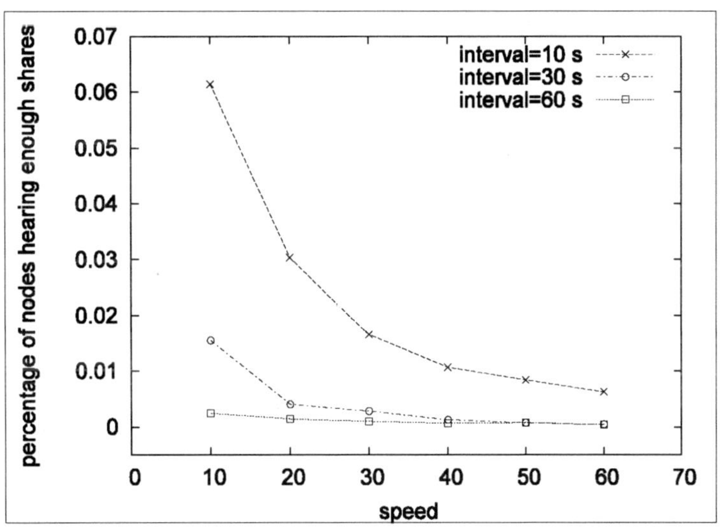

(Figs. 6 and 7). For Figure 8, when the sessions send fewer shares, the probabilities of hearing no share are higher than the ones sending more shares, e.g., the case of sessions having only 5 packets. In correspondence, if a session has more shares, the probabilities of leaking one or more shares are higher as expected. However, considering that the (k, n) scheme may also set a higher k for reconstructing the original message, these probabilities do not necessarily suggest a higher success ratio for an attacker. As an opposite example, the probability of hearing 4 packets in the case of session length being 5 is not zero, though it becomes zero for leaking all the 5 packets. For other session lengths, the percentages of nodes hearing more than 6 shares are all zeros. In fact, none of the figures show data beyond 6 shares because they are all zeros.

In the second set of simulation, we investigate how the sending interval and the speed influence the ability of an arbitrary node hearing enough shares. We pick the parameters of threshold secret sharing as (k, n) = (3, 5), i.e., a message is divided into 5 shares and hearing 3 or more shares will be able to recover the original message. So here "enough" shares means 3 or more shares. These values are chosen only for the purpose of illustrating the changing trend of the probability.

The results of the second set of simulation are shown in Figure 9. There are 100 sessions in the network. The speed ranges are [10, 20] m/s, [20, 30] m/s, [30, 40] m/s, [40, 50] m/s, and [50, 60] m/s. Note that the labels on the x axis show only the minimum speeds of their corresponding speed ranges. The three curves in the figure are corresponding to the sending intervals of 10 seconds, 30 seconds and 60 seconds respectively. The figure shows that as node speed increases, the percentage of nodes who hear at least 3 shares drops. It also shows that as time interval increases, the percentage of hearing at least 3 shares drops too. These results are consistent with our expectation and also with the numerical solutions in the previous section.

In all, the simulation results suggest that the probability of hearing enough shares is low. In practice, in order to achieve better protection, a larger n (and k) should be used: it is evident from the first set of simulation (where n = 10) that when we increase n (and k), the hearing probability approaches zero.

CONCLUSION

We presented and analyzed a data transmission scheme that enhances message confidentiality for mobile ad hoc networks. The scheme explores mobility through delaying the transmissions of message shares over a long period of time, resembling the application scenarios of Delay Tolerant Networks. We presented probability analysis, and used the isotropic random walk mobility model as a case study. We further used simulation to evaluate the scheme with different parameters and network scenarios. The results suggest that the probability for an attacker to hear multiple shares at different locations is small under the mobility model. The analysis and evaluation validate that the delayed transmission scheme is a valid alternative to preserve the message confidentiality for delay tolerant applications and also for network scenarios where spatial multi-paths are not feasible. Work in progress includes analytical work on parameter selection and simulation comparisons with more mobility models.

REFERENCES

Balfanz, D., Smetters, D. K., Stewart, P., & Wong, H. C. (2002). *Talking to strangers: Authentication in ad-hoc wireless networks*. In NDSS.

Capkun, S., Hubaux, J.-P., & Buttyan, L. (2003). Mobility helps security in ad hoc networks. In *MobiHoc '03: Proceedings of the 4th ACM international symposium on Mobile ad hoc networking & computing*, 46–56. New York, NY: ACM Press.

Chan, H., Perrig, A., & Song, D. X. (2003). Random key predistribution schemes for sensor networks. In *IEEE Symposium on Security and Privacy*, p. 197.

Eschenauer, L., & Gligor, V. D. (2002). A key-management scheme for distributed sensor networks. In *ACM Conference on Computer and Communications Security*, 41–47. Hubaux, J.-P., Buttyan, L., & Capkun, S. The quest for security in mobile ad hoc networks. In *MobiHoc '01: Proceedings of the 2nd ACM international symposium on Mobile ad hoc networking & computing*, 146–155. New York, NY: ACM Press.

Hughes, B. D. (1995). *Random Walks and Random Environments, Volume 1: Random Walks*. Oxford Science Publications.

Jain, S., Fall, K., & Patra, R. (2004). Routing in a delay tolerant network. In *ACM Sigcomm*.

Johnson, D. B., & Maltz, D. A. (1996). Dynamic source routing in ad hoc wireless networks. In Imielinski and Korth, (Eds.), *Mobile Computing*. Kluwer Academic Publishers.

Knuth, D. (1969). The Art of Computer Programming: *Vol. 2. Seminumerical Algorithms*. Reading, MA: Addison-Wesley.

Lee, M. G. S-J. (2001). Split multipath routing with maximally disjoint paths in ad hoc networks. In.ICC'01.

Lou, W., Liu, W., & Fang, Y. (2004, March). Spread: enhancing data confidentiality in mobile ad hoc networks. In *INFOCOM 2004, 2404–2413*. Hong Kong, China: IEEE.

Luo, H., Kong, J., Zerfos, P., Lu, S., & Zhang, L. (2004). Ursa: ubiquitous and robust access control for mobile ad hoc networks. *IEEE/ACM Transactions on Networking, 12*(6), 1049–1063. doi:10.1109/TNET.2004.838598

Papadimitratos, P., & Haas, Z. J. (2003). Secure data transmission in mobile ad hoc networks. In *WiSe '03: Proceedings of the 2003 ACM workshop on Wireless security*, 41–50. New York, NY: ACM Press.

Perkins, C. E., & Royer, E. M. (1999, February). Ad-hoc on-demand distance vector routing. In *Proceedings of Second IEEE Workshop on Mobile Computing Systems and Applications (WMCSA '99)*, 90–100. IEEE. Qualnet. http://www.scalable-networks.com/.

Rabin, M. O. (1989). Efficient dispersal of information for security, load balancing, and fault tolerance. *Journal of the ACM, 36*(2), 335–348. doi:10.1145/62044.62050

Radunovic, B., Gkantsidis, C., Key, P., & Rodriguez, P. (2008, April). An optimization framework for opportunistic multipath routing in wireless mesh networks. In *27th IEEE International Conference on Computer Communications (INFOCOM 2008)*, AZ, USA.

Shamir, A. (1979). How to share a secret. *Communications of the ACM, 22*(11), 612–613. doi:10.1145/359168.359176

Stajano, F., & Anderson, R. J. (1999). The resurrecting duckling: Security issues for ad-hoc wireless networks. In *Security Protocols Workshop*, 172–194.

Tsirigos, Z. H. A. (2001, November). Multipath routing in the presence of frequent topological changes. In *IEEE Communication Magazine*.

Wu, J. H. K. (2001). Performance study of a multipath routing method for wireless mobile ad hoc networks. In *the 9th international symposium on modeling, analysis and simulation of computer and telecommunication system*.

Yoon, J., Liu, M., & Noble, B. (2003). Random waypoint considered harmful. In *Infocom 2003: Proceedings of the Twenty-Second Annual Joint Conference of the IEEE Computer and Communications Societies*, 1312–1321. IEEE Press.

Zhang, Y. S. C., & Fang, Y. (2008, April). Modeling secure connectivity of self-organized wireless ad hoc networks. In *27th IEEE International Conference on Computer Communications (INFOCOM 2008)*. AZ, USA.

Zhou, L., Schneider, F. B., & Renesse, R. V. (2002). Coca: A secure distributed online certification authority. *ACM Transactions on Computer Systems, 20*(4), 329–368. doi:10.1145/571637.571638

This work was previously published in International Journal of Interdisciplinary Telecommunications and Networking, Volume 1, Issue 1, edited by Michael R. Bartolacci, and Steven R. Powell, pp. 80-94, copyright 2009 by IGI Publishing (an imprint of IGI Global).

Chapter 6
Hybrid Trust Structure in Self-Organizing Networks

Tong Zhou
University of Missouri at Kansas City, USA

Lein Harn
University of Missouri at Kansas City, USA

ABSTRACT

A traditional service provider of telecommunications is recognized as an authority which is trusted by the subscribers and the public. Ad hoc and Peer to Peer (P2P) networks have demonstrated advantages that service provider controlled networks lack, and they also exhibit self-organizing behaviors. A pure self-organizing network does not rely on any hierarchical management. Instead, it utilizes a web of trust for security. Its trust management is complicated and varies from node to node. In this article, we discuss a hybrid trust structure that leverages the involvement of an authority in a self-organizing network to increase trust levels between disconnected small-worlds. The new model will help service providers design more robust and innovative solutions for next generation networks and applications.

INTRODUCTION

Wireless and internetworking technologies (i.e. IEEE 802.11 and the Internet) have provided opportunities for user equipment (UE) to directly communicate among themselves, bypassing traditional service providers' physical and logical controls. Today, UE (i.e. wireless handset, Personal Digital Assistant (PDA) or laptop computer) is a multifunction and multipurpose device. Not only does it provide a connection channel, makes a phone call and browse the Internet, but also stores personal data, makes electronic payments, determines its location, and so on. The traditional way of offering services is through telecommunications service providers. Service providers control the admissions to their network infrastructures, including access networks, such as Worldwide Interoperability for Microwave Access (WiMAX), Transmission Control Protocol/Internet Protocol

DOI: 10.4018/978-1-60960-505-6.ch006

(TCP/IP) transport networks and service networks, such as IP Multimedia Subsystems (IMS) through Authentication, Authorization and Accounting (AAA). Agreements may exist among different service providers for roaming and service peering purposes. A subscriber either shares a secret with the service provider or uses a digital certificate issued by the Public Key Infrastructures (PKI) of an authority for security. The issue of a service provider's complete control is that all users' service requests must be backhauled to a control point at a national or regional data center or the edge of the service provider's network. In an emergency, such as natural disaster or terrorist attack, this infrastructure centric control model is not robust enough to handle larger than the normal bursts traffic. Even during normal operations it is inefficient for a user to transmit large amounts of data (i.e. file sharing and streaming video) to another user through a server provider's network infrastructure when a self-organizing network provides a direct channel or a shorter path.

A self-organizing network, which can be a Mobile Ad hoc NETwork (MANET), a Peer-to-Peer (P2P) network, a mesh network or a wireless sensor network, is a promising approach for providing flexibilities for users to form a network and control applications by themselves. It can potentially reduce the burden on a service provider's network, increase service availability and reliability, and drive innovations. However, the challenge of a self-organizing network is the lack of a centralized control of authority. Without this it is difficult to establish secure communications. A pure self-organizing network does not assume any authority for managing communications. A user makes their own decision. A reputation system can be used to improve the performance of a self-organizing network. It helps users identify trusted nodes.

Network security plays a crucial role for service providers. Popular applications are often targeted by hackers. Various security attacks, such as Distributed Denial of Services (DDOS), Man-in-the-Middle and SPAM can negatively impact service performance. Implementing strong authentication and diverting unknown traffic can effectively avoid attacks. In cryptograph, a digital certificate is used to bind the public key and the identity of the owner using a digital signature of a Certification Authority (CA) to prevent impersonation attack. Both hierarchical, such as ITU X.509, and nonhierarchical, such as Pretty Good Privacy (PGP), certification structures can be used to secure communications between two nodes. Hierarchical PKI require a root CA, which may not exist in cross-domain scenarios. A nonhierarchical structure, which is also known as web of trust, has the flexibility to allow any user to be a CA. However, it is very challenging to manage the trust relationships between CAs. PGP defines trust levels and allows a user to assign three levels of trustworthiness (complete trust, marginal trust and no trust) to another user's certification capability. In PGP, a user only accepts a stranger's certificate if it is issued by a CA that is completely trusted, or two CAs that are marginally trusted by the user. Trust is based on context and subjective. Li, Li and Kato (2008) define trust as a belief level that one node can put on another node for a specific action according to previous direct or indirect information from the observation of behaviors. In this article, trust refers to the belief of certification capability of a user. Studying how people trust each other will help improve the design of self-organizing networks.

Small world phenomenon reveals the fact that people are connected through six or less acquaintances (six degrees of separation) (Watt, 1999). Figure 1 shows two planes in a self-organizing network. The physical plane includes a variety of UE that is capable of forming self-organizing networks using existing (i.e. WiFi and TCP/IP) or future networking technologies. The social plane reflects the trust relationships among users. It controls the self-organizing networks on the physical plane through equipment ownerships. The discovery of six degrees of separations sup-

Figure 1. Self-organizing network layers

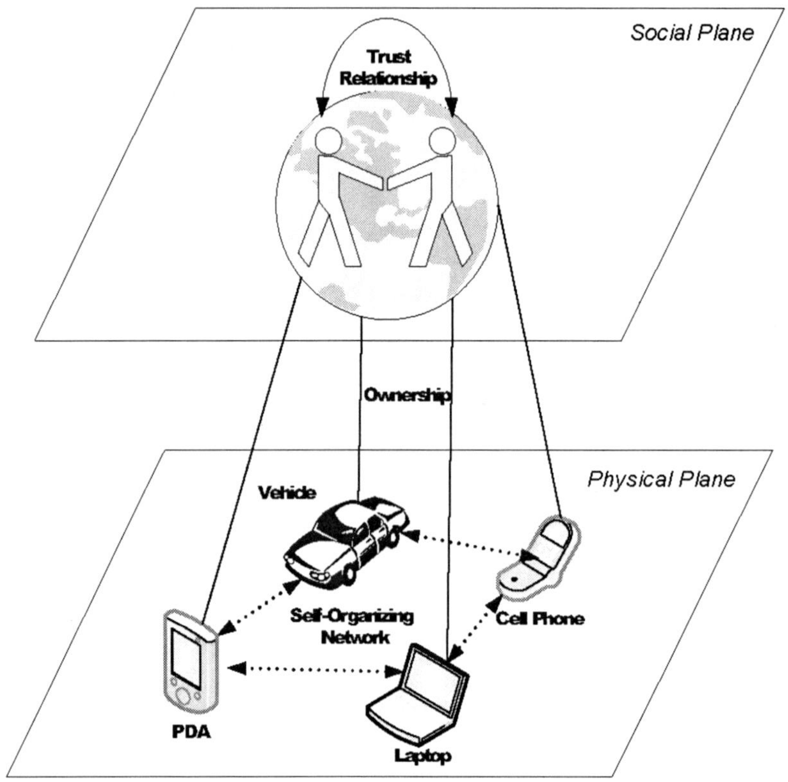

ports the feasibility of establishing self-organizing networks as people are all connected through a small number of intermediate hops. However, it does not mean people on the connection path all trust each other. Additionally, there are different levels of trusts assessed by each individual. Therefore, from a trust perspective, the whole world is viewed as a collection of loosely connected or isolated groups. The services offered by Tier 1 service providers in the United States (U.S.) are often targeted for 99.999% availability. A self-organizing network will not be highly available and scalable if there are too many loosely connected or isolated groups in it.

To solve the trust problem, we propose a novel method to utilize a hybrid trust structure for certification in order to connect groups where there is no strong trust between them. Within a group, the trust structure web can still be used for trust establishment. Between groups, a hierarchical trust structure is applied to bridge them. This solution will benefit both service providers and self-organizing networks. In human life, P2P services and mass collaboration, such as the online community for lending money (www.prosper.com) and wikipedia (www.wikipedia.com) are future trends, where trust is a fundamental issue that needs to be further studied.

The rest of the article is organized as follows: after a review of related works, we will explain the terminologies used in this article. Then the hybrid certification structure is described, followed by a section that provides a security protocol for exchanging private information between peers. The next section analyzes the effectiveness of endorsements, followed by the simulation results, with the final section summarizing the key contributions and introducing future work.

RELATED WORKS

A self-organizing network exhibits the structure of decentralization in many aspects, including signaling, data transportation, security, reputation management and charging. This section discusses prior art that is related to self-organizing security. Public-key and symmetric-key techniques can be used in self-organizing networks to establish security associations (Capkun, Hubaux, & Buttyan, 2006). Pairwise symmetric keys lack scalability (Zhang, Liu & Fang, 2006). In a P2P environment, public keys can be certified by an authority or individuals. In the scope of pure self-organizing networks, we do not assume any hierarchical certification structure. ID-based cryptography (IBC) is an alternative to certificate-based cryptography (CBC). Zhang et al. (2006) devise an ID-based key management scheme called IKM, which still requires a single authority. Suryanarayana, Erenkrantz and Taylor (2005) summarize different types of security threats. Sanzgirl, LaFlamme, Dahill, Levine, Shields and Belding-Royer (2005) design a protocol named 'authenticated routing' for ad hoc networks (ARAN), which uses public-key cryptography to defeat the identified attacks. Sun, Osborne, Xiao and Guizani (2007) present an overview of intrusion detection techniques and future research directions in MANET and wireless sensor networks.

There are many proposals on the web of trust based public-key certifications. Sun, Yu, Han and Liu (2006) propose entropy-based and probability based trust models. Caronni (2000) discusses the algorithms for computing multiple interconnected trust paths. Theodorakopoulos and Baras (2006) view the trust evaluation as a generalized shortest path problem on a weighted directed graph and propose a trust computation scheme. Li et al. (2008) design a robust and attack-resistant framework, which is called the objective trust management framework (OTMF) based on a modified Bayesian approach by which various weights are put on different information related to the observations of behaviors according to their occurrence time and providers. In previous research (Zhou & Harn, 2007), we discuss the method of utilizing the public key trust and validity levels to obtain an end-to-end trust, and summarize different thresholds for making a trust decision. When no connection with a newly joined node is found, and to mitigate the risks of blind trust, we introduce the macro and micro methods of analyzing the trust relationships among the CAs which issue certificates to the node (Zhou & Harn, 2008). More discussions on security and trust are found in Xiong and Liu, 2004; Josang, Hayward and Pope, 2006; Huynh, Jennings and Shadbolt, 2006; Papadimitratos and Hass, 2006; Yang, Shu, Meng and Lu, 2006; Ma and Orgun, 2006; Lin, Lu, Zhang, Zhu, Ho and Shen, 2008.

However, to the best of our knowledge, there is no comparison on the service quality of a carrier's network and a self-organizing network. In self-organizing networks, low and no trust are common obstacles to achieving high availability and reliability while maintaining security. Without quality and security, the applications of self-organizing networks will be very limited to specific environments and only provide better than nothing services. To improve the performance of self-organizing networks, we propose a framework that allows an authority to bridge the trust gaps between loosely connected or isolated groups by leveraging its authority-endorsements on selected nodes, which become super-nodes. The introduction of super-nodes can help users establish a new certification path or short-cut a low trust path. We expect the confidence on the public key ownership can be increased. Therefore, self-organizing networks will be more practical and popular. The hybrid certification structure, which is detailed in a later section, inherits the advantages of hierarchical and non-hierarchical certifications.

DEFINITIONS AND BACKGROUND

Authority

In this article, an authority is a truly accountable administration in communications. Its judgments are completely or strongly trusted by the public. An authority is trusted is because it has proven records of being responsible for its behaviors. It must have AAA functions and is supposed to have a strict process for making sound decisions. Most authorities have AAA functions. As discussed in the following sections, an authority is expected to identify super-nodes and issue endorsement certificates properly to improve the performance of self-organizing networks.

Endorsement

An authority may provide a node in self-organizing networks with an endorsement certificate that not only binds the node's public key to its identity, but also includes the node's certification capability that the authority recognizes. In this article, this behavior is called endorsement. When compared to a PGP certificate, an endorsement certificate makes the trust level (i.e. gold, silver, or bronze) public and recognized. An example of an endorsement certificate is as follows:

Version: 10
Serial Number: 8390
Algorithm/Parameter: MD5, RSA
Issuer: Service Provider Company Name
Validity Period: mm/dd/yyyy – mm/dd/yyyy
Subject Identity: Bob.Rose@company.com
Subject's Public Key:
00:c6:a9:d3:1b:25 …(RSA, 1024 bits)
Subject's Certification Capability: Gold
Signature: 45:9a:3b:12:8d …

Super-Node

An authority-endorsed node is referred as a super-node in this article. The certification capability of a super-node is described in the endorsement certificate example above. To become a super-node, a regular node must meet certain criteria, such as an excellent certification reputation and history, hardware and software capabilities, I/O throughput, and a willingness to assist other nodes in self-organizing networks. Through a common authority, a super-node trusts the other super-nodes. A similar concept exists in P2P VoIP (Voice over Internet Protocol) communications (i.e. Skype), where a super-node assumes the responsibility to assist other nodes to connect to each other as a routing hub. The super-node defined in this section functions as a reputable CA. The two types of super-nodes can coexist but are not necessarily in the same UE.

HYBRID CERTIFICATION STRUCTURE

In a service provider controlled network, a user can either share a secret with the service provider or use a service provider accepted certificate for security. As it is expensive for a service provider to build PKI, in most cases, an asymmetric key is not provided for the UE. In self-organizing networks, there is no centralized authority. A user stores the others' public keys in their key ring (local directory) and assigns a value of validity and trust to each key. Furthermore, a user can issue digital certificates to attest to the integrity and the ownership of the public keys. The certificates can be used by others who trust the issuer to verify the public keys and the identities.

The advantage of hierarchical certification systems is that users completely trust the CAs' certifications capability. However, due to different reasons, such as Return on Investment (ROI), an authority may not build a hierarchical structure when users need it. Then an individual user must obtain the certificate from the authority. In contrast, a self-organizing network may be voluntarily formed by individual users to overcome the absence of authority control. High trust can

Figure 2. Hybrid structure: a hypothetical example

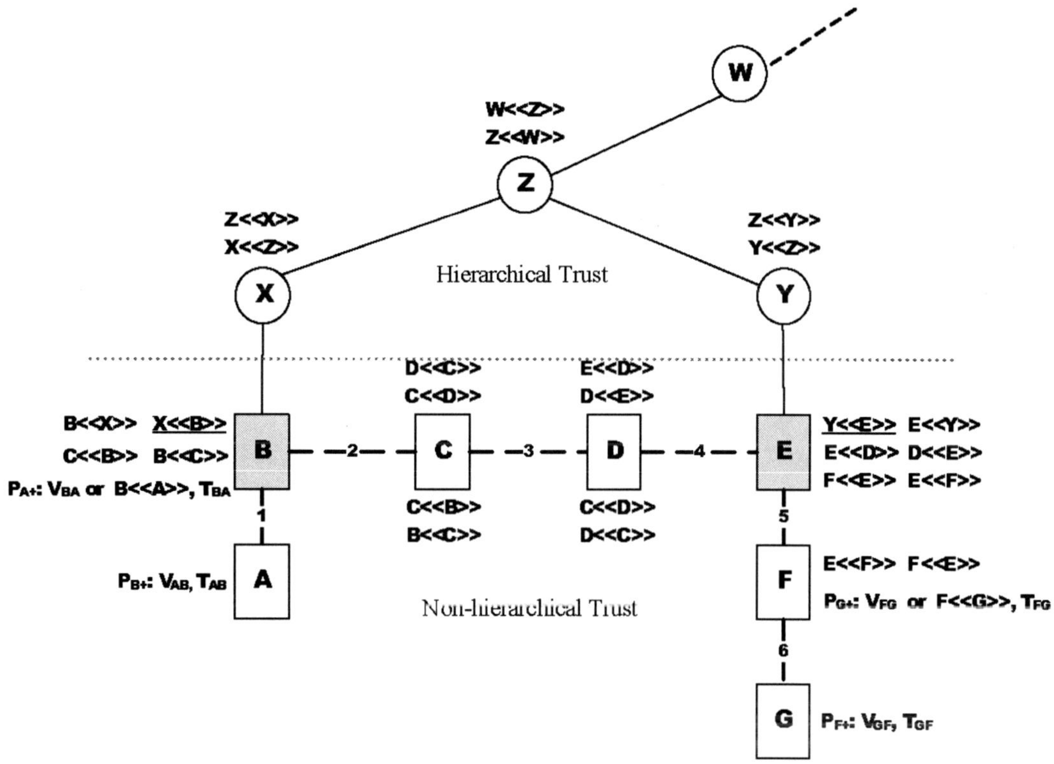

possibly exist in a small group. In the real world, not all groups are strongly connected. Therefore, a self-organizing network is hard to scale. The hybrid trust structure we propose in this article combines the advantages of hierarchical and non-hierarchical trust structures to solve the no trust or low trust issue in self-organizing networks. The trusts among authority-endorsed super-nodes can be increased such that two super-nodes can establish a strong path to replace a weak link. This benefit does not exist in a pure non-hierarchical structure. An example of hybrid structure is a self-organizing community WiFi network. A resident may establish a strong trust with the neighbors in the community, but may not trust other residents in a different community. Without the involvement of an authority, the person can only be served within the neighborhood due to the lack of strong trust outside the community. In a hybrid structure, super-nodes bridge disconnected communities through the endorsements of an authority. Thus, residents in different communities can establish strong trusts, which results in improved service performance.

Figure 2 illustrates a self-organizing network with authority endorsements. Nodes X, Y, Z and W represent CAs in a hierarchical structure. X and Y are leave nodes and are under Z. The root of the tree is not shown. M<<N>> represents a certificate that M issues to N. V_{MN} and T_{MN} are the validity and trust levels that M assigns to N's public key, P_{N+} respectively. B and E are super-nodes endorsed by an authority through its local CAs, X and Y, respectively. B and E also sign the public keys of X and Y, respectively. Through certifications, trusts can be propagated in the network.

Nodes A, B, C, D, F and G participate in a self-organizing network. Based on the small-world phenomenon, any nodes in a self-organizing network are connected by six or less intermediate nodes. Assume A does not know G directly and needs to communicate with it. A requests his or her friends to find connections to G. Among them, B's public key ownership (V_{AB}) and certification capability (T_{AB}) are trusted by A. B issues B<<C>> to C, and believes C's certification capability is T_{BC} (not shown); C issues C<<D>> to D, and believes B's certification capability is T_{CD} (not shown); D issues D<<E>> to E, and believes B's certification capability is T_{DE} (not shown); E issues E<<F>> to F, and believes B's certification capability is T_{EF} (not shown); F either issues F<<G>> to G or just stores the public key of G (P_{G+}) in its public key ring and assigns the validity of the key to V_{FG}.

Let the trust distance between two super-nodes be *d*. In the endorsement certificates that the authority issued to B and E, their certification capabilities (T_B and T_E) are included. There are different ways to calculate the end-to-end trust level (Sun et al., 2006; Zhou & Harn, 2007). For simplicity, we only discuss a single certification path in the following sections. The distance between two super nodes before endorsements can be different. Figure 2 shows B and E are three hops apart. For A to authenticate G, A needs to obtain the certificates on the certification path, which are B<<C>>, C<<D>>, D<<E>>, E<<F>> and F<<G>> or VFG, as well as the trust levels of certification capabilities, T_{BC}, T_{CD}, T_{DE} and T_{EF}. The following section discusses how to securely transmit trust levels and validity levels. Additionally, we assume A has securely obtained B's public key (P_{B+}), and can retrieve V_{AB} and T_{AB} from its key ring. With all the information received, A can verify the binding of G's public key and identity (Zhou & Harn, 2007). Similarly, G can assess the validity of A's public key for the purpose of mutual authentication.

With an authority's endorsements on nodes B and E, the hybrid certification path from A to G is B<<X>>, X<<Z>>, Z<<Y>>, Y<<E>>, E<<F>> and F<<G>> or V_{FG}. And the path from G to A is F<<E>>, E<<Y>>, Y<<Z>>, Z<<X>>, X<<B>> and B<<A>> or V_{BA}. The certification capabilities of super-nodes B and E are described in the endorsement certificates. As X, Y and Z are CAs in a hierarchical structure, they completely trust each other's certification capability. The hybrid certification path should have a higher trust than the self-organizing path. It is expected that the hybrid certification structure will improve the successful rate of communications in self-organizing networks. The quantitative studies are discussed in a later section.

PRIVATE INFORMATION EXCHANGE (PIE) PROTOCOL

This section describes a protocol for two nodes to securely exchange their personal information, such as trust levels and validity levels, in a self-organizing network. The following security requirements are identified for the design of the protocol (Sun et al., 2006):

- Mutual authentication – the two nodes involved in communications must authenticate each other.
- Data confidentiality – private information, such as trust level, transmitted between intermediate hops cannot be decoded by eavesdroppers.
- Data authenticity (integrity) – ensures data is from the intended source and not altered during transmission.
- Privacy protection – a receiver cannot prove to a third party that the received private information is from the sender.

To meet the requirements, we propose the PIE protocol, which is based on Internet Key Exchange - Secure Key Exchange MEchanism (IKE-SKEME) (Krawczyk, 1996) for the hop-by-hop private information exchange. In the following example, we assume that A stores the trust level (T_{AB}) and validity level (V_{AB}) for B's public key

(P_{B+}) in its public key ring. A sends a SEARCH request to B to obtain B's confidence on C's public key (V_{BC}). In the message, A indicates that it accepts both direct and indirect trusts. For security, A encrypts its identity (A) and a random number (N_A) it generated using B's public key. Based on six degrees of separations, A chooses six as the maximum number of intermediate hops from it.

REQEUST SEARCH C <carol@example.com>, PIE/1.0
From: A <alice@example.com>
To: B <bob@example.com>
Search-ID: 49dks0230slwla
Information: Trust Level
Direct Trust: Yes
Indirect Trust: Yes
Max-Hops: 6 (maximum number of intermediate nodes)
Security: IKE-SKEME
Content: 87:6d:c2:fe:a4:45...
($E_B[A, N_A]$, where A = alice@example.com)

B sends back a provisional response (100 Trying) to inform A that B does not know C directly, and will forward the request to a number of trusted nodes. To avoid impersonation attacks, B only forwards the request to the nodes whose identities it trusts. If B is a super-node, it can also broadcast the request to other super-nodes. B reduces the Max-Hops it received by one and includes the new value in the forwarded message.

RESPONSE 100 Trying PIE/1.0
From: B <bob@example.com>
To: A <alice@example.com>
Search-ID: 49dks0230slwla
Direct Trust: Not found
Indirect Trust: Forwarding to n node(s), where $n^{3} 1$.

Assume one of the trusted nodes, J responds to B and provides V_{JC}, which J believes is the validity of C's public key, using the same PIE protocol. B encrypts its identity (B), V_{JC}, B<<J>> (or V_{BJ}), T_{BJ} and a random number it generated (N_B) using A's public key and sends a 200 Found message to A.

B also attaches a Message Authentication Code (MAC) for data authenticity (secret key $K_0 = H(N_A, N_B)$). Therefore, A is assured that the data is from B. However, A cannot prove it to a 3rd party because the message from B does not include B's digital signature, and A can possibly forge the message by itself. The communication between B and J is protected in the same way.

RESPONSE 200 Found PIE/1.0
From: B <bob@example.com>
To: A <alice@example.com>
Search-ID: 49dks0230slwla
Indirect Trust: Found
Content: 32:1d:b2:5c:6a:f0 ...
($E_A[Data, B, N_B]$, $MAC_{K0}(Data, B, A)$,
where Data = $T_{BJ} \| V_{BJ}$ or B<<J>> $\| V_{JC}$ or J<<C>>,
$K_0 = H(N_A, N_B)$,
B = bob@example.com, and A = alice@example.com.)

A decrypts the received message to obtain the data using A's private key and sends back an ACKnowledgement to B. A uses B<<J>> to verify J's identity and calculates the trust distance to J by calculating $T_{AB} \times T_{BJ}$. If J<<C>> is received, A concludes that the validity level of C's public key is $T_{AB} \times T_{BJ}$. If J does not issue a certificate to C, A verifies that a friend's friend, J trusts C's public key at the V_{JC} level. So the validity of C's public key is $T_{AB} \times T_{BJ} \times V_{JC}$. When A uses multiple certification paths for public key validity evaluation, it is important to identify if the paths are independent or interconnected.

ACK <carol@example.com>, PIE/1.0
From: A <alice@example.com>
To: B <bob@example.com>
Search-ID: 49dks0230slwla
Security: IKE-SKEME
Content: 14:7c:a2:df:58:93... ($MAC_{K0}(A, B)$)

QUANTITY ANALYSIS

Trust Gain

In this section, we discuss the effectiveness of introducing two super-nodes on a certification path (see Figure 2). The super-node at the beginning of a directional path is called the upstream super-node, and the super-node at the end is called the downstream super-node. The end-to-end trust on G's public key from A is in Equation 1. After the endorsements on B and E, the trust is increased to Equation 2. The trust gain is defined in Equation 3.

$$T_{AG} = T_{AB} \times T_{BC} \times T_{CD} \times T_{DE} \times T_{EF} \times V_{FG} \quad (1)$$

$$T_{AG}' = T_{AB} \times T_{E} \times T_{EF} \times V_{FG} \quad (2)$$

$$T_{gain} = \frac{T_{AG}'}{T_{AG}} - 1 = \frac{T_{super-node}}{T_{SON-path}} - 1, \quad (3)$$

where $T_{super-node}$ is the certification capability of the downstream super-node recognized by an authority in the endorsement certificate, and $T_{SON-path}$ is the trust level between the upstream and downstream super-nodes. In Figure 2, $T_{super-node}$ and $T_{SON-path}$ are T_E and $T_{BC} \times T_{CD} \times T_{DE}$, respectively. Equation 3 indicates that the endorsement will be more effective if the downstream super-node, E improves its certification capability, $T_{super-node}$ and/or $T_{SON-path}$ is a weak link (low trust or no trust).

Assume a super-node obtains a gold level certification from an authority, the weak link length is l, and the average trust per hop is T_0. The lower limit of trust gain can be obtained from Equation 4. This is because the arithmetic mean is greater than the geometric mean (see Equation 5).

$$T_{gain} \geq \frac{T_{Gold}}{T_0^{\,l}} - 1 \quad (4)$$

$$\frac{1}{n}\sum_{i=1}^{n} x_i \geq \left(\prod_{i=1}^{n} x_i\right)^{1/n}, \quad (5)$$

where x_i ³ 0

Figure 3 shows the increases of trusts for different average trusts (0.1, 0.5 and 0.9) and path lengths (1, 2, 3 and 4). The X-axis is the distance, *d* between two super-nodes in terms of the number of intermediate hops, and the Y-axis is the trust increase in a logarithmic scale. In the case of low trust (avg. trust = 0.1), the super-nodes improve the trusts significantly.

Figure 3. Trust gains

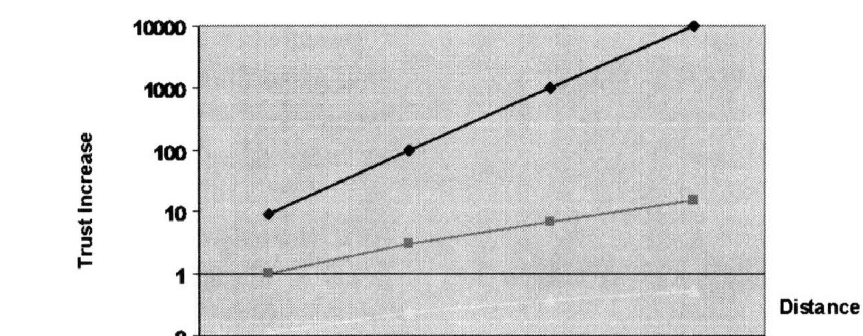

	1	2	3	4
Avg. Trust = 0.1	900%	9900%	99900%	999900%
Avg. Trust = 0.5	100%	300%	700%	1500%
Avg. Trust = 0.9	11%	23%	37%	52%

End-to-End Trust with Authority Endorsements

In a self-organizing network, a node needs to determine the end-to-end trust level (Z) to another node it encounters. Z is the product of an upstream trust level, X, and a downstream trust level, Y (see Equation 6).

$$Z = X \times Y \tag{6}$$

Assume the Probability Density Functions (PDF) of upstream and downstream trusts are $f_X(x)$ and $f_Y(y)$, respectively (see Figure 4). The PDF varies from group to group. Figure 4 illustrates a hypothetical example, where the upstream group has strong trusts within it while, on average, the downstream group has marginal trusts.

The PDF of Z is calculated in Equation 7, where X and Y are independent trust variables.

$$f_Z(z) = \int_0^1 f(z / y, y)dy = \int_0^1 f(z / x, x)dx \tag{7}$$

As $f(x,y) = f_X(x)f_Y(y)$, $f_Y(y)$ can be obtained from Equation 8.

$$f_Z(z) = \int_0^1 f_X(x)f_Y(z / x)dx = \int_0^1 f_X(z / y)f_Y(y)dy \tag{8}$$

If a node selects a trust threshold T ($0 \le T \le 1$), the possibility that the end-to-end trust is higher than T is obtained from Equation 9. The higher the trust threshold used, the stronger the security achieved. The tradeoff is that the successful rate of inter-group communications becomes lower. The goal of the hybrid trust structure is to connect isolated groups which have high or medium internal trusts through endorsing super-nodes. With this approach, it is expected that the successful rate of communications is higher than 80% if a medium trust threshold is selected. The following section uses simulation to explain this in detail.

$$P(Z > T) = 1 - \int_0^T f_Z(z)dz \tag{9}$$

Caveman Model

Watts (1999) describes a caveman world, "Everybody you know knows everybody else you know and no on else." However, this model does not include how well a node knows other nodes. This missing part is important for people to make trust decisions. In this section, we discuss three

Figure 4. Upstream and downstream trust distribution example

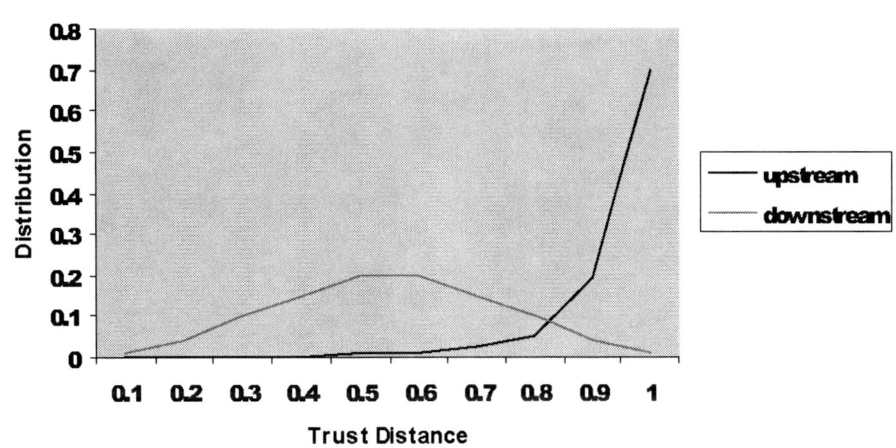

types of caves, where the internal trusts are High, Medium and Low (see Table 1). In each cave, we assume the trusts among the nodes are identical.

An authority should focus on connecting isolated high trust and medium trust caves (see Figure 5). Although super-nodes serve as hubs connecting isolated groups, the end-to-end trust still includes the intra-group trust.

When comparing connecting two high trust groups and connecting two low trust groups, the end-to-end trust of the former is higher than that of the latter. It is unlikely that two low trust groups can establish trust unless the acceptable trust threshold is very low, which increases the risk of trusting a malicious node. Figure 6 exhibits the results of connecting H, M and L groups. There are six different combinations, HH, HM, HL, MM, ML and LL, which have the trust values of 0.81, 0.45, 0.09, 0.25, 0.05 and 0.01, respectively.

SIMULATION

We created two groups A and B, each of which has 10 nodes (see Figure 7). The weights on the lines represent mutual trust levels. Initially, the two groups were disconnected. Group A includes Nodes 1 to 10 and Group B includes Nodes 11 to 20. The average intra-group trust is the arithmetic mean of inter-nodes direct trusts (see Equation 10).

Table 1. H/M/L trusts

Intra-cave Trust Level	Quantification
High (H)	0.9
Medium (M)	0.5
Low (L)	0.1

$$T_{avg} = \sum_{i,j=1,n,i\neq j} T_{ij} / N_{MAX} \qquad (10)$$

where N_{MAX} is equal to $n^2 - n$. n is the number of nodes in a group. For Groups A and B, the average intra-group trusts are 0.767 and 0.770 respectively (see Figure 10). Nodes 10 and 20 were identified and then endorsed by an authority to become super-nodes. Thus, inter-group communications are possible. For example, we can find a trust path from Node 3 to Node 12, which is 3 → 9 → 10 → 20 → 18 → 12. The end-to-end trust level from Node 3 to Node 9 is 0.66. If we divide the trust into 10 equal levels, 0.66 falls into the 0.6 to 0.7 range. The PDFs of the trust levels from regular nodes to the super-nodes in group A and B have the same distribution (see Figure 8). After

Figure 5. Connecting caves

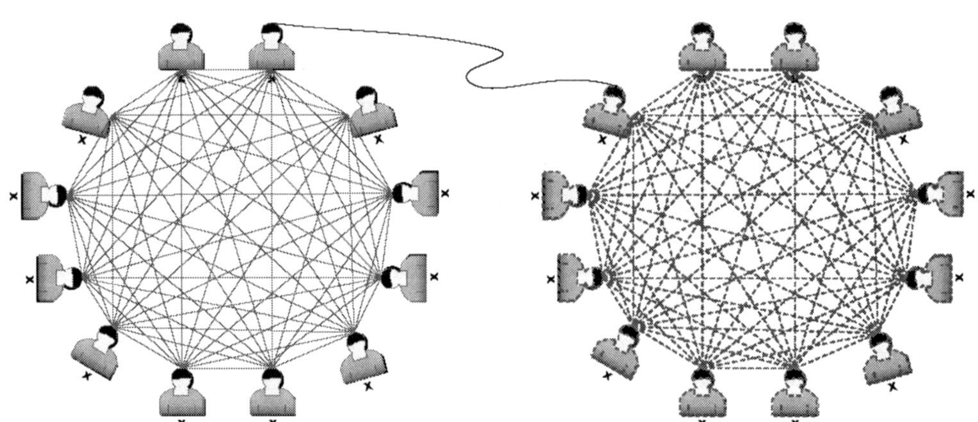

Figure 6. Connecting groups with high (H), medium (M) and low (L) trusts

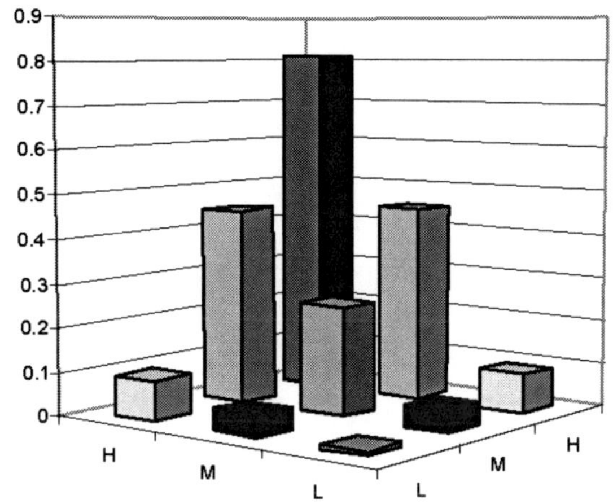

Figure 7. Two isolated groups with one connection

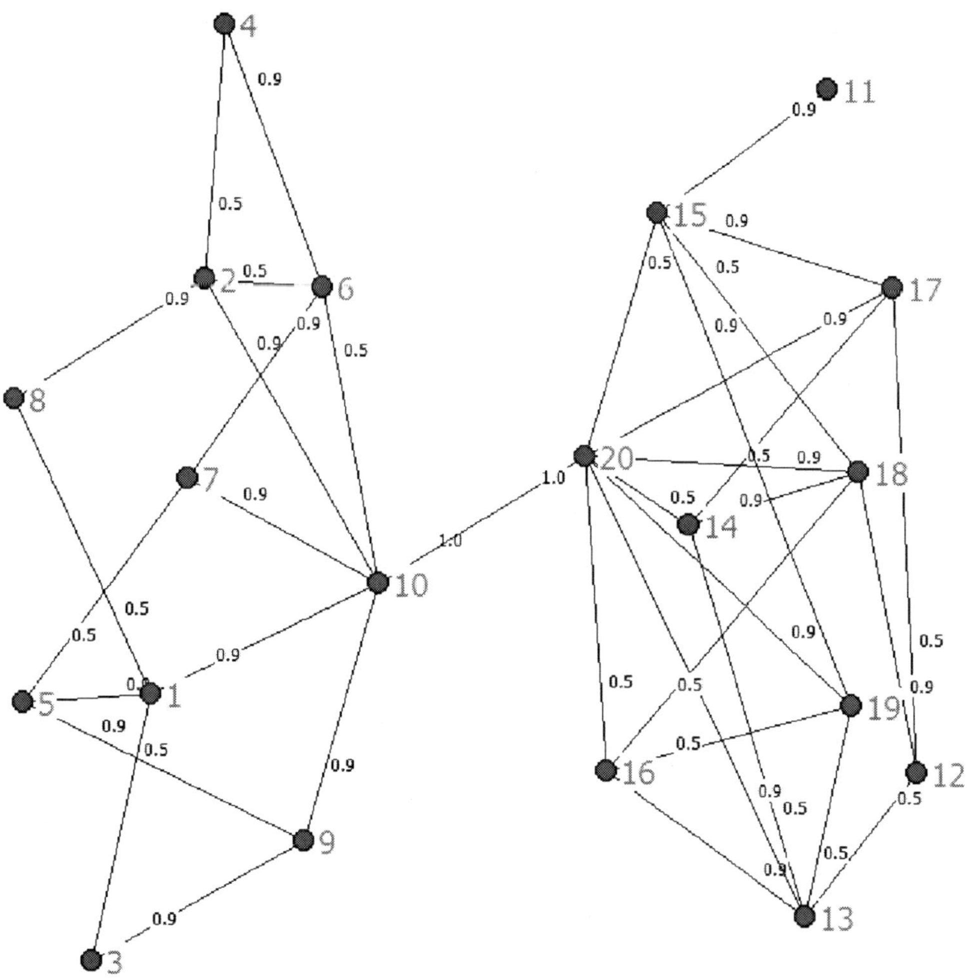

Figure 8. Simulation of upstream and downstream PDFs - $f_X(x)$ and $f_Y(y)$

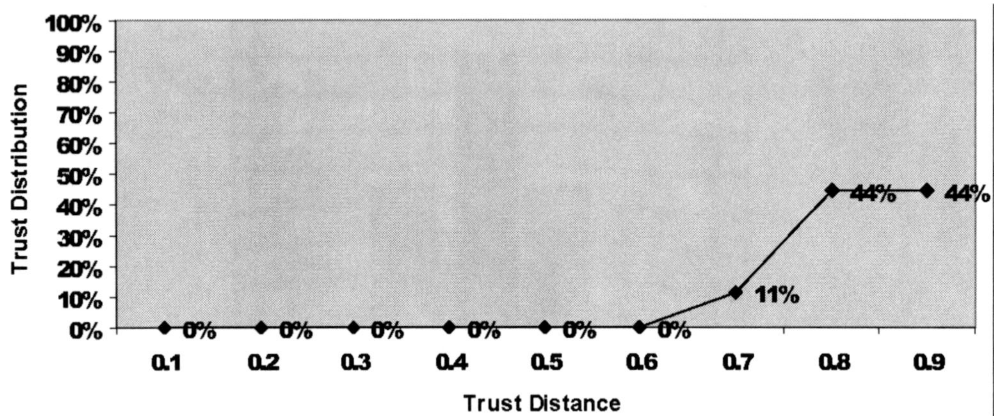

Figure 9. End-to-end trust PDF - fZ(z)

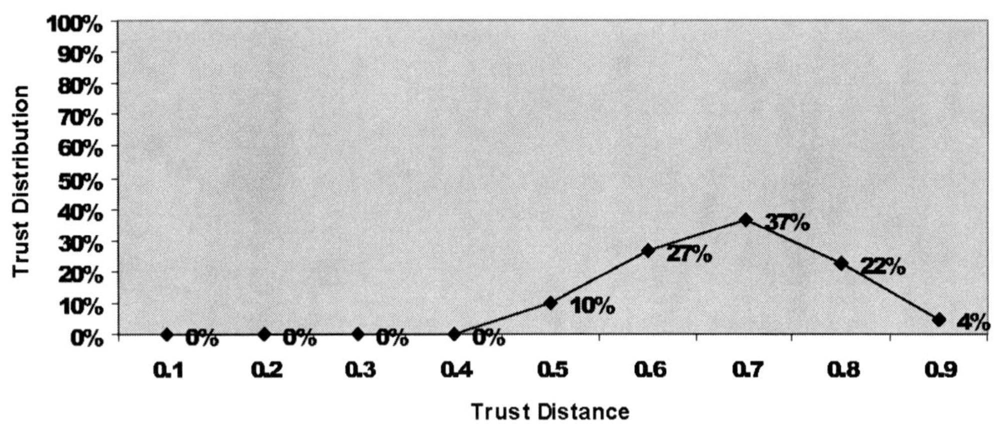

the introduction of super-nodes, the inter-group trust distribution is shown in Figure 9.

In our study, we used UCINET Version 6 software (UCINET, 2007) for the social network analysis to find connections between Groups A and B. (Under the main manual, select Network, Cohesion, Reachability, and then Probabilities in the "Type of data" field.) The probability of a path is equal to the product of the probabilities of its edges. The algorithm finds the probability of the most probable path between each pair of nodes (UCINET, 2007).

It is the user's decision to select a trust threshold. Different thresholds incur different successful rates and risks of communication. Figure 11 shows when the thresholds are 0.6 and 0.8 the communication successful rates are 90.0% and 26.7%, respectively. Trust is a dynamic learning process in self-organizing networks. After a positive experience is obtained, the trust on the path will be increased. A UE can cache the certificates and the trust data. So next time when the same path is found, the trust will be increased to $T_{current} = T_{past} + \Delta T$, where $T_{current}$ is the current trust level to be determined, T_{past} is the trust level used in the last event, and ΔT is recommended to be a fraction of T_{past} until $T_{current}$ reaches the maximum trust value (1.0). With this approach, self-organizing networks

Figure 10. Average trust

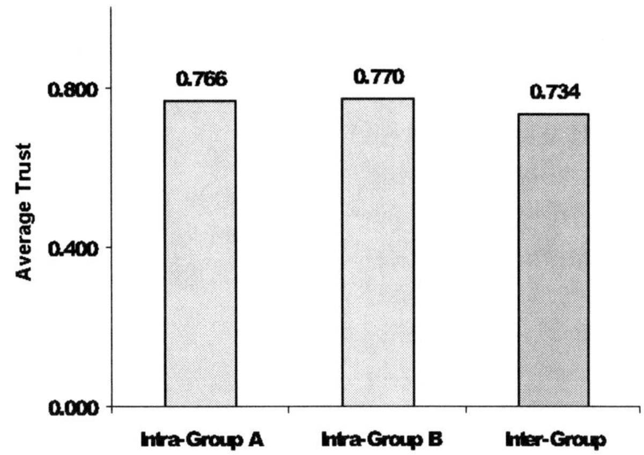

Figure 11. Communication successful rate

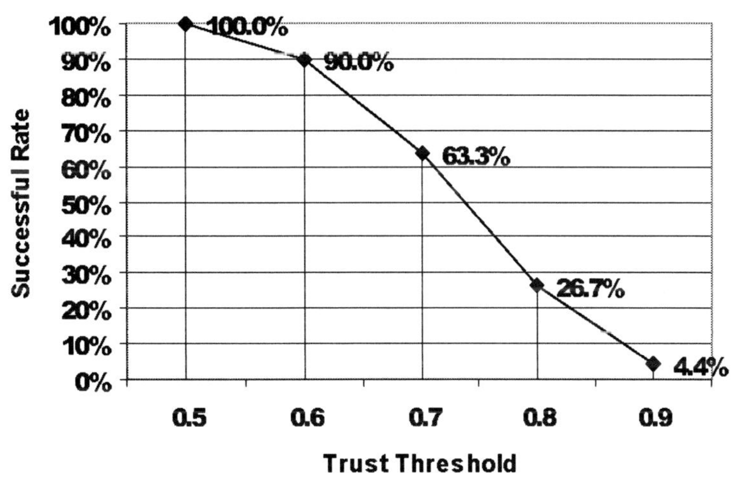

will become more successful with increasing trusts from past experiences.

CONCLUSION AND FUTURE WORKS

In contrast to traditional communication services provided by government authorized service providers, self-organizing networks are formed by individual users using their own equipment. Each of the two models has advantages and disadvantages. The service providers provide so-called carrier grade services with strong authentication control which should yield high quality and security. Self-organizing networks can be applied to P2P and ad hoc environments. Due to the lack of a central authority in self-organizing networks, it is difficult to verify the identity of the participating users. The studies on the web of trust and small world phenomenon have revealed the possibility of establishing trusts among users who do not directly know each other. However, the end-to-end trust level decreases when the path length of certification gets longer, which results in low success rates of communications. To solve the problem, we propose a hybrid trust structure for certification by combining the strengths of both models, and introducing the concept of authority endorsement and super-node for public key certification. In other words, an authority is involved in self-organizing networks to improve their performance. Through quantity analysis and simulations, we have shown the trust gain and the successful rate against different trust thresholds. In addition, the proposed PIE protocol provides security and privacy protection for transporting user sensitive information over open networks. In summary, the hybrid trust structure for certification will achieve improved security and performance. For our future research, we plan to focus on studying the trust development process (from a stranger to a friend) and cryptographic processing times.

REFERENCES

Capkun, S., Hubaux, J.-P., & Buttyan, L. (2006). Mobility Helps Peer-to-Peer Security. *IEEE Transactions on Mobile Computing*, *5*, 43–51. doi:10.1109/TMC.2006.12

Caronni, G. 2000. Walking the Web of Trust. *Proceedings of IEEE 9th International Workshops on Enabling Technologies: Infrastructure for Collaborative Enterprises* (pp. 153-158).

Huynh, T., Jennings, N., & Shadbolt, N. 2006. Certified Reputation: How an Agent Can Trust a Stranger. *Proceedings of the Fifth International Joint Conference on Autonomous Agents and Multiagent Systems.*

Josang, A., Hayward, R., & Pope, S. 2006. Trust Network Analysis with Subjective Logic. *Proceedings of the Twenty-Ninth Australian Computer Science Conference* (pp. 85-94).

Krawczyk, H. 1996. SKEME: A Versatile Secure Key Exchange Mechanism for Internet. *Proceedings of the 1996 Internet Society Symposium on Network and Distributed Systems Security* (pp. 114-127).

Li, J., Li, R., & Kato, J. (2008). Future Trust Management Framework for Mobile Ad Hoc Networks. *IEEE Communications Magazine*, *46*, 108–114. doi:10.1109/MCOM.2008.4481349

Lin, X., Lu, R., Zhang, C., Zhu, H., Ho, P., & Shen, X. (2008). Security in Vehicular Ad Hoc Networks. *IEEE Communications Magazine*, *46*, 88–95. doi:10.1109/MCOM.2008.4481346

Ma, J., & Orgun, M. (2006). Trust Management and Trust Theory Revision. *IEEE Transactions on Systems, Man, and Cybernetics. Part A, Systems and Humans*, *36*, 451–460. doi:10.1109/TSMCA.2006.871628

Papadimitratos, P., & Hass, Z. (2006). Secure Data Communication in Mobile Ad Hoc Networks. *IEEE Journal on Selected Areas in Communications*, *24*, 343–356. doi:10.1109/JSAC.2005.861392

Sanzgirl, K., LaFlamme, D., Dahill, B., Levine, B., Shields, C., & Belding-Royer, E. (2005). Authenticated Routing for Ad Hoc Networks. *IEEE Journal on Selected Areas in Communications*, *23*, 598–610. doi:10.1109/JSAC.2004.842547

Sun, B., Osborne, L., Xiao, Y., & Guizani, S. (2007). Intrusion Detection Techniques in Mobile Ad Hoc and Wireless Sensor Networks. *IEEE Wireless Communications*, *14*, 56–63. doi:10.1109/MWC.2007.4396943

Sun, Y., Yu, W., Han, Z., & Liu, K. J. (2006). Information Theoretic Framework of Trust Modeling and Evaluation for Ad Hoc Networks. *IEEE Journal on Selected Areas in Communications*, *24*, 305–317. doi:10.1109/JSAC.2005.861389

Suryanarayana, G., Erenkrantz, J., & Taylor, R. (2005). An Architectural Approach for Decentralized Trust Management. *IEEE Internet Computing*, *9*, 16–23. doi:10.1109/MIC.2005.119

Theodorakopoulos, G., & Baras, J. S. (2006). On Trust Models and Trust Evaluation Metrics for Ad Hoc Networks. *IEEE Journal on Selected Areas in Communications*, *24*, 318–328. doi:10.1109/JSAC.2005.861390

UCINET. 2007. Received September 25, 2007 from http://www.analytictech.com/

Watts, D. (1999). *Small Worlds: The Dynamics of Networks between Order and Randomness*. Princeton, NJ: Princeton University Press.

Xiong, L., & Liu, L. (2004). PeerTrust: Supporting Reputation-Based Trust for Peer-to-Peer Electronic Communities. *IEEE Transactions on Knowledge and Data Engineering*, *16*, 843–857. doi:10.1109/TKDE.2004.1318566

Yang, H., Shu, J., Meng, X., & Lu, S. (2006). SCAN: Self-Organized Network Layer Security in Mobile Ad Hoc Networks. *IEEE Journal on Selected Areas in Communications*, *24*, 261–273. doi:10.1109/JSAC.2005.861384

Zhang, Y., Liu, W., & Fang, Y. (2006). Securing Mobile Ad Hoc Networks with Certificateless Public Keys. *IEEE Transactions on Dependable and Secure Computing*, *3*, 386–399. doi:10.1109/TDSC.2006.58

Zhou, T., & Harn, L. 2007. Security in User-Assisted Communications. *Proceedings of IEEE Wireless Telecommunications Symposium (WTS)*.

Zhou, T., & Harn, L. 2008. Risk Management of Digital Certification in Ad Hoc and P2P Networks. *Proceedings of the 21st Canadian Conference on Electrical and Computer Engineering*.

This work was previously published in International Journal of Interdisciplinary Telecommunications and Networking, Volume 1, Issue 2, edited by Michael R. Bartolacci, and Steven R. Powell, pp. 1-15, copyright 2010 by IGI Publishing (an imprint of IGI Global).

Chapter 7
On Peer-to-Peer Location Management in Vehicular Ad Hoc Networks

Zhaomin Mo
Florida International University, USA

Hao Zhu
Florida International University, USA

Kia Makki
Florida International University, USA

Niki Pissinou
Florida International University, USA

Masoumeh Karimi
Florida International University, USA

ABSTRACT

Vehicular ad-hoc networks (VANETs) have been gained importance for the inter-vehicle communication that supports local communication between vehicles without any expensive infrastructure and considerable configuration efforts. How to provide light-weight and scalable location management service which facilitates geographic routing in VANETs remains a fundamental issue. In this paper we will present a novel peer-to-peer location management protocol, called PLM, to provide location management service in VANETs. PLM makes use of high mobility in VANETs to disseminate vehicles' historical location information over the network. A vehicle is able to predict current location of other vehicles with Kalman filtering technique. Our theoretical analysis shows that PLM is able to achieve high location information availability with a low protocol overhead and latency. The simulation results indicate that PLM can provide fairly accurate location information with quite low communication overhead in VANETs.

DOI: 10.4018/978-1-60960-505-6.ch007

INTRODUCTION

Vehicular Ad-hoc Networks (VANETs) have received considerable attention in recent years due to its potential in supporting vast value-added and customized applications. By leveraging low cost and high bandwidth wireless interface installed on each vehicle, a VANET can be deployed quickly and economically. It provides high data rates, and is more cost-effective than systems that need pre-installed infrastructures (e.g. cellular networks). Hence a wide-range of applications such as increasing road safety, providing passengers with business information and entertainment, and improving traffic flow and efficiency on the roads, can be provided using VANETs as the communication infrastructure (Riva, O., Nadeem, T., Borcea, C., & Iftode, L., 2007; Morris, R., Jannotti, J., Kaashoek, F., Li, J., & Decouto, D., 2000; Dikaiakos, M., Florides, A., Nadeem, T., & Iftode, L., 2007; Festag, A., Fler, H., Hartenstein, H., Sarma, A., & Schmitz, R., 2004). Because of the short communication range of each network interface, multi-hop data transmission may be needed for end-to-end data delivery in VANETs. Recent study in multi-hop routing in VANETs (Zhao, J., & Cao, G., 2006; Mo, Z., Zhu, H., Makki, K., & Pissinou, N., 2006; Naumov, V., Baumann, R., & Gross, T., 2006; Sun, W., Yamaguchi, H., Yukimasa, K., & Kusumoto, S., 2000; Lochert, C., Hartenstein, H., Tian, J., Fler, H., Herrmann, D., & Mauve, M., 2003) has shown that, with the global position system (GPS) and digital map (e.g. Map Mechanics, 2005), geographic routing, in which data packets are forwarded from the source to the destination with the aid of nodes' location information, has high end-to-end packet delivery ratio, low end-to-end delay and low control overhead. In Zhao et. Al studied the carry-and-forward scheme, named VADD, which delivers packets in sparse VANETs where disconnection happens frequently. In Mo, Z. et. Al (2006) a multi-hop routing protocol, called MURU, was proposed to set up robust end-to-end path for urban VANETs. An advanced greedy forwarding (AFG) (Naumov, V., et. Al, 2006) based on GPSR was proposed to increase packet delivery ratio in VANETs. All these protocols assume an efficient location management service is available to provide the source node with the destination's location. Therefore, it is fundamentally important to design a good location management scheme in VANETs to support geographic routing and other location-based applications.

Up to date some location management schemes have been proposed in mobile wireless ad hoc networks (MANETs) (Li, J., Jannotti, J., Decouto, D., Karger, D., & Morris, R., 2000; Xue, Y., Li, B., & Nahrstedt, K., 2001; Kiess, W., Fler, H., Widmer, J., & Mauve, M., 2004; Basagni, S., Chalamtac, I., Syrotiuk, V. R., & Woodward, B. A., 1998; Sasson, Y., Cavin, D., & Schiper, A., 2005) and most of them are grid-based. That is: they divide the network area into ordered grids and select location servers based on a particular grid mapping algorithm. As all these schemes adopt the client-server architecture, they usually involve two phases: location update and location query. In the location update process, mobile nodes periodically send their up-to-date location information to one or more location servers to update their locations. In the location query process, a node queries target node's location servers for target node's location information.

However, in a VANET where vehicles are highly dynamic, how to efficiently select proper location servers for a node turns out to be extremely difficult (if possible). Moreover, updating a node's location frequently and explicitly makes the client-server architecture inefficient in terms of communication overhead. Therefore an efficient location management protocol in VANETs should have the following characteristics. First, it is able to be adaptive to high mobility in VANETs. Second, since the diameter of a VANET can be large, the location management protocol should be scalable. To satisfy these two requirements, in this article we propose a novel location management protocol, called PLM (Peer-to-peer Location Management), which achieves the design goal by combining local information exchange and intelligent information processing technique. The basic idea of the PLM protocol is to disseminate a node's historical location information to other nodes in a peer-to-peer manner which does not require any pre-specified servers. By employing a carry-and-disseminate method, a node is able to disseminate its location information to other nodes far away from it through multi-relayed dissemination with the help of node mobility in VANETs. Then each node independently applies Kalman filter (Welch, G., & Bishop, G.) to process received location information. As a result, a node is able to predict other nodes' location information without querying location servers. An on-demand query method is also proposed to supplement the dissemination scheme to improve the performance of location management by reducing the location acquisition delay. The performance of PLM is extensively evaluated via simulations and the results show that PLM can provide location information with a low control overhead and a low latency.

The rest of this article is organized as follows. In Section II, we summarize the related work. We describe the details of PLM and justify the feasibility of PLM through theoretical analysis in Section III. Section IV gives the performance evaluation results, and Section V concludes the article.

RELATED WORK

A number of Location management schemes (Camp, T., Boleng, J., & Wilcox, L., 2002; Chen, C., Lemberg, H., Philips, S., den Berg, E. V., & Zhang, T., 2002; Yu, Y., Lu, G.-H., & Zhang, Z.-L., 2004; Flury, R., & Wattenhofer, R., 2006; Haas, Z. J., & Liang, B., 1999; Ko, Y-B., & Vaidya, N. H., 2000; Xue, Y., Li, B., & Nahrstedt, K., 2001; Kiess, W., et. al, 2004; Basagni, S., et. al, 1998; Sasson, Y., et al. 2205; Rezende, J., Ziviani, A., Fdida S., & Duarte, O.; Philip, S. J., & Qiao, C., 2003) have been proposed to facilitate location-based routing in MANETs. Das et. al (2005) summarized a taxonomy of location management schemes and conducted performance comparison of scalable distributed location management schemes. In general, the location management schemes can be divided into flooding-based and quorum-based approaches. In the proactive flooding-based approach, destination nodes periodically flood its location to other nodes in the network, each of which records the most recent locations of other nodes. The Simple Location Service (SLS) (Camp, T., Boleng, J., & Wilcox, L., 2002) and the DREAM location management scheme (Basagni, S., et al., 1998) are two examples of proactive flooding-based approach. In SLS, nodes periodically send tables containing the location information of a few nodes in the system to their neighbors. In DREAM, a node transmits its own location information to nearby nodes at a particular rate and to faraway nodes at another lower rate. A node updates its location table if it either receives a location information packet or if it overhears it. In DREAM, the farther two nodes are apart, the less often they exchange location information. With the existing proactive flooding-based approach, higher network mobility leads to more frequent location updates, which indicates that it is not suitable for highly dynamic VANETs. In reactive (on-demand) flooding-based approach (e.g. LAR [2000]), if a node cannot find a recent location of a destination to which it is trying to send data packets, it initiates a scoped query in the network in search of the destination. However, due to the dynamic of VANETs, it is not easy to efficiently adapt the search scope to strike the balance between the bandwidth overhead and the success rate of search.

Most of quorum-based location management schemes are grid-based and employ the client-server architecture. In quorum-based approaches, all modes in the network will agree on a mapping that creates a virtual grid in the network, in which each node's ID will be mapped to one or more nodes called *location servers*. A target node periodically updates its location servers with its up-to-date location. When a node needs the target node's location, it launches a query message which is routed through the network to one of the location servers of the target node. The location server can either answer the query message or forward it to the target node.

In the explicit quorum-based approach (Haas, Z. J., et al., 2003) each location update of a node is sent to an update quorum of available nodes, and a location query is sent to query quorum. As the two quorums overlap, the location query will be serviced by some node in the update quorum. In theory, each node's location is disseminated to $O(\sqrt{N})$ location servers. Different from the explicit quorum-based approach, the implicit quorum-based approach, location servers are selected based on a hashing function which maps the node's ID or location to a virtual hierarchical or flat grid. In hierarchical hashing-based protocols (Li, J., et al., 2000; Xue, Y., 2001), the network is recursively divided into a hierarchy of subareas. Location updates and queries are routed up and down the hierarchy. In flat hashing-based protocols (Woo, S-C. M., & Singh, S., 2001), each node's ID is mapped to a home region consisting of one or more nodes within a fixed network area. All nodes in the home region maintain the location information for the node and can service the location queries of that node. Since both the explicit quorum-based approach and the flat hashing-based approach scale poorly with the network size, most of previous research effort has been put towards the hierarchical hashing-based location management schemes in MANETS.

One example scheme can be found in (Li, J., et al., 2000), where a location management scheme named GLS was proposed. In GLS, each node is assigned with a unique ID. A small set of nodes having the IDs "closest" to a node's ID are chosen as its location servers. The node periodically updates its location servers with its up-to-date location information. GLS works well when network mobility is low and becomes much less efficient when network mobility grows. This is because a node needs to scan the entire network to find nodes with the "closest" ID as its appropriate servers. In a network with high mobility, a node needs to scan the entire network frequently in order to keep the "closest ID" property, as a result a lot of overhead is introduced. Besides, high network mobility in VANETs makes querying of location information more difficult.

Xue et al. (2001) proposed a distributed location management scheme named DLM. It makes use of hash function to map a node's ID to a set of sub-domains in the network. Several nodes in the sub-domains are picked as location servers. When a node crosses the borders of certain region or periodic updates for maintaining a soft state are transmitted, all location servers of that node have to be updated. Similar to DLM, Yu et al. (2004) proposed the scheme called HIGH-GRADE that tries to reduce the location update cost by a multilevel hierarchical location server structure and a multi-grained location information organization. In both DLM and HIGH-GRADE, when the location servers are evenly distributed in the network, the location update overhead is similar to that of network-wide flooding, which may cause considerable bandwidth overhead as node mobility increases.

S. Philip et al. (2003) proposed a hierarchical grid location management scheme called HGRID. In HGRID, the topography is divided into logical grids, and a multi-level hierarchy is established by selecting nodes located within specific locales in the topography as leaders. HGRID can achieve a slow protocol overhead increase as the number of nodes becomes larger. However, mobility is not considered in the design of HGRID.

Flury and Wattenhofer (2006) proposed a routing framework called MLS in which each node can send messages to any other node without knowing the position of the destination node. The routing (lookup) algorithm works hand in hand with a publish algorithm, through which moving nodes publish their current location on a hierarchical grid in the network. In contrast to previous work, they considered a concurrent setup where nodes are truly mobile and move even while messages are being routed towards them. The correctness and efficiency of MLS was proven and the maximum speed at which the nodes might move is upper bounded by up to 1/15 of the routing speed. However, MLS does not focus on the efficiency of location management with high node mobility.

As discussed above, the hieratical client-server architecture is not feasible for VANETs due to the high mobility in VANETs. In this article we study a peer-to-peer location management scheme (PLM) to address the impact of high mobility on the location service in VANETs. With PLM, nodes are equal peers with no centered servers. A carry-and-disseminate scheme is used to take advantage of the high mobility of VANETs to disseminate a node's location to other nodes far away from it. With target node's historical location, a node independently predicts target node's current location without querying any location servers. When target node's historical location is not available, a node can initiate an on-demand query by sending a query message to its all one-hop neighbors for the target node's location. PLM works in a distributed manner and is adaptive to the high mobility in VANETs by tuning the system parameters.

THE PLM PROTOCOL

In this section we give the details of the PLM protocol, which consists of two procedures: the mobility-assisted location information dissemination and the Kalman-filter based location information prediction. These two schemes work together to perform location management function in the network. In particular, the location information of each node is efficiently disseminated by the mobility-assisted location dissemination scheme.

Figure 1. An illustration of mobility-assisted dissemination

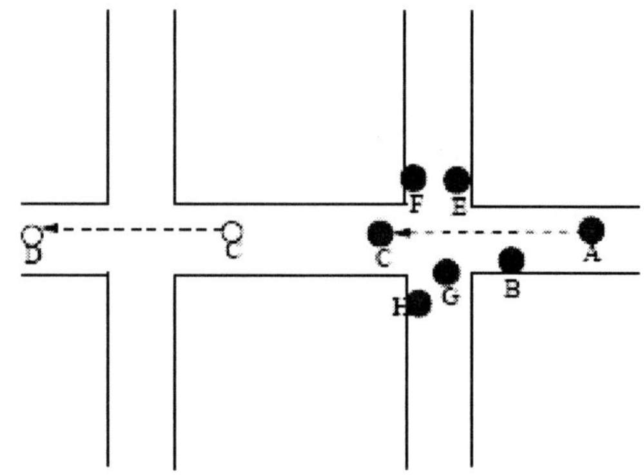

With the disseminated location information, a node is able to predict the current location of the target node by applying the Kalman-filter based location prediction scheme.

The Mobility-Assisted Location Information Dissemination

Since in VANETs nodes[1] move fast along roadway trajectory, PLM takes advantage of such characteristic and employs a carry-and-disseminate method to help disseminate a node's location information to other nodes efficiently. The basic idea is shown in Figure 1. At time t nodes are in locations depicted by dark circle, and node *A* sends its location update message, including its up-to-date location information, to all its one-hop neighbors.

Node *C* overhears node A's location information from the received message. After a certain time period, all nodes reach their new locations depicted by empty circles. Then node *C* sends its location update message, including node A's historical location information, to its neighbors, including node *D*. Hence node *D* that is not able to directly receive node *A's* historical location information can indirectly get that information with the aid of node *C's* mobility.

Following the basic idea aforementioned, we design our dissemination mechanism as follows.

In PLM each node maintains two tables: *location table* (*LT*), which stores the location update information overheard from other nodes; and *dissemination table* (*DT*), which stores the location information to be disseminated. A location update message includes a number of *Location Information Units* (*LIUs*). As shown in Table 1, an *LIU* includes node ID, node's location (represented as X coordinate and Y coordinate), direction, hop number and timestamp (the time associated with the node's location and direction). Initially the hop number is set to zero. When a node (say, node B) receives a location update message from another node (say, node A), it saves all *LIUs* in the location update message into its *LT* for future prediction usage. Meanwhile, node B compares the *LIU*s in the received location update message with the *LIUs* in its *DT*.

If an *LIU* that is in the location update message but not in its *DT* or the timestamp of the *LIU* in location update message is newer than that of the *LIU* in its *DT*, node B adds this *LIU* into its *DT*. In this way, only most recent location information will be disseminated further. To save the storage consumption, a node may only need to store the location information of other nodes which it is interested in. For example, the location information of the nodes in the contact list which is generated from the historical data. Since only recent historical information is needed when applying Kalman filter, to save storage, a node only keeps the location information within a certain freshness threshold in its *LT* and *DT*. Note that the freshness thresholds of *LT* and *DT* can be different. The pseudocode of the dissemination mechanism is shown in Figure 2.

The Hop Number Threshold Selection

After being disseminated by several other nodes, a node's *LIU* can reach other nodes at different locations. The number of times that the *LIU* is disseminated is controlled by the hop number threshold. Hence, there is a tradeoff between the availability of *LIUs* and the network bandwidth consumption. For example, if an *LIU* is only allowed to be disseminated one time, only one-hop neighbors of this node are able to receive this *LIU*. On the other hand, if an *LIU* is allowed to be disseminated many times, the corresponding bandwidth consumption becomes considerable and may hurt the overall performance of the network. In the following we will first show the calculation of the availability of an *LIU*, then we study the impacts of network conditions to the balance of the tradeoff.

We consider an area divided into m segments (a section of a street between two intersections) and with n nodes. For simplicity, we assume that every node moves in the same speed and the time that a node needs to across a segment is T[2]. Initially nodes are randomly deployed in the network. The direction selection of a node at any intersection is approximated by a first order Markov chain, in which a state in the Markov chain corresponds to a segment. A node is associated with a transition matrix *P* with size of m×m, which describes the probability of this node moving from one segment to another. The sum of each row in *P* equals to one, which describes all possible directions that a node can take at an intersection. By generating a node's transition matrix, we can get the stable state probabilities[3] of a node on each segment at any time instance.

Table 1. The format of a node's location information

X Coordinate	Y Coordinate	Dirct
Node ID	Timestamp	Hop No

Figure 2. The pseudocode of the dissemination mechanism

```
Notations:
  n_i: Node i;
  n_j: Node j;
  CT: Current time;
  FT: Freshness threshold;
  HT: Hop number threshold;
  L: Length threshold of dissemination table;
  p(i): The location updating packet of Node i;
  LT(i): The location table of Node i;
  LT(j): The location table of Node j;
  DT(i): The dissemination table of Node i;
  DT(j): The dissemination table of Node j;
Upon n_j receiving p(i) from n_i:
for each LIU(i) in p(i)
  if no LIU(j) from LT(j) has the same timestamp and node ID as those of LIU(i)
    add LIU(i) into LT(j);
  if (timestamp of LIU(i) > (CT-FT)) /* LIU(i) is fresh enough*/ and (the hop number of LIU(i) < HT)
    if LIU(i)'s node ID is not found in all LIUs in DT(j)
      increase hop number of LIU(i) by one and add LIU(i) into DT(j);
    if LIU(i) and LIU(j) in DT(j) has the same node ID and the timestamp of LIU(i) > the timestamp of LIU(j)
      increase hop number of LIU(i) by one, add LIU(i) into DT(j) and remove LIU(j) from DT(j);
    if there exists an LIU(k) in DT(j) that has the same timestamp and node ID as those of LIU(i)
      remove LIU(k) from DT(j);
Upon n_i sending location updating message:
for each LIU(i) in DT(i)
  if (timestamp of LIU(i) > (CT-FT) and hop number of LIU(i) < HT)
    add LIU(i) into p(i);
if number of LIU in p(i) > L
  remove the LIUs with oldest timestamps in p(i) to fit L;
disseminate p(i);
```

An example of generating a row of transition matrix at an intersection is shown in Figure 3. A node on segment 1 moving towards east arrives at an intersection. The probabilities that this node moves to segment 2, segment 3, segment 1 and segment 4 are 0.70, 0.13, 0.03 and 0.14, respectively; and the probability that this node moves to any other segments is 0. Thus we can get the following elements in the transition matrix for this node as follows: $P[1][1] = 0.03$; $P[1][2] = 0.7$; $P[1][3] = 0.13$ and $P[1][4] = 0.14$. A more realistic mobility model can be simulated by letting a node having different transition matrix at different intersections. A node's probability vector, with size of $1 \times m$, which describes this node's possible location at any time interval, can be calculated with this node's transition matrix. For instance, node i is initially located on segment $s(i)$ and moves according to its transition matrixes. At time t, the probability vector of node i's location is $s(i) * \Pi_{r=1}^{\lceil t/T \rceil} P_i(r)$ where $P_i(r)$ is node i's r^{th} transition matrix at the rth time interval. Since the time interval T is much longer than the time needed to transmit a packet (queuing delay plus transmission delay), it is reasonable to assume that two nodes are able to exchange packets when they are on the same segment (or, equivalently, stay in the same state). As a result, after n time intervals, a node (say, node A) can receive another node's (say, node B) location information generated at time t_0 if one of the following events Since we assume each node moves within the area, the Markov chain of each node is recurrent occurs:

1. Node A and node B meet with each other in a segment during n time intervals, and exchange location update messages. Such situation is called zero - relay - dissemination.
2. Node B disseminates location update message to another node (say, node C); and then node C meets with node A and sends node B0s location information to node A. Such situation is called one-relay-dissemination.
3. Node B0s location information is indirectly disseminated to node A by more than one relays during n time intervals. Such situation is called multiple-relay-dissemination.

Thus, the probability that B0s location information is available for node A during n time intervals is the combination of the probabilities of the three cases mentioned above. We will show the calculation of probabilities of the first two cases. The probability that zero relay dissemination occurs at least once during n time intervals, denoted by $\Pr^0_{A,B}(1,n)$, is calculated as following (See Box 1) where $s(A)$ and $s(B)$ are the initial location of node A and B respectively, and $P_A(i)$ is the transition matrix for node A at i^{th} time interval. The operation of $P(i) \otimes P(j)$ is defined as following (See Box 2) where m is the number of segments in the area.

Figure 3. An example of how to generate a row in transition matrix

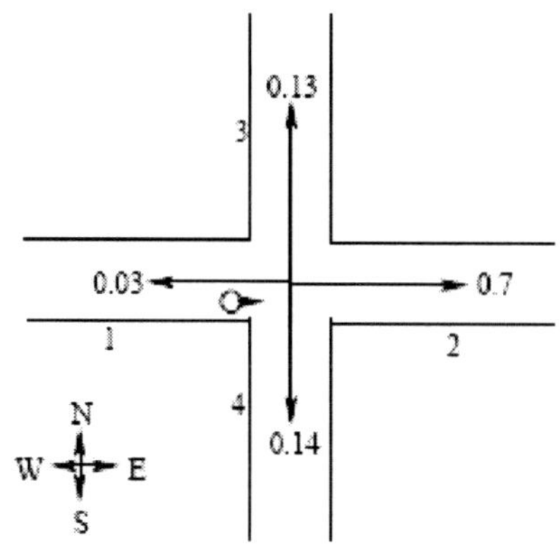

The probability of one relay dissemination through node C during n time intervals, denoted by $\Pr^1_{A,B,C}(1,n)$, is calculated as following (See Box 3).

Then, the probability of one relay dissemination through any node is shown as following:

$$\Pr^1_{A,B}(1,n) = 1 - \prod_{k \in H}(1 - \Pr^1_{A,B,h}(1,n)) \qquad (4)$$

Box 1.

$$\Pr^0_{A,B}(1,n) = 1 - \prod_{q=1}^{q=n}(1 - (s(A) * \prod_{r=1}^{q} P_A(r)) \otimes (s(B) * \prod_{r=1}^{q} P_B(r))) \qquad (1)$$

Box 2.

$$P(i) \otimes P(j) = 1 - \prod_{k=1}^{m}\prod_{l=1}^{n}(1 - P(i)[k]\,[l] * P(j)[k]\,[l]) \quad (2)$$

Box 3.

$$\Pr^{1}_{A,B,C}(1,n) = 1 - \prod_{q=1}^{n}(1 - \prod_{x=1}^{q}(1 - \Pr^{0}_{A,C}(1,x)))(1 - \prod_{y=q}^{n}(1 - \Pr^{0}_{C,B}(q,y))) \quad (3)$$

where h can be any node except node A and B and $\Pr^{1}_{A,B}(1,n)$ is defined in Equation (3). By using similar method, the probability of two or more relay dissemination can be calculated. To save space, we do not show the detailed calculation.

A higher availability of *LIU* can be achieved with more relayed disseminations, whereas more bandwidth consumption would be consumed. Therefore in order to save bandwidth consumption, a *hop threshold* is used to limit the maximum hop number that an *LIU* is allowed to be disseminated. An example of location information availability in a network with 200 nodes and 60 segments is shown in Figure 4. We observe that the availability when the hop threshold is set to 1 is much lower than the availability when the hop threshold is set to 2 or 3. Compared with the hop threshold of 3, the availability when the hop threshold is set to 2 is just a little lower, while much overhead can be saved with the hop threshold of 2.

Figure 4. Location information availabile

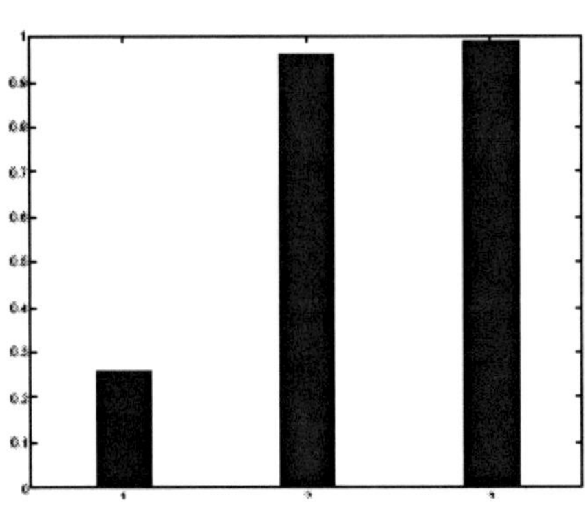

Our study indicates that network mobility and density also affect the availability of LIU. As a node moves faster, it is able to meet more nodes, therefore there are more chances for other nodes to get its location update messages. An simulation of 200 nodes moving with different average speeds in a network was conducted and the simulation results are shown in Figure 5 (a). It shows that the location information availability increases as the average node velocity increases. Such property indicates our dissemination mechanism can exploit high mobility of VANETs. In addition, the location information availability increases as the node density becomes higher. The reason is that higher density brings more relay nodes participating in disseminating location update messages; and higher location information availability can be achieved. A simulation with different node densities was conducted and the simulation results are shown in Figure 5(b). We can see that higher node density does bring higher location information availability.

Figure 5. Impact of the node velocity and node density on the location information availability

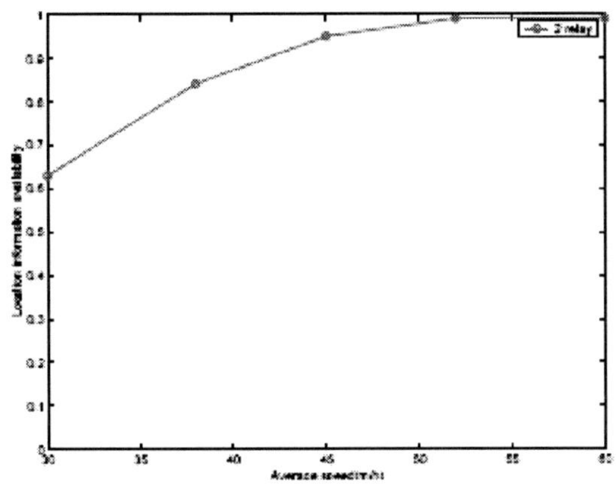

The On-Demand Query Mechanism

Following the scenario in the previous section, when node *B's* historical location information is not available for node *A*, node *A* needs to defer the prediction until it gets enough historical location information of node *B* from newly received location update messages. Although this passive approach is simple, and does not incur extra communication overhead, a considerable delay may be introduced when node *A* is waiting for node *B's* historical location information. To reduce the waiting time, in this article we propose an on-demand mechanism, called on-demand query, as

Figure 6. The pseudocode of the on-demand query

Notations:

n_i: Node i;
n_j: Node j;
n_k: Node k;

Upon n_i in need of n_j's current location

if n_i fails in predicting n_j's location
  n_i broadcasts query message with a unique sequence number to its one-hop neighbors asking for n_j's current location;

Upon n_k receiving n_i's query message

if n_k succeeds in predicting n_j's current location
  n_k replies n_i including n_j's predicted location and the sequence number in the query message;
else
  n_k drops the query message;

Upon n_i receiving n_k's reply message

if n_i never receives a reply message with the sequence number included in the reply message before
  n_i processes the reply message and gets n_j's up-to-date location;
else
  n_i drops the reply message;

a supplement component of *PLM*. When the on-demand query is initiated, node *A* sends a query message to its one-hop neighbors asking for node *B's* location information. After receiving the query message, a node (say, node *C*) checks its LT to see if node B0s historical location information is available. If yes, node *C* replies node *A* with the message containing node *B's* historical location information. As long as at least one of node A0s neighbors replies the query message, node A is able to get node *B's* location information. Thereafter node A has a shorter acquisition delay of node *B's* location information.

In a reply message, several *LIUs* may be included in the message to carry node *B's* historical locations with different timestamps. In order to reduce the size of reply messages, an alternative is that node *C* predicts node *B's* location and includes only the predicted location, which only occupies one *LIU*, instead of multiple *LIUs* in the reply message. If multiple replies from different neighbors are received by node *A*, to save waiting time, node *A* only uses the one that is received earliest, and other replies are ignored. If node *C* doesn't have node *B's* historical location information, it just drops the query message.

By using the on-demand query, a node is able to get another node's historical (or predicted) location information which is not available to it under controlled disseminations. Since the query is on-demand, only very limited overhead is involved. The simulation results shown in Section IV indicate that on-demand query only brings little extra overhead, while the location acquisition delay can be reduced greatly.

The Kalman-Filter Based Location Prediction

Kalman filter and extended Kalman filter are well-known tools used to estimate a system's past, present or future state by using a time sequence of measurements of the system state and a statistical model that characterizes the system and measurement errors, along with initial conditions (Simon, D., 2001 Kalman and extended Kalman filters: Concept, Derivation and Properties). In cellular networks, Kalman filter and extended Kalman filter have been widely used for user locating and tracking (Fan, D., Zhang, R., & Jin, L., 2003; Tung, T., & Jamalipour, A., 2003; Pathirana, P., Savkin, A., & Jha, S., 2004). In PLM, with node *B's* historical location information stored in node *A's LT*, node A is able to predict node *B's* current location information by implementing Kalman filter or extended Kalman filter. Compared with Kalman filter, extended Kalman filter can provide higher prediction accuracy, but much more calculation overhead could be involved. To save calculation overhead, in this article we will focus on Kalman filter.

Since location update messages are disseminated periodically (e.g., every 5 seconds), any two *LIUs* of node *B* have at least 5 seconds' time difference. In order to fill up the time slots between two historical location updates, interpolation is implemented. In our current work, liner interpolation (Interpolation) technique is adopted. After interpolation, a series of node B0s historical location information are generated as the input for node *A's* Kalman filter to predict node *B's* current location. In order to describe a node's location and mobility status in Kalman filter, four system states are needed: the node's location at x-dimension and y-dimension, and the node's velocity projected to x-dimension and y-dimension. The state vector *x* with time index *k* is shown as below:

$$x_k = (px_k, py_k, vx_k, vy_k)^T \quad (5)$$

Where px_k and py_k are the location at time index *k*, vx_k and vy_k are the projected velocity at time index *k*. We assume a node's speed changes linearly, thus the state at time *k* +1 can be calculated as following:

$$x_{k+1} = Dx_k + w_k \quad (6)$$

and the input equation is shown in the following:

$$y_k = Cx_k + z_k \quad (7)$$

where D and C are matrix index, y_k is the measurement based on the input x_k, w is process noise, and z is GPS measurement noise. Matrix D relates the state at the previous time step to the state at the current step. Matrix C relates the state to the input x_k, which is from the disseminated location information. Nodes moving towards different directions may implement different matrix D. An example of the matrix D when nodes move along X-dimension is shown as following:

$$D = \begin{pmatrix} 1 & t & 0 & 0 \\ 0 & 1 & 0 & 0 \\ 0 & 0 & 1 & 0 \\ 0 & 0 & 0 & 0 \end{pmatrix} \quad (8)$$

where t is the time interval between two Kalman filter's inputs. The matrix C is shown as following:

$$C = \begin{pmatrix} 1 & 0 & 0 & 0 \\ 0 & 0 & 1 & 0 \end{pmatrix} \quad (9)$$

With all these parameters, Kalman filter takes recurrent steps to calculate a node's current or future states by using the following equations [27].

$$K_k = DP_kC^T (CP_kC^T + S_z)^{-1} \quad (10)$$

$$x_{k+1} = Dx_k + K_k(y_k+1 - Cx_k) \quad (11)$$

$$P_{k+1} = DP_kD^T + S_w - DP_kC^T S^{-1}_z CP_kD^T \quad (12)$$

where S_z is GPS measurement noise covariance, S_w is process noise covariance, K matrix is Kalman gain and P matrix is estimation error covariance.

DISCUSSIONS

Compared with the client-server architecture, which is commonly adopted by traditional location management schemes in MANETs, PLM has the following merits. First, PLM works in a peer-to-peer manner without location updates and queries on location servers, and is distributed and scalable. Second, different from traditional location management schemes where before sending out data packets, a node has to explicitly query for the target node's location and wait for the reply from location servers, in PLM a node can get target node's location information in real-time by applying Kalman filter to do location prediction by itself. Third, in PLM, a node's up-to-date location information is achieved by executing a prediction algorithm; thereafter the requirement for the frequency of location update is relieved. While with traditional location management scheme, a node may have to frequently update its location servers to keep them with its up-to-date location information. Fourth, with client-server architecture, the query overhead may increase proportionally as the number of query increases. Thus traditional location management scheme may not be practical in dynamic network with high node mobility. In contrast, in PLM such problem can be avoided since there is no need of location servers that handle location update and query for certain nodes in a centralized manner.

PERFORMANCE EVALUATION

We build the simulation using the ns-2 simulator (Vint Project), which supports for simulations of routing protocols in wireless networks. Vehicles are equipped with 802.11 interface, global position system (GPS), and digital maps (MapMechanics, 2005). Thus a vehicle is able to get its location (imprecise) and street-level road geometry at any instance. IEEE 802.11 DCF protocol is used as the MAC layer transmission protocol, and the

Figure 7. Illustration of vehicle traces in Zurich

shadowing propagation model is used to reflect the dynamic channel conditions in VANETs.

SIMULATIONS RESULTS BASED ON REALISTIC TRACES

To evaluate the performance of our approach in the context of vehicular networks, we simulate our protocol based on the traffic traces generated by a multi-agent microscopic traffic simulator (MMTS). It was developed by K. Nagel at ETH Zurich (2006). These traces contain mobility patterns of 260,000 vehicles on real road maps in the canton of Zurich within 24 hours (Figure 7 ETH Zurich Vehicular Network Traces). In our simulation, we extracted a smaller area from the 260km x 260km area of the traces. This is because simulation of 260,000 vehicles makes the simulation extremely slow. The area that we selected is 12km × 10km. 1200 vehicles are randomly picked from the area and the average speed of vehicles are linearly adjusted to measure the protocol performance in different network motilities.

In order to better evaluate the performance of PLM, we compare the performance of PLM with a location management protocol based on the client-server architecture. It is similar to GLS (Li, J. et al., 2000), except that all location servers are static and GPSR (Karp, B., & Kung, H. T., 2000) is employed in network layer to support location update and location query. The performance of this protocol is evaluated under different number of location servers to observe the impact of number of location servers. In order to evaluate the efficiency of on-demand query, two versions of PLM are evaluated: PLM with and without on-demand query. We measure three performance metrics: the location information availability rate, the average delay, and the protocol overhead. The availability rate presents the probability that the destination's historical *LIUs* are available for prediction usage at the source. The availability rate of client-server based location management protocol measures the probability that the source successfully gets the destination's location by location query. The availability of PLM with or without on-demand query measures the probability that destination's LIUs is available for the source. The delay of client-server based location management protocol comes from the time spent on querying location servers. The delay of PLM with on-demand query comes from the time on sending location query message and waiting for reply from one hop neighbors of the querying node. In PLM without on-demand query, the delay comes from the time that a node passively waits for target node's historical location informa-

Figure 8. Performance comparison in realistic scenario

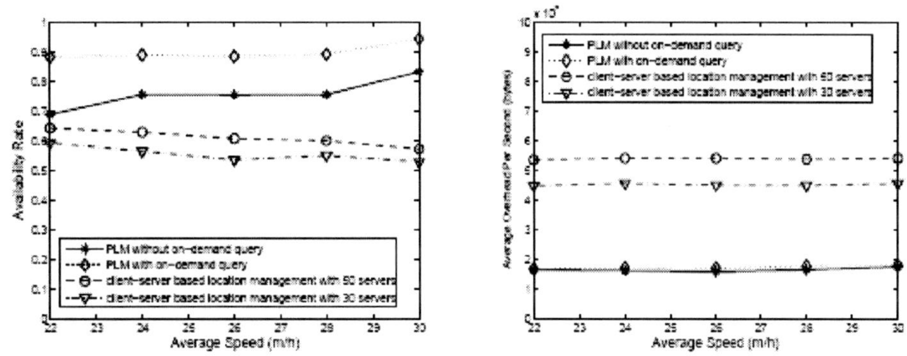

(a) Location information availability

(b) Protocol overhead

tion from location update messages when target node's historical location information is not available.

Figure 8(a) shows the availability rate of the four protocols. In both versions of PLM (with and without on-demand query), the *hop threshold* is set to a small number of 2 to save protocol overhead. As a result, an *LIU* can only be disseminated no more than 2 hops. We observe that PLM without the help of the on-demand query can achieve the availability greater than 70%. With the on-demand query, the availability rate increases about 10%, this proves the effectiveness of on-demand query. The availability rate of both versions of PLM goes up as node speed increases, which verifies that node mobility does help location information dissemination in PLM. In client-server based location management service, location update and location query rely on successful routing, which is still a challenging issue in VANETs due to the high mobility, dynamic topology, and large area of VANETs. From Figure 8(a) we can see that the client-server based location management provides a lower availability rate than that of PLM. After carefully studying the simulation traces, we find that most failures are due to the loss of location updates or location queries at location servers. From Figure 8, we can see that higher number of location servers leads to higher availability rate, while higher protocol overhead is involved.

Figure 8(b) shows the average protocol overhead. Compared with PLM without on-demand query, on-demand query involves query messages and reply messages, thereafter increases the protocol overhead. As shown in Figure 8(b), on-demand query only involved very limited overhead. This is because query and reply messages are in an on-demand mode, so a lot of overhead can be saved. The overhead of client-server based location management is a lot higher than that of both versions of PLM. The higher overhead comes from frequent location update and query due to high node mobility in VANETs.

The delay comparison is shown in Figure 9. PLM with on-demand query has an average delay of lower than 0.3 seconds, which is the lowest average delay among all the protocols. It shows the efficiency of our approach. The average delay of PLM without on-demand query involved passive waiting time, thus has a higher delay than the delay of PLM with on-demand query. While in client-server based location management, query messages have to be sent to servers that may be multi-hops away from the source node. Thus the delay of client-server based location management is much higher than that of PLM.

Coupled with Routing Protocol

In this section we evaluate the performance of PLM coupled with MURU[20], a location based routing protocol in VANETs. MURU has been shown to be an efficient geographical routing for urban VANETs. To evaluate the efficiency of PLM, we evaluate the performance of MURU with the help of PLM and MURU working with ideal location service (i.e., the source node is assumed to always have the destination's current location information). The performance metrics are packet delivery ratio and end-to-end packet delay. Since the ideal location service doesn't involve any overhead, we don't compare the overhead of these two scenarios. The data traffic is generated by five source nodes randomly selected from networks and is sent to five randomly selected destinations. The number of nodes in the simulation varies from 60 to 150 to evaluate the impact of network density to our protocol performance. The simulation area is 700m£700m with block length of 100m. Simulation runs 300 seconds and each case is repeated 20 times to achieve a high confidence of the results.

Figure 10(a) shows the data delivery ratio of MURU working with PLM and ideal location service. In both scenarios, MURU can provide high data delivery ratio. The data delivery ratio increases as network density increases, this shows that MURU makes good use of network density. Compared with ideal location service, MURU with PLM only delivers less than about 10 percent of data. It indicates that PLM provides MURU with effective location service in VANETs. Figure 10(b) shows the average end-to-end delay of MURU with PLM and ideal location service. The end-to-end delay of MURU with ideal location service comes from MURU's routing delay. The end-to-end delay of MURU with PLM mainly comes from PLM's location acquisition delay.

Figure 9. Average delay comparison in realistic scenario

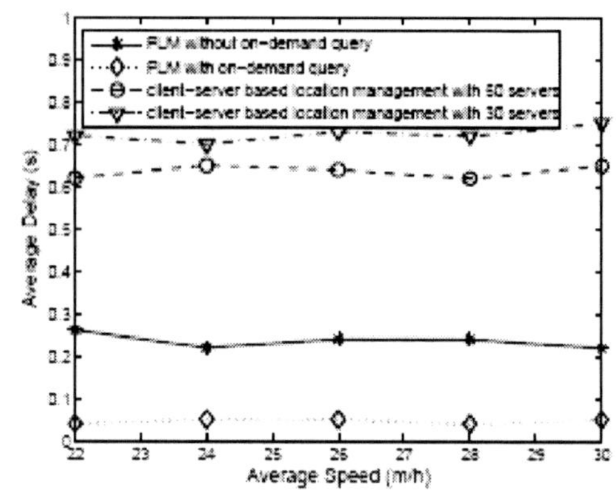

Figure 10. Performance comparison for different network densities

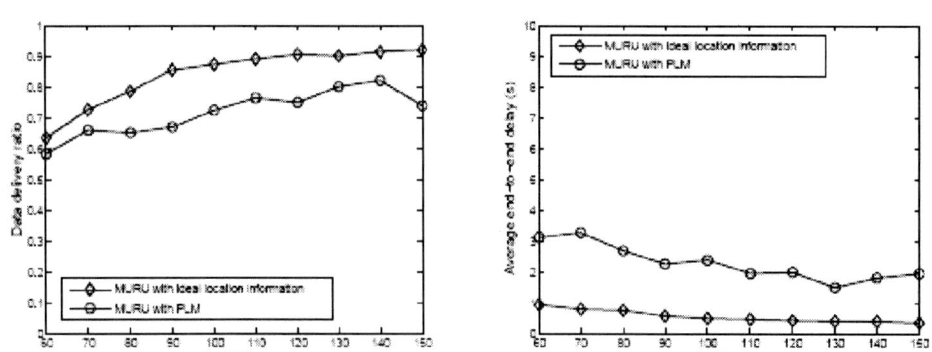

(a) Data delivery ratio

(b) Average end-to-end delay

In both scenarios, the average end-to-end delay decreases with the increase of networks density since the performance of MURU improves as network density increases. The average delay of MURU with PLM decreases faster than that of MURU with ideal location service. The reason is that higher network density improves the performance of PLM.

CONCLUSION

In this article we proposed a novel location management protocol, called PLM, to facilitate location-based applications and geographic routing protocols in VANETs. PLM utilizes the high mobility of VANETs to help disseminate location information updates in the network. An on-demand query is used to improve the location information's availability rate with the cost of very limited extra communication overhead. With other nodes' historical location information, a node is able to predict other nodes' current or future location by using Kalman filter or extended Kalman filter. PLM is designed to be distributed and scalable and doesn't need to help of any pre-install infrastructures. Compared with traditional client-server based location management schemes, PLM doesn't require concentrative location update and location query, and is more adaptable to the dynamic of VANETs. We performed theoretically analysis to show that PLM is able to achieve high location information availability with small protocol overhead.

The performance of PLM is extensively evaluated through simulations. Simulation results show PLM outperforms client-server based location management scheme in terms of availability rate, protocol overhead and average delay. Simulation results also show that PLM works well with the geographic routing protocol.

REFERENCES

Basagni, S., Chalamtac, I., Syrotiuk, V. R., & Woodward, B. A. (1998, October). A Distance Routing Effect Algorithm for Mobility. In ACM MOBICOM'98. DREAM.

Camp, T., Boleng, J., & Wilcox, L. (2002). *Location information services in mobile ad hoc networks* (pp. 3318–3324).

Chen, C., Lemberg, H., Philips, S., den Berg, E. V., & Zhang, T. (2002, March). *Slalom: A scalable location management scheme for large mobile ad-hoc networks.*

Das, S. M., Pucha, H., & Hu, Y. C. (2005, March). *Performance comparison of scalable location services for geographic ad hoc routing.*

Dikaiakos, M., Florides, A., Nadeem, T., & Iftode, L. (2007, October). Location-aware Services over Vehicular Ad-Hoc Networks using Car-to-Car Communication. *IEEE Journal on Selected Areas in Communications, 15*(8), October 2007.

ETH Zurich Vehicular Network Traces. (n.d.). http://lst.inf.ethz.ch/ad-hoc/car-traces/.

Fan, D., Zhang, R., & Jin, L. (2007). *2007.* Extended Kalman Filter Location and Tracking Method Using Single Base Station in Cellular Network. In Wireless Communications, Networking and Mobile Computing.

Festag, A., Fler, H., Hartenstein, H., Sarma, A., & Schmitz, R. (2004). FleetNet: Bringing Car-to-Car Communication into the Real World. *In 11th World Congress on ITS, Nagoya, Japan, 2004.*

Flury, R., & Wattenhofer, R. (2006). *Mls: An efficient location service for mobile ad hoc networks.*

Haas, Z. J., & Liang, B. (1999). Ad hoc mobility management with uniform quorum systems. *IEEE/ACM Transactions on Networking, 7*(2), 228–240. doi:10.1109/90.769770

Interpolation (n.d.). http://www.mathworks.com/moler/interp.pdf.

Kalman and extended Kalman filters: Concept, derivation and properties. http://users.isr.ist.utl.pt/ mir/pub/kalman.pdf.

Karp, B., & Kung, H. T. (2000, August). Greedy perimeter stateless routing for wireless networks. In ACM MOBICOM'00. GPSR.

Kiess, W., Fler, H., Widmer, J., & Mauve, M. (2004). Hierarchical Location Service for Mobile Ad-Hoc Networks. In *ACM SIGMOBILE Mobile Computing and Communications Review, 2004*.

Ko, Y.-B., & Vaidya, N. H. (2000). Location-aided routing (lar) in mobile ad hoc networks. *Wireless Networks*, *6*(4), 307–321. doi:10.1023/A:1019106118419

Li, J., Jannotti, J., Decouto, D., Karger, D., & Morris, R. (2000). A Scalable Location Service for Geographic Ad-hoc Routing. In *ACM MobiCom'00*.

Lochert, C., Hartenstein, H., Tian, J., Fler, H., Herrmann, D., & Mauve, M. (2003). A Routing Strategy for Vehicular Ad Hoc Networks in City Environments. In *IEEE Intelligent Vehicles Symposium, June 2003*.

MapMechanics. (2005, May). *GB Traffic Volumes*. www.mapmechanics.com.

Mo, Z., Zhu, H., Makki, K., & Pissinou, N. (2006). MURU: A Multi-Hop Routing protocol for Urban Vehicular Ad Hoc Networks. In *Int'l Conf. on Mobile and Ubiquitous Systems: Networking and Services (MobiQuitous 2006)*.

Morris, R., Jannotti, J., Kaashoek, F., Li, J., & Decouto, D. (2000). CarNet: A Scalable Ad Hoc Wireless Network System. In *9th ACM SIGOPS European Workshop, 2000*.

Naumov, V., Baumann, R., & Gross, T. (2006, May). An Evaluation of Inter-Vehicle Ad Hoc Networks Based on Realistic Vehicular Traces. In *ACM MOBIHOC*.

Pathirana, P., Savkin, A., & Jha, S. (2004). Robust extended Kalman filter based technique for location management in PCS networks. *Computer Communications*.

Philip, S. J., & Qiao, C. (2003). *Hierarchical grid location management for large wireless ad hoc networks*. ACM SIGMOBILE Mobile Computing and Communications Review.

Rezende, J., Ziviani, A., Fdida, S., & Duarte, O. (n.d.). *Enhancing Location Management in Mobile Ad Hoc Networks*.

Riva, O., Nadeem, T., Borcea, C., & Iftode, L. (2007). Context-aware Migratory Services in Ad Hoc Networks. *IEEE Transactions on Mobile Computing*, 2007.

Sasson, Y., Cavin, D., & Schiper, A. (2005). A Location Service Mechanism for Position-Based Multicasting in Wireless Mobile Ad hoc Networks. In *Hawaii International Conference on System Sciences (HICSS'05)*.

Simon, D. (2001). *Kalman Filtering*. Embedded System Programming.

Sun, W., Yamaguchi, H., Yukimasa, K., & Kusumoto, S. (2000). GVGrid: A QoS Routing Protocol for Vehicular Ad Hoc Networks. In *Fourteenth IEEE International Workshop on Quality of Service (IWQoS 2006)*.

Tung, T., & Jamalipour, A. (2003). A Kalman-Filter Based Paging Strategy for Cellular Networks. In *36th Annual Hawaii International Conference on System Sciences (HICSS'03)*. Vint project (n.d). http://www.isi.edu/nsnam/ns/.

Welch, G., & Bishop, G. (n.d.). *The kalman filter*. http://www.cs.unc.edu/ welch/kalman/.

Woo, S.-C. M., & Singh, S. (2001). Scalable routing protocol for ad hoc networks. *Wireless Networks*, *7*(5), 513–529. doi:10.1023/A:1016726711167

Xue, Y., Li, B., & Nahrstedt, K. (2001). A scalable location management scheme in mobile ad-hoc networks. In *IEEE Conference on Local Computer Networks*.

Yu, Y., Lu, G.-H., & Zhang, Z.-L. (2004). Enhancing location service scalability with high-grade.

Zhao, J., & Cao, G. (2006). Vehicle-Assisted Data Delivery in Vehicular Ad Hoc Networks. In *IEEE INFOCOM'06*. VADD.

ENDNOTES

1. For convenience, we use vehicle and node interchangeable in the rest of article.
2. By applying different value of T, different speeds of nodes can be simulated.
3. Since we assume each node moves within the area, the Markov chain of each node is recurrent.

This work was previously published in International Journal of Interdisciplinary Telecommunications and Networking, Volume 1, Issue 2, edited by Michael R. Bartolacci, and Steven R. Powell, pp. 28-45, copyright 2009 by IGI Publishing (an imprint of IGI Global).

Chapter 8
Multi-Layer Network Performance and Reliability Analysis

Kostas N. Oikonomou
AT&T Labs Research, USA

Rakesh K. Sinha
AT&T Labs Research, USA

Robert D. Doverspike
AT&T Labs Research, USA

ABSTRACT

The authors describe a methodology for evaluating the performability (combined performance and reliability) of large communications networks. Networks are represented by a 4-level hierarchical model, consisting of traffic matrix, network graph, "components" representing failure modes, and reliability information. Network states are assignments of modes to the network components, which usually represent network elements and their key modules, although they can be more abstract. The components can be binary or multi-modal, and each of their failure modes may change a set of attributes of the graph (e.g. the capacity or cost of a link). Their methodology also captures the effect of automatic restoration against network failures by including two common rerouting methods. To compute network performability measures, including upper and lower bounds on their cumulative distribution functions, we augment existing probabilistic state-space generation algorithms with our new "hybrid" algorithm. To characterize the network failures of highest impact, we compute the Pareto boundaries of the network's risk space. The authors have developed a network analysis tool called nperf that embodies this methodology. To illustrate the methodology and the practicality of the tool, they describe a performability analysis of three design alternatives for a large commercial IP backbone network.

DOI: 10.4018/978-1-60960-505-6.ch008

INTRODUCTION

Network service providers usually guarantee a certain level of service to their enterprise customers by specifying limits on network measurements such as down time, restoration delay, and packet delay. An example of such a guarantee, also called a "service-level agreement" (SLA), is "with 99.99% probability, at most 5% of the network traffic will be unavailable". Violation of these agreements results in a penalty to the service providers. In turn, the enterprise customers incur business losses if their networks do not perform as expected[1]. Some aspects of the service level involve only reliability, such as downtime, while others are pure performance measures, such as packet delay. Each can be evaluated or predicted in isolation by an appropriate model and technique. However, as networks become larger and more complex, it is less realistic to consider only pure reliability measures or to evaluate performance measures as if the network were always in its perfect (no-failure) state[2].

Even though the complexity of real networks often requires that performance and reliability evaluations be treated separately from each other, a combined, simultaneous evaluation, known as *performability* analysis, is more useful (Meyer, 1995). Performability analysis characterizes failures probabilistically and evaluates a performance measure (or measures) over a large set of network states, one of which is the no-failure state and the rest represent combinations of failures. Then one can estimate the expected value of a measure or estimate the probability that the measure does not exceed a threshold. These results often expose the inadequacy of separate performance and reliability analyses, but the simultaneous analysis is much more difficult. For example, if the effects of *restoration* after network failure are included, performability computation becomes significantly more complex. See Colbourn (1999) for some more on network reliability vs. performability evaluation.

Network performability, as well as stand-alone reliability or performance models, fall into two basic categories. One approach is to consider the effect of equipment failure and network restoration on the performance of a fixed "reference" connection, chosen to represent either the typical or the worst case. This allows very detailed and precise results to be obtained (Oikonomou, 2006), but ignores the interaction among connections in a network. The other approach, which we employ here, is a performability analysis of the entire network. There are two main points to note. First, the complexity of any realistic network necessitates a description in various *layers* and a *failure mapping* between these layers. Second, because the size of the state space is exponential in the number of failure modes, exploring all possible failure states is impractical. An approach proven successful in practice is to generate failures *in order of probability*, i.e. most likely first, and evaluate the performance measures on these states as they are generated.

Related Work

Colbourn (1999); Harms, Kraetzl, Colbourn, and Devitt (1999); and Sahner, Trivedi, and Puliafito (1996) provide background on reliability and performability evaluation. Algorithms for generating the states of a system in order of probability are in Yang and Kubat (1990) (1989); Gomes and Craveirinha (1998); and Jarvis and Shier (1996). Network performability studies have also appeared in the literature. Levy and Wirth (1989) study *response time* in a multiprocessor system and *call completion rate* in a communications network. Alvarez, Uysal, and Merchant (2001) study performability guarantees for the *time* required to satisfy a web request in a network with up to 50 nodes, where only nodes can fail, and without restoration. Levendovszky, Jereb, Elek, and Vesztergombi (2002) study the expected lost *traffic* in the Hungarian backbone SDH network with 52 nodes and 59 links, and no restoration.

Carlier, Li, and Lutton (1997) use a 3-level network model, and study expected lost *traffic* in a 111-node, 180-link network using *k*-shortest path restoration. Gomes and Craveirinha (1997) study a 46-node, 693-link representation of the Lisbon urban network with a 3-level performability model, and compute *blocking probabilities* for a Poisson model of the network traffic, with no restoration. Finally, layered specification of a network for the purposes of performability evaluation has been used in Gomes et al. (1997) and Carlier, Li, and Lutton (1997) which separate the network into a "physical" and a "functional" layer, and in Mendiratta (2001) which uses a special-purpose separation into "node cluster" and "call processing path" layers.

With regard to previous work, the contributions of this article are a more realistic and comprehensive multi-layer network model, representation of minimum-cost (shortest-path), near-optimal (multicommodity flow), and multicast routing during restoration, and performance measures that include lost traffic from lack of connectivity and from congestion. In particular, the methodology we describe is actively used to evaluate the performability of many types of AT&T networks (including IP backbone, fiber backbone, and local enterprise) in order to study trade-offs about their long-term evolution and design, and to help predict performance for enterprise customers.

Summary

The section *Network Performability Evaluation* of the article is a review of the network performability evaluation problem from a theoretical standpoint. *Hierarchical Network Model* describes our hierarchical, multi-layer model for a network. In *Performance Measures and Risk* we present the traffic performance measures that we use in our analyses, discuss the monotonicity property, describe how to incorporate network restoration time into the analysis, look at the risk space defined by the measures, and outline an efficient way to compute its Pareto boundary. In *Network Restoration In Performance Measures* we describe the two schemes we use for path selection in network restoration and present techniques that improve the efficiency of path computations when generating states in most-probable order. In *The* nperf *Tool* we give a short description of nperf, a network performability analysis tool that we have developed and which embodies the ideas presented in the preceding sections. Finally, in *Applications and Experimental Results* we present a case study of the performability analysis of a large commercial IP backbone network done with the nperf tool.

NETWORK PERFORMABILITY EVALUATION

We describe the problem of network performability evaluation in an abstract manner and, for clarity of exposition, restricted to networks whose components have only two modes (binary components). Models with multi-mode components are discussed in *A Hybrid State Generation Algorithm*. Useful background is in Harms, et al., (1995) and Shier (1991).

The Theoretical Problem

Let C = $\{c_1,..., c_n\}$ be a set of "components", each of which is in either a working or failed mode. Components associated with network elements may represent hardware or software that can fail. Other components may represent more complex combinations of events, such as a maintenance activity or a natural disaster, that bring down all network equipment in a certain geographical area. The main assumption about components is that c_i fails with probability q_i *independently* of any other components.

A network *state* is an assignment of a mode to every component and can be represented by a binary *n*-vector. The set of all network states *S*(C) has size 2^n, and the probability of a particular state is the product of the mode probabilities of the n components. Let $F(\cdot)$ be a vector-valued *performance measure* (function) defined on S(C), mapping each state to an m-tuple of real numbers; examples are in *Performance Measures and Risk*.

The most basic characteristic of F is its expected value over the space of network states:

$$\bar{F} = \sum_{s \in S(C)} F(s)\Pr(s). \qquad (1)$$

Our main purpose is to compute *performability guarantees* which are more sophisticated characteristics of F than just its expectation. For example, "with 99.9% probability, at most 5% of the total traffic is down" or "with 99.9% probability at most 5% of the total traffic is down, and with 99.99% probability at most 10% is down". Formally, these guarantees are statements of the type

$$\Pr(F < x_1) \geq p_1, \quad \Pr(F < x_2) \geq p_2, \ldots, \quad or$$
$$\Pr(F > y_1) \leq q_1, \quad \Pr(F > y_2) \leq q_2, \ldots \qquad (2)$$

that apply to the entire space S. These guarantees, referred to as "multi-part performability specifications" in Alvarez et al. (2001), are much stronger and more useful than just the expectation (1). For instance, they can be used to set, or to check, service-level agreements (SLAs) for the network. There is also an alternative assessment, which uses *multi-attribute risk*, examined in *Risk and the Pareto Frontier*.

Note on Terminology

In this article we make a distinction between reliability and performability, in accordance with the traditional meaning of the term "reliability" and with the usage of, e.g., Colbourn (1999); Harms et al. (1995); Sahner, Trivedi, & Puliafito (1996); Bolch, Greiner, de Meer, and Trivedi, (2006); and Levy et al. (1989). Nevertheless, some authors (e.g. Rocco, & Muselli, 2005; Ramirez-Marquez, & Gebre, 2007; and Lisnianski & Levitin, 2005; Aven & Jensen, 1999) still use "reliability" to mean what we call performability. These authors also use the terms "structure function" for performance measure, "multistate system" for system with multi-mode components, "coherent structure function" for monotone measure, and "reliability to level x" for $\Pr(F \geq x)$.

Time Interpretation

If a component is represented by a 2-state ergodic Markov process, then S is the state space of a large Markov chain, defined by the parallel composition of the n independent 2-state processes. Then it is casy to see that the limiting average of F over time as the network moves through its states is precisely the $\bar{F}$ of (1), if we take the failure probability of c_i to be the steady-state probability that the ith Markov chain is in its failed state. Since the individual Markov processes are ergodic the probabilities in (2) have similar interpretations as limiting time averages, i.e. yield the long-term *fraction of time* over which the corresponding statements are true.

Approximate Evaluation

Since the size of S(C) is normally very large, the straightforward computation of (1) or (2) is generally prohibitive. The complexity of the Performability Evaluation problem Given a set and a polynomial-time computable function F($\cdot$) defined on S(C), compute the expectation $\bar{F}$ defined by (1), depends on F. Even for very simple F this problem is computationally difficult. For example, it contains as a special case the #P-complete problem Two-terminal Network Reliability evaluation, which asks for the probability of there being a path between vertices s and t of a graph whose edges are subject to failure with some probability; see Shier (1991) and Garey and Johnson (1978) problem ND20.

Various ways of approximating $\bar{F}$ are known in the literature. Without attempting to be all-inclusive, they can be categorized as

1. Lower and upper bounds for certain F such as connectivity or maximum flow, or for special network/graph structures (Harms et al., 1995;Shier, 1991; Lisnianski et al., 2003). These bounds are based on the notions of cut and path sets, and their multi-mode generalizations, and on the principle of inclusion-exclusion (Aven et al., 1999; Lisnianski et al., 2003).
2. Monte Carlo sampling approaches (Harms et al., 1995; Carlier et al., 1997; Rocco et al., 2005).
3. Mixed Monte-Carlo and cut/path set methods, such as Ramirez-Marquez et al. (2007).
4. Probabilistic approximation algorithms for simple F, see Karger (1995).
5. "Most probable states" methods (Shier, 1991; Harms et al., 1995; Gomes et al., 1997; Yang et al., 1990; Gomes et al., 2002; Shier et al., 1990).

These methods can be classified according to the *bounds* that they produce on $\bar{F}$, i.e. algebraic or statistical, and by whether they require that F have special properties, such as monotonicity (defined below). Methods of of type 1 and 3 work for monotone F and produce algebraic bounds on $\bar{F}$. Type 2 methods work for any F and result in statistical bounds. Type 5 methods work for arbitrary F and produce algebraic bounds.

Here we adopt the "most probable states" methods, which are based on the observation that if the component failure probabilities are small, most of the probability mass of S will be concentrated on a relatively small fraction of the state space (equivalently, there will be many failure states of low probability). Let ω and α be the smallest and largest values of F over S(C), and suppose we generate the k highest-probability elements of S(C). If these have total probability P, we have algebraic lower and upper bounds on $\bar{F}$

$$\begin{aligned} \bar{F}_l &= \sum_{i=1}^{k} F(s_i)\Pr(s_i) + (1-P)\omega, \\ \bar{F}_u &= \sum_{i=1}^{k} F(s_i)\Pr(s_i) + (1-P)\alpha. \end{aligned} \tag{3}$$

(These were first pointed out in Li & Silvester, 1984)

By viewing states as sets of failed components, F is *monotone increasing* if $F(s) \le F(s')$ whenever $s \subseteq s'$, and *monotone decreasing* iff. $F(s) \ge F(s')$ The importance of monotonicity is that it is possible to improve the bounds (3) if F is monotone; some recent results are in Oikonomou (2009). The particular performance measures we use in this article are defined in *Performance Measures and Risk* and we show that some, but not all, are monotone.

A generalization of most probable states is the notion of "most important states", i.e. those which contribute the most to the value of $\bar{F}$; see Colbourn (1999) and Jarvis (1996), and the references therein. We did not adopt this – in principle more powerful – approach, because determining the states on which the function F takes large values (a) is difficult for the F and types of networks that are of interest here (cf. *Performance Measures and Risk*), and (b) depends strongly on the network design, and so does not generalize to the degree we require. Furthermore, our experience with practical networks shows that most-probable-states methods usually generate states in most-important order as well.

If one thinks of the value of F on a state as a *reward* associated with that state, the performability evaluation problem appears similar to that of evaluating a Markov reward model, a problem for which there exists a large body of knowledge, and which is still an active area of research (see, e.g. *ComputationalProbability,* 2000, and *Formal Methods for Performance Evaluation,* 2008). The key difference between the two problems is not so much the size of the state space, as the fact that in the performability evaluation problem the values of the rewards are in general not known a priori, they can be computed only *after* a state is generated.

State Generation Algorithms

There are two fundamentally different types of algorithms for generating states in order of probability, exemplified by the algorithms in Gomes (1998) and in Yang (1990). The algorithm of Gomes (1998) views the space S as a partially-ordered set (lattice), while the algorithm of Yang (1990) treats it as a binary tree. Both these algorithms work only on systems of binary components. Also, each is best suited to a different problem domain. For a monotone F, the algorithm of Yang (1990) produces better bounds on $\overline{F}$ than those of (3). Equivalently, this algorithm will need to explore fewer network states to achieve a given accuracy on $\overline{F}$. However, the algorithm works only for monotone F, and it consumes a large amount of memory, which depends strongly on the size of C; this tends to render the algorithm impractical on large networks. When we have non-monotone measures and large networks with binary components, we use the algorithm of Gomes et al. (1998) that is applicable to any measure, monotone or not, at the expense of the looser bounds (3), but whose memory usage is a very weak function of the size of C. Finally, for network models most of whose components are binary but some have more than two modes, we present a new *hybrid* algorithm in *A Hybrid State Generation Algorithm.*

Bounds on the CDF

So far we have discussed bounds only on the expectation of a measure F. The same techniques allow us to derive bounds on the *performability guarantee* $\Pr(F \leq x)$, (cdf of F) for a given x. Let the notation [C] yield 1 if the condition C is true and 0 otherwise. Then

$$\Pr(F \leq x) = \sum_{s \in S} [F(s) \leq x] \Pr(s)$$

is a sum of the form (1), so bounds of the type (3) can be found for it as well

$$\sum_{i=1}^{k} [F(s_i) \leq x] \Pr(s_i) \leq \Pr(F \leq x) \leq \sum_{i=1}^{k} [F(s_i) \leq x] \Pr(s_i) + 1 - P, \quad (4)$$

where P is the probability of the already-generated k highest-probability states as in (3).

A Hybrid State Generation Algorithm

Our experience indicates that many practical telecommunications networks can be adequately modelled by binary components together with a relatively small number of components with multiple failure modes. To our knowledge, the GC algorithm of Gomes (1998) is the most efficient for systems of purely binary components, but cannot handle any non-binary components. On the other hand, the algorithms of (Jarvis et al., (1996; Shier et al., 1990; Gomes, 2002) work with multi-mode components, including binary, but they are not nearly as efficient as the GC algorithm in the purely binary case. A newer algorithm (Rauzy, 2005), using a transformation to a shortest-path problem, is an improvement over Gomes (2002).

Oikonomou et al. (2006) provides a method for combining an *arbitrary* algorithm B for binary components and an *arbitrary* algorithm M for multi-mode components into a *hybrid* algorithm which works for combinations of binary and non-

Figure 1. Hybrid state generation scheme, combining a binary algorithm B and a (modified) multi-mode algorithm M'

```
HYBRID MOST PROBABLE STATES ALGORITHM

Set up data structures needed by algorithms B and M'
Put the most probable state in the priority queue Q
repeat {
  remove the most probable state s from Q
  exit if the accumulated probability is ≥ 1 − ε
  if s is a 1-state then
    apply the rules of B to generate its next 1-states and add them to Q
  if s contains non-binary components then
    apply the rules of M' to generate its next non 1-states and add them to Q
}
```

binary components, but is particularly efficient for mostly binary systems of components. By "mostly binary" we mean such that the number of binary failure modes is at least as large as the total number of multi-mode failure modes. All state generation algorithms that we are aware of use a priority queue which stores network states according to probability, and a set of rules for generating next states from a current state. The hybrid algorithm follows the same pattern, and is described in Figure 1; the concept of *1-states* is central to the algorithm, and will be defined shortly.

Note that the multi-mode algorithm M chosen for incorporation into the hybrid scheme needs a slight generic modification, described in Oikonomou et al. (2006); we denote the modified algorithm by M ′.

Efficiency

We now present a more complete discussion of the efficiency of the hybrid algorithm than was given in Oikonomou et al. (2006). Given the components $C = \{c_1,..., c_n\}$, we define a network state to be an n-tuple of (component id, mode) pairs. If a component has μ failure modes, we denote their probabilities by $p_0, p_1,..., p_\mu$, where p_0 is the probability of the working (non-failure) mode. We assume that each component's modes are arranged in order of decreasing probability, so $p_0 > p_1 \geq p_2 \geq$.... A *1-state* is a state in which all components are in mode 1, their most probable failure mode.

Notice that for a given set C and ε > 0 *any* algorithm, hybrid or not, for generating states in order of probability, will produce exactly the same sequence of states (ignoring the ordering of states with equal probabilities). So the algorithms M and (B,M ′) will generate the same sequence s_0, s_1, s_2,... when run on C. The gain of the hybrid (B,M ′) over M comes from lines 6 and 7 in Figure 1, and is based on the assumption that B generates 1-states more efficiently than M. The gain is quantified by the fraction of states among s_0, s_1,... that are 1-states. This fraction is difficult to calculate, but, intuitively, it can be claimed to be close to 1 as follows: Let S_k be the set of states consisting of k failed components. Then

Claim 1

A 1-state in S_k involving a particular set of k components is more probable than any other state in S_k involving these components.

Figure 2. Maximum queue size as a function of the number of generated states until P = 0.998 is achieved. This example uses the binary algorithm of Gomes et al. (1998) and the multi-mode algorithm called POSE2 in Shier et al. (1990).

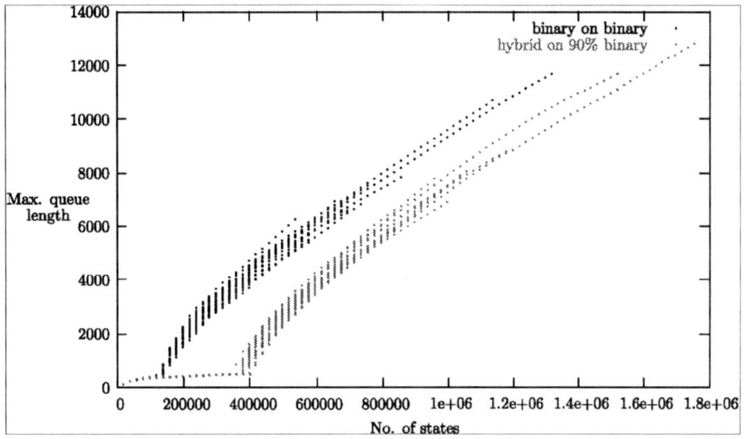

Claim 2

For small k, a lot of the states of S_k are 1-states.

Claim 1 follows from the assumed ordering of the mode probabilities. To substantiate Claim 2, suppose C contains b binary components and m multi-mode components, which have a total of M failure modes. 1-states with k components can be formed by choosing k out of $b + m$ failure modes. Any state with k components can be formed by choosing k out of $b + M$ failure modes, but choices involving two or more modes among the M that belong to the same component will be invalid. Therefore

$$\frac{\text{no. of } 1-\text{states in } S_k}{\text{no. of all states in } S_k} > \frac{\begin{pmatrix} b+m \\ k \end{pmatrix}}{\begin{pmatrix} b+M \\ k \end{pmatrix}} > \left(\frac{b+m-k+1}{b+M-k+1} \right)^k . \qquad (5)$$

In a "mostly binary" network as defined above we have $b > M$, and for relatively small k the ratio in (5) is comparable to 1, approaching 1 as b becomes larger with respect to M. E.g., for $k \leq 3$, if $M = b/10$, the ratio is ≥ 0.75 for $b \geq 100$.

Now if M is comparable to b, the fraction of 1-states in S_k is going to be smaller, but in any practical situation, we only generate a small subset of states. Because the 1-states have high probabilities, we can make an intuitive argument that 1-states form a large percentage of the *generated* states $s_0, s_1, \ldots$

Suppose we sort all $b + M$ failure modes in the decreasing order of probability. A network state corresponds to a subset of these modes and its probability is proportional to the product of the corresponding mode probabilities. Let us consider the states in S_3 in a system with $b + m = 500$ components. There are more than $5 \cdot 10^6$ triplets formed out of the first 250 modes, and in any practical situation we expect most of the states to be formed out of these first 250 modes. This can be seen as follows: the mode probabilities of any c_i are in decreasing order, so a large number of the first 250 modes will be mode 1. If, for example, 80% of them are mode 1, then the fraction of 1-states among all triplets formed out of the first 250 modes is about $(0.8)^3 \geq 1/2$ (think of selecting a mode at random out of 250). In other words, more than half of the states of S_3 that are generated will be 1-states.

Figure 2 supports this argument by showing the experimental maximum queue length as a function of the number of generated states in a network that is not mostly binary. The black points show the behavior of the binary algorithm on a system with 500 purely binary components. The red points (lower) show the behavior of the hybrid algorithm on the same system, but where 50 of the components have been made to have 10 modes each. Each algorithm is run with 50 randomly-selected values of p_0 in [0.999, 1), and the failure modes of the multi-mode components have equal probabilities. It can be seen that the efficiency of the hybrid algorithm is competitive with that of the binary algorithm.

HIERARCHICAL NETWORK MODEL

Figure 3 shows two views of a network: a view of the "real" network, and the view adopted by the performability model. Each view consists of a number of layers/levels, and the figure indicates the correspondence between them. The real network view is divided into *traffic*, *transport*, and *physical* layers. Both "transport" and "physical" are commonly-used terms in the networking field; we clarify later the sense in which we use them. The l.h.s. of Figure 3 shows an IP network carried over an optical transport layer, but the "real" network could also be of an entirely different type, such as a set of satellites that communicate among themselves and a set of ground stations via microwave links, whereas the ground stations themselves are connected by a terrestrial network.

The performability model view has a *demand* level, a *graph* level (multigraph, in general), a *component* level, and a *reliability* level (network: layers, model: levels). The demand level specifies the amount, and possibly type, of traffic flowing from one node of the graph to another via the graph's edges. The graph is the level at which the network routing and restoration algorithms operate, so graph edges have associated *capacities* and (routing) *costs*.

At the component level, each component corresponds to an independent failure mechanism, as explained in *Network Performability Evaluation*; failure of a component may affect a whole set of graph-level elements. The components may, in fact, have an arbitrary number of failure modes; see *Attributes for Graph Elements, Multi-Mode Components* and *Markov Models for Components*. In general, the effect on a graph element of a component entering one of its failure modes is to *change some of the element's attributes*. For example, the capacity of an edge may decrease, or a node may become unable to perform routing. Depending on the final values of the attributes, e.g. edge capacity = 0, the graph element may be considered "failed". Finally, at the reliability level, a failure mode of a component is characterized simply by its mean time between failures (MTBF) and mean time to repair (MTTR).

The model view in the right side of Figure 3 implicitly assumes that network restoration happens at only one level. However, multi-mode components afford the capability to model restoration occurring at more than one network layer; for an example, see Oikonomou, Ramakrishnan, Doverspike, Chiu, Martinez-Heath, and Sinha (2007).

Example for an IP-Over-Optical Network

Today's typical commercial packet network consists of IP routers whose links, connecting the routers, are transported by an underlying optical network. For such a network we illustrate how we model the layers on the left side of Figure 3 and then map them to the model on the right side. Our discussion is motivated by Phillips, Reingold, and Doverspike (2002).

Traffic Layer

Based on an estimate of the peak or average traffic pattern, we create a matrix giving the demand (typically called "flows") between each pair of routers. A demand has a rate, and possibly a type associated with it.

Figure 3. The layers of an IP-over-optical network (left) and the corresponding model levels (right). We use the term "layer" for the actual network, and the term "level" for the model.

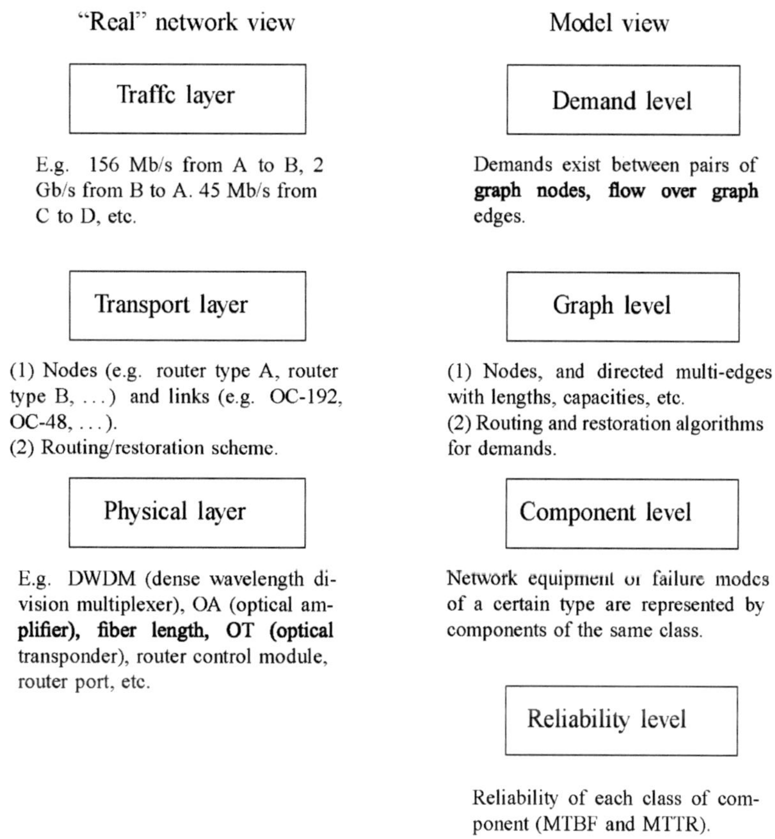

Transport Layer Nodes

A network node represents an IP router. At the component level this node explodes into one data plane, one control plane, one hardware and software upgrade component, and a number of networking interfaces (line cards/ports). The data plane, or switching fabric, is responsible for routing packets, while the control plane computes routing tables and processes other network signalling protocols (e.g., OSPF, BGP). When a data plane component fails, all the links incident to its router fail. When a control plane component fails, the router continues to switch packets, but cannot participate in rerouting, including restoration. Failure of a port component fails the corresponding edge(s). Finally, at the reliability level we specify an MTBF and an MTTR for each class of component.

Transport Layer Links

A link between routers fails if either one of its endpoint-port components fail, or the node data plane fails or a lower-layer component over which it is routed fails, such as a network span failure, discussed next. Two network nodes may be connected by multiple parallel links. These parallel links may be grouped into a type of virtual link called a *composite link* (also "aggregate" or "bundled" link), whose capacity is the sum of the capacities of its constituent links. For the purposes of IP routing, the routers see only the one composite link. When a constituent link fails, the capacity of the composite link is reduced. A composite link fails (or more precisely is "taken out of service") when the aggregate capacity of its non-failed constituent links falls below a specified threshold. An edge in the nperf graph level has a vector capacity, and so can directly represent a composite link.

Physical Layer Spans

We use the term "span" to refer to the network equipment and media (such as optical fiber) at the physical layer that carries the transport-layer links. Failure of a span component affects all transport-layer links which are routed over this span. When modeling an IP-over-optical layered network, the physical layer usually uses Dense Wavelength Division Multiplexing (DWDM) and a span consists of a concatenation of point-to-point DWDM systems called Optical Transport Systems (OTS)[3]. In turn, an OTS is composed of many elements, such as optical multiplexers/demultiplexers, optical amplifiers, and optical transponders. Also, a basic constraint in commercial transport networks is that a span is considered to be working only if *both* of its directions are working. With this assumption, it is not difficult to compute the failure probability of a span based on the failure probabilities of its individual elements in both directions. Thus, for simplicity, we generally represent a network span by a single aggregate component whose MTBF and MTTR are calculated as explained in *Markov Models for Components*.

Other Components

A bundle of fibers that is likely to fail together because they are contained in a single conduit/bundle can be represented by a *fiber cut* component that brings down all network spans (hence higher IP-layer links) that include this fiber bundle. Other types of catastrophic failures of sets of graph nodes and edges may be similarly represented.

Attributes for Graph Elements, Multi-Mode Components

A distinguishing feature of our model is that it allows graph nodes or edges to have multiple (possibly vector-valued) *attributes*, whose values can be set by the modes of the components. Having such attributes at the graph level greatly increases the representational convenience of the model.

Attributes

As a simple example, suppose we want to model a composite link, as defined earlier, that consists of two member links e_1, e_2, each of the same capacity c. In the real network e_1 can fail because of a failed "port" component e_1, p_2 because of a port p_2, and both edges can fail because of a "span" component S, as shown in Figure 4, left.

Now we can calculate a measure F for a network including such a link in a model with binary components and with ordinary graph elements, by examining the states of p_1, p_2 and S and implementing the truth table of Figure 4, right. However, if the network includes many such instances of composite links, this requires writing significant custom code in the computation of the measure F (as will become clearer in *Performance Measures and Risk* and *Network Restoration in Performance Measures*).

Now consider allowing graph edges to have a *multi-element capacity* attribute, **cap[i]**, $i = 1, 2, \ldots$, with the convention that the edge's capacity is the sum of the **cap[i]**. Then modelling our composite link becomes simple: with two capacity elements, when p_i fails, $i = 1, 2$, it sets **cap[i]** to 0, whereas when S fails, it sets both **cap[1]** and **cap[2]** to 0. The point is that this association between component modes and element attributes can be defined, and changed, entirely independently of the performance measure.

Figure 4. A composite line and its capacity table

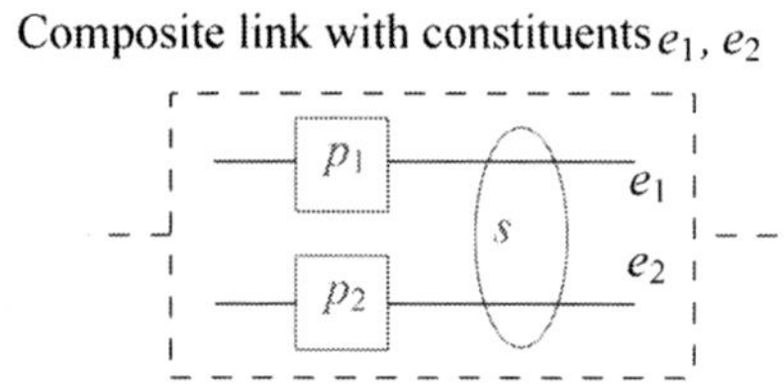

s	p_1	p_2	Link capacity
0	-	-	0
1	0	0	0
1	0	1	c
1	1	0	c
1	1	1	$2c$

Multi-Mode Components

Components with multiple (more than 1) failure modes are not new to our work; see, e.g., Jarvis et al. (1996) for some examples. One application to our network setting is in modelling router software upgrades or maintenance, whose impacts on the network are similar to brief router failures. To minimize these impacts, the maintenance/upgrade policy in most commercial networks is that only a single router is maintained at a time. Representing these mutually-exclusive events with binary components would require implementing truth tables similar to the one in Figure 4, but, for a network with n routers, it is much more convenient to define a $(n + 1)$-mode component, with mode i corresponding to maintaining the ith router, and the last mode representing no router in maintenance mode. A more sophisticated example is in Oikonomou et al. (2007), where a component with a few hundred modes is used to abstract an entire network restoration layer.

Combination

The nperf tool combines multi-mode components and graph element attributes in a general way, by allowing any component to change an arbitrary set of graph-level attributes upon entering one of its failure modes. Taking this one step farther, the *traffic matrix* may be thought of as an attribute of the graph as a whole, and a component may sometimes change the traffic matrix. As an example of where this may be useful, consider a refinement of the above network maintenance policy, where the router upgrade happens only when the network traffic is at a minimum.

Markov Models for Components

Each component of the network model is represented by a simple continuous-time Markov process with states G and B ("good" and "bad"), and parameters λ and μ (failure and repair rates), as shown in Figure 5.

Note that the multi-mode model assumes that every failure is followed by a return to the working state. We use these models for components that correspond to a single piece of equipment, as well for components that represent a connection of several pieces of equipment in series, e.g. the spans of *Example for an IP-Over-Optical Network*. In the first case λ and μ are directly available from the MTBF and MTTR of each mode, but in the second they are calculated, as follows.

If we have a series connection of k independent binary blocks, the corresponding Markov model has a single "up" state and 2^k-1 "down" states. We approximate this model by a 2-state model of the form shown in Figure 5a as follows. The 2-state model has states G and B. G corresponds to the single working state of the 2^k-state model, and B to the set of failure states. Let $\pi(G)$ be the steady-state probability of state G and $\pi(B)$ the sum of the steady-state probabilities of the states in B. Then the parameters of the 2-state model are

Figure 5. Markov models for binary and multi-mode components

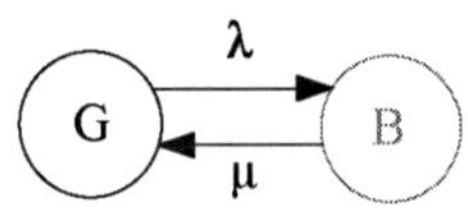

(a) Binary component (one failure mode)

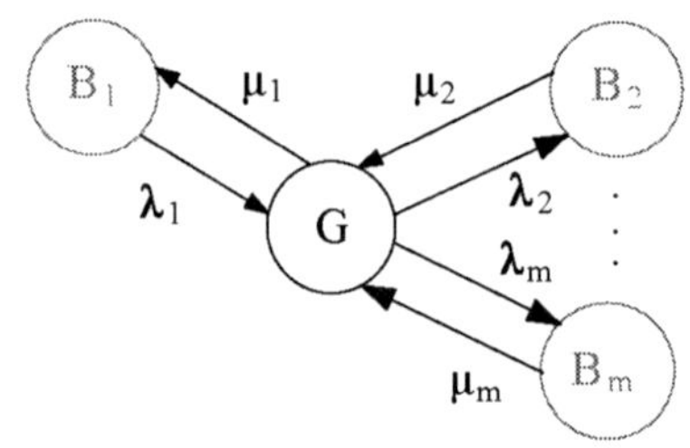

(b) Multi-mode "star" component with m failure modes

$$\Lambda = \sum_i \lambda_i, \quad M = \Lambda \frac{\pi(G)}{\pi(B)}, \quad \text{where } \pi(G) = \prod_i \frac{\mu_i}{\lambda_i + \mu_i}, \tag{6}$$

and λ_i and μ_i are the failure and repair rates of components 1,..., *k*. When $\lambda_i << \mu_i$, the above yields $\pi(B) = q \approx \sum_i q_i$, i.e., as expected, the failure probability of the series system is close to the sum of the failure probabilities of the *k* blocks. This approximation has two desirable properties: (a) the corresponding steady-state probabilities of the original and aggregate models are equal, and (b) the exponential sojourn time of the 2-state model in each of its states equals the approximately exponential sojourn time of the original model in each of the sets *G* and *B*. (The derivation of [6] involves the details of the 2-state approximation, which is based on Keilson, 1979.)

PERFORMANCE MEASURES AND RISK

Traffic Measures

Given a network specified by the hierarchical model of *Hierarchical Network Model*, we define the following performance measures on every network state *S*:

$t_{aff}(s)$ = total traffic affected in state *s*,

$t_{fail}(s)$ = total traffic over failed edges in state *s*,

$t_{lrt}(s)$ = total traffic lost because of no route in state *s*,

$t_{lcg}(s)$ = total traffic lost because of congested edges in state *s*, (7)

In our IP-optical-layer example, a *demand* corresponds to a router source-destination pair and we use *traffic* to mean size (volume) of a demand or set of demands. A demand is *affected* when an edge on its route fails. A demand *fails* when a link (multiedge) on its route fails. A failed demand is *lost*, and remains so if no path for it has been found after the restoration process completes.

If the network routing allows congestion, the total traffic lost because of congestion can be computed by finding the maximum total flow F that can be carried without congestion and subtracting it from the total traffic in the network. Let *D* be the set of all subdemands (a demand may be split and routed over multiple paths; we refer to each of these as a "subdemand"), and D_e the set of subdemands that route over edge *e*. Also let f_d be the carried flow corresponding to subdemand *d*. Then F is computed by solving the linear program

$$\mathcal{F} = \max \sum_{d \in D} f_d$$
$$\text{subject to}$$
$$\forall e \sum_{d \in D_e} f_d \leq c_e U_c, \qquad (8)$$
$$f_d \leq v_d,$$

where c_e is the capacity of edge e, $U_c \in (0, 1]$ is the congestion threshold, and v_d is the requested flow of subdemand d. Finally, $t_{lcg} = \sum_{d \in D} v_d - F$. However, for IP networks, in general this provides a crude estimate of packet loss, because packet loss depends on a model for the traffic, a queuing model for the nodes, and details of the protocol. E.g., TCP will "throttle down" traffic as soon as it detects congestion, so that packet loss is mitigated (see Padhye, Firoiu, Towsley, & Kurose, 2000 for details); in that case it is more accurate to call the t_{lcg} measure as computed above "loss of network capacity".

Monotonicity

Whether a traffic measure is monotone (as defined in *Network Performability Evaluation*) depends both on the measure itself and on the routing/restoration method used by the network. It is clear that the measure t_{aff} of (7) is monotone increasing, and so are t_{fail} and t_{lrt}. This holds for both of the minimum-cost routing (shortest-path) and optimal routing schemes considered in *Network Restoration In Performance Measures*. However, in general the determination of monotonicity needs to be done carefully because intuition can be misleading. For example, the natural measure t_{res}, the amount of traffic that is affected and successfully restored in state *S*, is not monotone: consider the example in Figure 6 with a unit demand on each of the heavy arcs. The demand is defined to be restored if there is a path between the source and destination after restoration; the exact restoration method is irrelevant. The figure also shows that congested traffic is non-monotone, again irrespective of the restoration scheme. Hence the same holds for the measure t_{lcg} of (6).

Incorporating Time

When computing the measures in (6) we also want to take into account the duration τ of a failure as well as the time τ_r needed for restoration to complete. To do this, we will use the generic term "bad" traffic for the measure of interest. Let $D_{bb}(s)$ be the set of demands in S that are bad after the failure but before restoration, and $D_{ba}(s)$ the set of bad demands after restoration. These two sets may intersect. If $t_{bb}(s)$ and $t_{ba}(s)$ are the traffic rates corresponding to $D_{bb}(s)$ and $D_{ba}(s)$, the bad traffic as a function of time is depicted in Figure 7. In practice, τ_r is usually much less than τ. The duration τ of state S is distributed exponentially with mean $\overline{w}_s$, the mean waiting time of the Markov process in S (recall *Markov Models for Components*). It can be shown that the expectation of the bad traffic in this state with respect to τ is

$$\overline{t}_b(s) = t_{bb}(s) + (t_{ba}(s) - t_{bb}(s))e^{-\tau_r/\overline{w}_s}. \qquad (9)$$

For example, when "lost because of congestion" is substituted for "bad" and the appropriate "before" and "after" sets of demands are computed, the result of (8) is to be taken as the value of $t_{lcg}(s)$ in (6).

The analysis just given clearly involves some idealizations. For example, the set of demands $D_{bb}(s)$ actually changes during the restoration process, hence t_{bb} also changes and is not constant as implied by Figure 7.

Risk and the Pareto Frontier

Given a set of measures such as the ones in (6), one way to assess the performability of a network is to check if their cdf's satisfy certain requirements, the performability guarantees of (2). An alternative to looking at performability guarantees

Figure 6. Non-monotone behavior of restored traffic and congested traffic

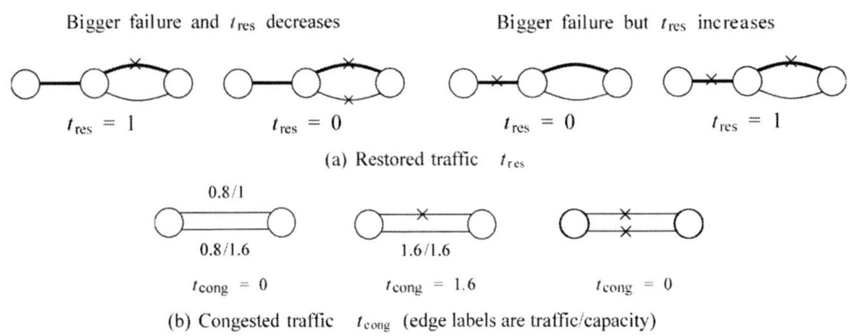

is to examine the space of explored failures from the viewpoint of *risk*.

Given a performance measure, let the "risk" of a failure with respect to this measure be the product of the probability of the failure and the value of the measure. With an m-dimensional measure, the risk associated with a failure has m aspects, or attributes. An immediate question is "what are the riskiest failures for the network?". Because of the multiplicity of attributes, the set of network states cannot be totally ordered with respect to risk, and there will usually not exist a failure that is worse than all the others. But we can say that a failure state s *dominates* (is worse than) another state s' if the risk attributes of s are $\geq$ the corresponding attributes of s', with strict inequality for at least one attribute. Given a set of network states, we can eliminate from consideration all states whose risk is dominated by that of some other state. What remains is known as the *Pareto boundary* or *Pareto frontier* of the risk set. The frontier can be extracted from the set by what is known as a "Pareto filter". We define the filtering problem and discuss solutions in the Appendix.

Network Restoration in Performance Measures

This section describes the algorithms we implemented to model the routing in commercial IP-over-optical layered networks. Recall that routing methods are important in the context of performability analysis because the IP layer restores network failures by rerouting affected demands. We model two major routing schemes: *minimum-cost* (uncapacitated shortest-path) routing and *optimal* routing. The nperf tool, discussed in *The* nperf *Tool*, also models other routing methods, such as multicast, e.g., IETF Protocol Independent Multicast-Single Source Mode (PIM-SSM), but we will not discuss these other routing methods in this article.

Uncapacitated Minimum-Cost Routing

This represents routing by the OSPF (Open Shortest Path First) protocol. Link costs correspond to OSPF *administrative weights*. Note that OSPF path computation does *not* take into account the capacities or utilizations of the links. Also, the other main IP-layer routing protocol, Intermediate System—Intermediate System (IS-IS), would behave similarly in our context.

Figure 7. "Bad" traffic in state S up to time τ

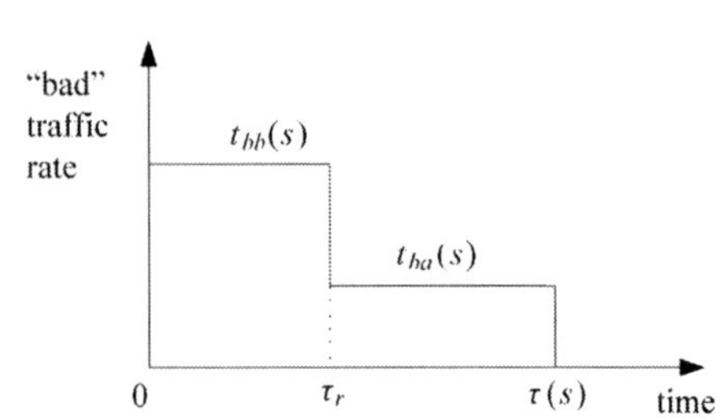

If there is more than one minimum-cost path between a pair of nodes, we balance the traffic as follows: for each source and destination pair (s, d) we compute **NextHops**(s, d), the set of neighbors of s that are on at least one minimum-cost path from s to d. Then we distribute all packets from s to d equally among all nodes in **NextHops**(s, d), and repeat recursively. To compute the all-pairs minimum-cost matrix we use the Floyd-Warshall algorithm (Ahuja, R., Magnanti, T., & Orlin, J., 1998). For a graph with n nodes and m edges, the complexity of this algorithm is $O(n^3)$, but we describe some improvements in *Speeding up the Routing Computation for Large State Spaces*.

Optimal Routing

In network flow terminology, this is an *integer min-cost max-multicommodity flow* problem, i.e. among the maximal flows find one with least total cost (i.e. Σ_i flow over edge $i \times$ cost of edge i). See Ahuja et al (1998). and Bertsekas (1998) for background on multicommodity flows. This models a network implementation where a centralized server has complete knowledge of the entire network and computes optimal paths. Capacity-sensitive extensions of routing protocols, such as OSPF-TE (Traffic Engineering) or Multiprotocol Label Switching-TE (MPLS-TE), sometimes can be used to provide approximate solutions to this problem. In the context of performability analysis, the network attempts to restore by solving the optimal routing problem on the affected network graph.

Even very simple versions of this problem are hard. For example, Garey and Johnson (1978) problem ND38 Directed Two-Commodity Integral Flow, shows that deciding whether just two demands can be routed in a graph with edges all of the same capacity and no costs is NP-complete. But if the integrality (indivisibility) requirement on the flows is removed, minimum-cost multicommodity flow is solvable in polynomial time via linear programming.

We find an approximate solution to the optimal restoration problem by using a greedy heuristic suggested in Phillips et al. (1999), which reduces the problem to a set of 1-commodity flow problems, and at the same time takes care of the integrality requirement. For example, to speed-up the computation as we are generating network states, our greedy heuristic stops the algorithm if a solution that routes all of the offered demand is found, no matter what its cost. Also, the heuristic lends itself well to speedup by parallelization and we have implemented a version using the Parallel Virtual Machine infrastructure, Geist, Beguelin, and Dongarra (1994).

Speeding up the Routing Computation for Large State Spaces

We describe some improvements in the overall efficiency of the path computation that exploit the fact that when states are generated in order of probability, the state sequence exhibits a certain *locality*: typically, successive network states differ in only one component. In the case of minimum-cost routing, this implies that the all-pairs minimum-cost (distance) matrix typically changes little between adjacent states. One way to take advantage of this is via a dynamic graph algorithm. There are several such algorithms with strong worst-case asymptotic guarantees (see Demetrescu & Italiano, 2001, and the references therein), but they are not easy to implement. In contrast, the improvements we describe below are very easy to implement, and result in near-optimal running time for our application.

Improvement 1: Affected Node Pairs

Let D_0 be the all-pairs minimum-cost (distance) matrix for the network in the perfect state. We want to reuse D_0 as much as possible in any state by computing the set of node pairs for which the minimum cost has changed. Because there is no easy way of computing this set we use a filter

Figure 8. A dynamic algorithm for computing all-pairs minimum-cost paths

```
DYNAMIC ALL-PAIRS MINIMUM-COST PATH

Initialize D o D_0, the all-pairs minimum-cost path matrix for the graph with no failed edges
for every (s,d) ∈ AffectedNodePairs {
      % EdgeCost[s,d] is ∞ if (s,d) is undefined or failed
      D[s,d] := EdgeCost[s,d]
}

for k := 1 to n {
      for every (s,d) ∈ AffectedNodePairs {
         D[s,d] := min(D[s,d], D[s,k] + D[k,d])
      }
}
return D
```

to compute a superset. For any node pair, if the minimum-cost path between these nodes contains at least one failed edge, we put the node pair in the set **AffectedNodePairs**. If a node pair is not in this set, its minimum-cost path does not change (all of the edges in the minimum-cost paths are there, and no new edges have been created). Then we run the modified Floyd-Warshall algorithm of Figure 8, where the two innermost loops range over the nodes in **AffectedNodePairs**. To establish the correctness of this algorithm, it remains to show that the minimum-cost for node pairs in **AffectedNodePairs**, which has either stayed the same or increased, is correctly computed. The only difficult part of this proof is to argue that the initialization in line 1 of D to D_0 (instead of initialization to edge costs) does not affect the correctness of the Floyd-Warshall algorithm. We establish this by a general proposition:

Proposition For any graph G = (V, E), construct a graph G' = (V, E') as follows. Initialize E' to E. Then pick an arbitrary pair of nodes u, v and add an edge (u, v) to E' with cost equal to the minimum-cost path between u and v in G. Repeat this step for an arbitrary number of node pairs. Then the minimum-cost matrices for G and G' are identical.

This can be proved by induction on the number of extra edges added to E'. For each extra edge, a path of the same cost existed in the graph and thus none of the costs change because of the addition of this edge. Because the effect of initialization is the same as adding an extra edge of cost equal to the minimum-cost path, this also completes our proof of the dynamic all-pairs minimum-cost algorithm.

Given the matrix D_0, the set of **AffectedNodePairs** can be computed in $O(n^2 \cdot$ (number of failed edges)) time by cycling over each pair of failed edge (u, v) and node pair (s, d) and checking if $D_0[s, u] + EdgeCost[u, v] + D_0[v, d] = D_0[s, d]$. It takes an additional $O(n^2 \cdot |AffectedNodePairs|)$ steps to run the modified Floyd-Warshall algorithm. Thus the overall complexity is $O(n^2 \cdot$ (nbr of failed edges) + |**AffectedNodePairs**|)

Improvement 2: Matrix Caching

We "reduce" the case of multiple component failures to the case of *single* component failures. This reduction is possible quite often because of the locality in the sequence of states produced by the algorithms of *Approximate Evaluation*. When the reduction is possible, its effect is that "number of failed edges + size of **AffectedNodePairs**" corresponds to a single component failure, sig-

*Figure 9. Block diagram of the core **nperf** network performability analyzer. Notation: iMCF = integral multicommodity flow, rMCF = (real) multicommodity flow, USP = un-capacitated shortest path*

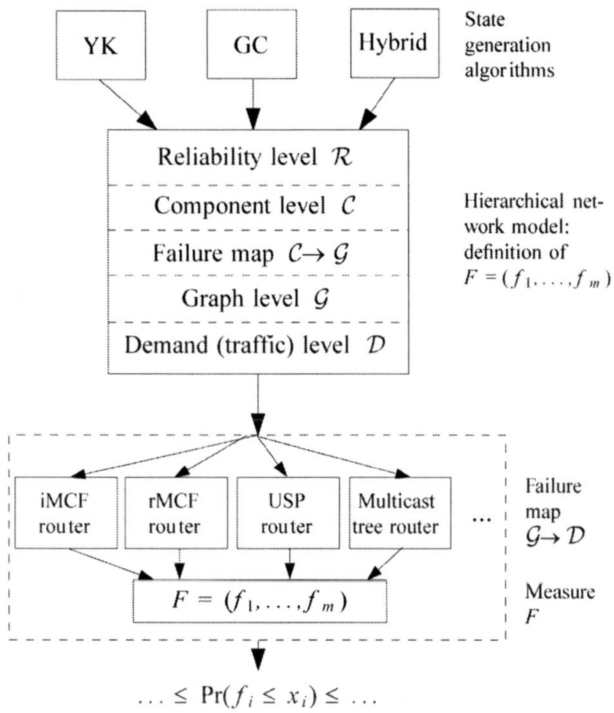

nificantly improving the speed of the algorithm in Figure 9.

Based on these observations, we modify the algorithm to maintain a cache of (set, matrix) pairs. In a state corresponding to a set of failed components C, we first check if the cache contains an entry (C_1, D_1) such that C_1 is missing exactly one element of C. If such an entry is found, the algorithm of Figure. 5.1 is run with $D_0 = D_1$ and as if a single component $c = C \backslash C_1$ has failed. If there is no match for C in the cache, we construct C_1 by copying C and removing the component which is least likely to fail from it. Then we compute the distance matrix D'_1 corresponding to C_1 and add (C_1, D'_1) to the cache.

In experiments with this scheme with a cache size of 25 and least-recently used replacement policy, we found a cache hit rate of nearly 99%. Moreover, as we searched the cache from top to bottom, about 80% of the time we had a match among the first two entries, and about 90% of the time a match among the first 5 entries. This makes the average complexity per state of the all-pairs minimum-cost paths algorithm be a small constant times n^2.

THE NPERF TOOL

We have incorporated the ideas of the preceding sections – hybrid state generation algorithm, hierarchical network model with multi-mode components and attributes for graph elements, optimized computation of traffic-related performance measures – into a network performability analysis tool called nperf. The tool's basic structure is depicted in Figure 9.

Some functionality of the core tool that is omitted from Figure 9 has to do with additional performance measures, such as various connectivity measures, measures of network latency, and a specialized computation of flows in stochastic networks (networks whose edges have stochastic capacities).

nperf is written in the very high-level language Unicon (Jeffery, Mohamed, Parlett, & Pereda, 2008), and uses C and C++ libraries for the numerically-intensive parts (performance measures), notably the Symphony library for mixed integer-linear programming from the COIN-OR project (Computational Infrastructure for Operations Research), and the Goblin graph and network flow library (GOBLIN graph library).

Finally, the core nperf is part of a larger system for network performability analysis, comprising interfaces to AT&T's databases for network topology data as well as traffic data, various network visualization tools, and a database back end which stores both network models and results of runs, and allows them to be analyzed together to answer queries such as "what failures have MTTR > 4 hrs. and result in more than 20% of the traffic originating at node *A* being lost because of congestion?"

nperf has been used in a variety of network studies. Some of these are the following: to compare the effects of different routing schemes on network performability (Agrawal, Oikonomou, & Sinha, 2007), to analyze the performability of an IPTV long-distance backbone networkd (Doverspike, Li, Oikonomou, Ramakrishnan, & Wang, 2007), to investigate 2-layer restoration in a satellite network (Oikonomou et al., 2008), and to set thresholds for composite links in an IP backbone (Oikonomou et al., 2008). In this article, we present a study of an IP backbone network design in section *Applications and Experimental Results*.

APPLICATIONS AND EXPERIMENTAL RESULTS

In this section we report our nperf analyses of three network designs, which are variations on the IP-over-optical backbone network of a large Internet Service Provider.

Access/Backbone Topology

Commercial IP networks are usually partitioned into "access" and "backbone" (or "core") geographical segments. A backbone location (building) houses one or two backbone routers (BRs). A set of access routers (ARs) connects to each backbone location. An AR can be co-located with its BRs and connected to them via intra-building fiber, which is assumed not to fail, or can be remotely located and connected to its BRs via the transport layer, which can experience failures as described earlier. Two BRs in the same building are connected via intra-building fiber. This basic structure is depicted in Figure 10.

To analyze the performability of the backbone segment, we first approximate the access segment as follows. At each backbone location we aggregate the demand (IP flows) from all the access routers that home onto that location into a single *virtual access router*. We then use a simple approximate method to distribute the demand across the ports of each virtual access router.

How the Networks were Designed

The network designs are generated using the "W/P" rule. This rule specifies that link utilization should not exceed W% in the no-failure state and not exceed P% in a failure state $s \in S_h \subset S(C)$. S_h corresponds to high-probability failure events, and when combined with the above W/P design rule, defines a simple mechanism for evaluating whether a network achieves a restoration objective. For our studies we set $W/P = 80/95$ for all three designs, but the definition of S_h differs:

1. The "fat" design. Each location has two BRs and each homing AR is connected to both BRs. S_h is the set of states corresponding to a *single* failure of either a network span, or a router port, or an entire BR.
2. The "thin" design. As with the fat design, each location has two BRs. S_h is the set of states corresponding to a single failure of

Figure 10. Three variations on an IP network design.

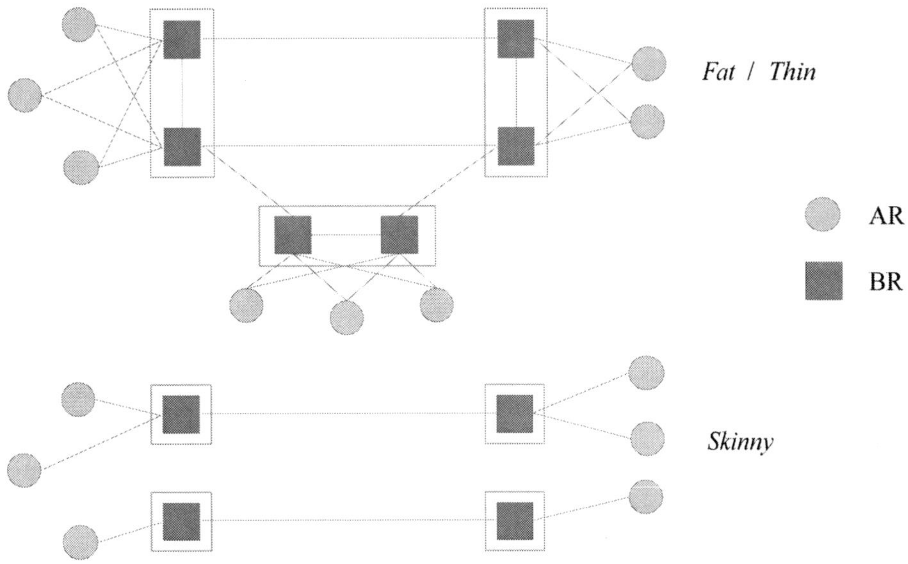

either a network span or a router port. It does not include states corresponding to failure of an entire BR.

3. The "skinny" design. Here we have only *one* BR per location, and each homing AR is connected only to this single BR. S_h is the same as in the "thin" case.

This approach is common among most large ISPs, although the utilization levels and network design specifics differ. The link capacities were determined by the network design algorithm of Phillips (1999). Note that during a network failure that is not in S_h, some links may be utilized over 95% and some demands may have no surviving path.

What We Evaluate

Our goal is to evaluate the effect of using reduced network capacity and fewer routers on the performability of the network. We also evaluate the effects of some improvements to routers, namely "hitless" software/hardware upgrades, and N:1 port protection.

Our component failure probabilities are derived from a combination of network data and specifications supplied by equipment manufacturers. Our traffic matrix is based on real network data at the time of study and we use min-cost routing (*Uncapacitated Minimum-Cost Routing*) to evaluate all designs. Table 1 lists some information on the nperf model for the "fat" variant of the network.

For each network in Figure 10 we compute the ccdf (complementary cdf) of the lost traffic, $t_{lost} = t_{lrt} + t_{lcg}$, as defined in (6). To protect proprietary information, we have multiplied the actual probabilities (y-axes) in all our plots by a fixed constant.

Lost Traffic

Figure 11a shows the bounds on t_{lost} for the fat design obtained by (4) after generating about 400,000 network states. The lower bound assumes that *none* of the traffic is lost in the unexplored state space, and the upper bound assumes that *all* of the traffic is lost in the unexplored state space [the "naive" bounds of eq. (2.3)]. In our runs, the unexplored failure space consists of some states

*Table 1. Information on the **nperf** model for the "fat" network. ARs are modelled at a high level of abstraction by a single component*

Demand level	
306 demands, 223223317 kb total demand	
400 sub-demands (path flows) after routing	
Graph level	
57 nodes, 1368 edges	
Edge utilization:	
average = 0.261, min = 0.000, max = 0.794	
Component level	
1055 components	
BR data plane	57
BR control plane	57
AR	24
span	66
port pair	684
port	167
upgrade	1

corresponding to double, triple, and quadruple failures and all of the states corresponding to more than four failures. If we weight these states according to their probabilities, most of the probability mass of the unexplored states is concentrated on states corresponding to double and triple failures. Our experience indicates that the exact lost traffic measure is much closer to the lower bound than to the upper bound. Therefore, for simplicity, in all subsequent plots we show only the lower bound curve. Figure 11b shows lost traffic for all three network designs.

In all these ccdf graphs, the plots are smoothed versions of staircase-like functions. The abrupt changes in slope of the ccdf curves are due to the discrete nature of the probability space and the complex mapping from this space to the value of the measure. In particular, the relatively flat segments are not caused by any particular set of failures, but by the *absence* of failures that, via the discrete mappings from the physical to the traffic layer, cause traffic loss of a certain magnitude.

One may note in Figure 11b that the skinny network gives a *better* guarantee than the thin network at low values of t_{lost}. The reason is that the skinny network has fewer components (e.g. router ports), and thus a higher probability of being in the no-failure state. As the fraction of lost traffic increases, the better restoration behavior of the thin network translates into better performance. On the other hand, the fat network is superior everywhere because its restoration performance is so much better than that of the thin and skinny networks.

Router Enhancements

Here we look at two possible improvements to the reliability of routers and assess the effect of these improvements on the entire network from the viewpoint of lost traffic.

Hitless Upgrades

Figure 11 assumes that whenever a router's software or hardware is upgraded, the router is taken out of service. We assume three 30-minute software upgrades and one 60-minute hardware upgrade per year for each router, and model them using multi-mode components such that that at most one BR from a given backbone location gets upgraded at a time (recall *Attributes for Graph Elements, Multi-Mode Components*). A new generation of commercial routers provides "hitless" software upgrades, and we assess the resulting performability improvement. Figure 13 shows that the improvement is dramatic. In fact, the performance of the thin and skinny designs becomes better than the performance of the fat design with normal (non-hitless) upgrades. These results assume a time-invariant traffic matrix. However, in reality router upgrades are often scheduled during low-traffic periods of the day, and in that case the difference in the performance

Figure 11. Lost traffic $t_{lost} = t_{lrt} + t_{lcg}$, see (Traffic Measures)

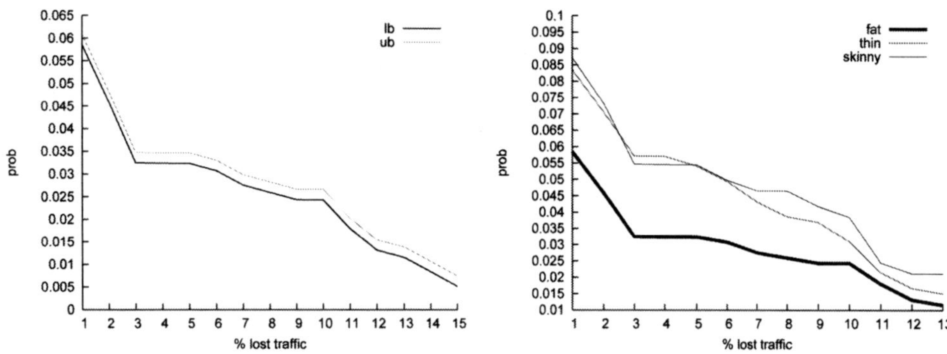

(a) Upper and lower bounds for the "fat" network

(b) Lower bounds for all three designs

of the three designs would be smaller than Figure 12 indicates.

Port Protection

Here we investigate the effect of adding N:1protection on router ports. Here, groups of N ports are protected by an $(N + 1)$th spare port. which takes over role of a failed port in the group; hence, traffic is lost only when multiple ports in the group fail. If an individual port fails with probability q, then we can compute Pr("a port is up or replacee by its protection port") = $(1-q) + q(1-q)^N \geq 1 - Nq^2$

Figure 12. Lost traffic for the "fat", "thin", and "skinny" network designs. The fat design has normal upgrades, and the thin and skinny designs have hitless upgrades

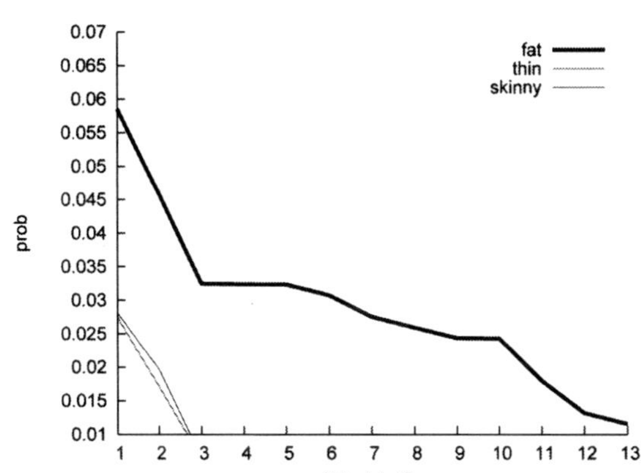

This shows that for typical (small) values of q, the effect of protection is, surprisingly, relatively independent of the protection group size N.

Figure 13a shows the results for the fat design. It can be seen for port protection to have a positive impact, hitless upgrade is also required. This is because all ports on a router are down when the entire router is down and the probability of a router being down (e.g., due to common component failure, software upgrade or maintenance) is an order of magnitude higher than q. However, router software upgrades are the major contributor to a router being down and, therefore, if we make the upgrades hitless, the individual port failures become significant, in which case Figure 13b shows that port protection improves performability. The benefits of port protection are concentrated near low percentages of lost traffic. The higher percentages of lost traffic result from failure states where port protection has insignificant impact (e.g., router hardware failures and fiber cuts).

Risk Assessment

In *Lost Traffic* and *Router Enhancements* we evaluated three network designs by looking at performability guarantees for the lost traffic measure. Another, complementary, assessment consists in examining the space of explored failures from the viewpoint of risk and the Pareto frontier.

As explained in *Risk and the Pareto Frontier*, risk is defined with respect to a particular traffic measure: at each network state the value of the measure is multiplied by the probability of the state. With the 4-dimensional traffic measure of (6), the risk associated with a failure has 4 attributes. Figure 14a shows a scatterplot of two of these attributes for the fat network design, t_{lcg} and t_{lrt} risk. There are about 400,000 points in the plot, and each point (x, y) denotes a failure with no-route risk x and congestion risk y. Evidently the risks vary by many orders of magnitude, but the upper right "corner" of the scatterplot, in this case above the point (10000,10000), contains the worst failures with respect to both risk attributes. These are the failures that one would want to address first when contemplating improvements to the network; they lie on Pareto frontier of the failure set.

Figure 14b shows the result of running the Pareto filter on the failures of Figure 14a three times in succession, each time eliminating the extracted frontier from the set. The first and second frontiers consist of a single point each, corresponding to the failure of a major virtual access router (recall *Example for an IP-Over-Optical Network*), while the third frontier consists of the failures of another two major virtual access routers.

CONCLUSION

Although non simulation-based performability analysis of multi-layer networks has proved to

Figure 13. "Fat" network: effect of port protection on lost traffic with and without hitless router upgrades

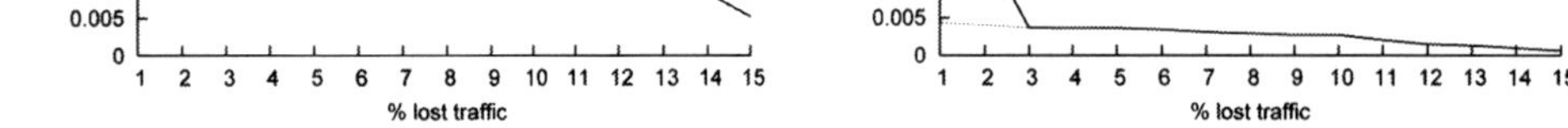

(a) With normal upgrades (b) With hitless upgrades

Figure 14. Congestion vs. no-route traffic risk for the "fat" network

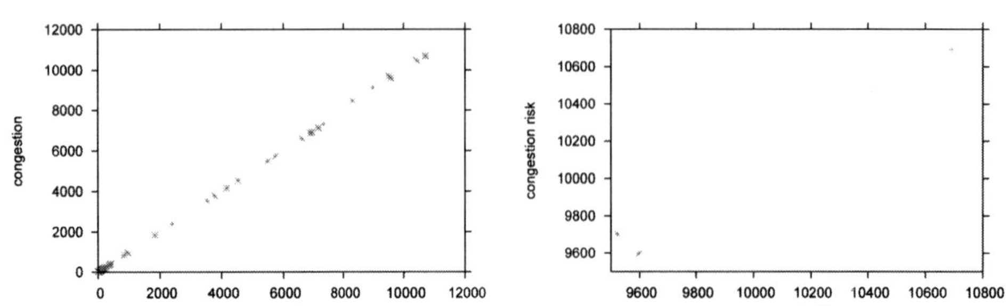

(a) Congestion vs. no-route risk (b) Three successive Pareto frontiers

be a daunting problem, our methodology of detailed multi-layer network modeling, multi-mode components, state-space generation methods, and efficient routing algorithms, all embodied in the nperf tool, makes the problem tractable for real networks. We applied this methodology to a case study of a multi-layered IP-over-Optical backbone network. We found that, based on the reliability data applicable at the time, the most significant performability driver for the more costly "dual backbone router" architecture employed by most ISPs is router software maintenance/upgrade events. If the upgrades could be made "hitless", port protection would further improve the network's performability, and there is even a promise of being able to reduce restoration capacity and still maintain the same performability levels. Our methodology has also been applied to many other types of commercial networks.

REFERENCES

Agrawal, G., Oikonomou, K. N., & Sinha, R. K. (2007). Network Performability Evaluation for Different Routing Schemes. *Proceedings of the OFC,* Anaheim, CA.

Ahuja, R., Magnanti, T., & Orlin, J. (1998). *Network Flows*. Prentice-Hall.

Alvarez, G., Uysal, M., & Merchant, A. (2001). Efficient verification of performability guarantees. *PMCCS-5: The Fifth International Workshop on Performability Modelling of Computer and Communication Systems*.

Aven, T., & Jensen, U. (1999). *Stochastic Models in Reliability*. Springer-Verlag, 1999.

Bertsekas, D. P. (1998). *Network Optimization: Continuous and Discrete Models*. Athena Scientific.

Bolch, G., Greiner, S., de Meer, H., & Trivedi, K. S. (2006). *Queueing Networks and Markov Chains*. Wiley. doi:10.1002/0471791571

Carlier, J., Li, Y., & Lutton, J. (1997). Reliability evaluation of large telecommunication networks. *Discrete Applied Mathematics*, 76.

Climaco, J. N., & Martins, E. Q. (1982). A Bicriterion Shortest Path Algorithm. *European Journal of Operational Research,* 11.

Colbourn, C. J. (1999). Reliability Issues in Telecommunications Network Planning. In B. Sansó, (Ed.), *Telecommunications Network Planning*. Kluwer Academic Press. Computational Infrastructure for Operations Research. http://www.coin-or.org/.

Demetrescu, C., & Italiano, G. (2001). Fully Dynamic All Pairs Shortest Paths with Real Edge Weights. *IEEE Symposium on Foundations of Computer Science*.

Doverspike, R. D., Li, G., Oikonomou, K. N., Ramakrishnan, K. K., & Wang, D. (2007). IP Backbone Design for Multimedia Distribution: Architecture and Performance. *Proceedings of the IEEE INFOCOM*

Formal Methods for Performance Evaluation of LNCS 4486. (2007). Springer.

Garey, M., & Johnson, D. (1978). *Computers and Intractability: A Guide to the Theory of NP-completeness*. W.H. Freeman.

Geist, A., Beguelin, A., & Dongarra, J. (1994). *PVM: Parallel Virtual Machine: A Users' Guide and Tutorial for Network Parallel Computing*. MIT Press. GOBLIN graph library. http://goblin2.sourceforge.net/.

Gomes, T., Craveirinha, J., & Martins, L. (2002). An efficient algorithm for sequential generation of failures in a network with multi-mode components. *Reliability Engineering & System Safety, 77*, 111–119. doi:10.1016/S0951-8320(02)00006-6

Gomes, T. M. S., & Craveirinha, J. M. F. (1997). A Case Study of Reliability Analysis of a Multi-exchange Telecommunication Network. In Soares, C. G. (Ed.), *Advances in Safety and Reliability*. Elsevier Science. doi:10.1016/B978-008042835-2/50269-1

Gomes, T. M. S., & Craveirinha, J. M. F. (1998). Algorithm for sequential generation of states in failure-prone communication network. *IEE Proceedings. Communications*, *145*(2). doi:10.1049/ip-com:19981838

Harms, D. D., Kraetzl, M., Colbourn, C. C., & Devitt, J. S. (1995). *Network Reliability: Experiments with a Symbolic Algebra Environment*. CRC Press.

Jarvis, J. P., & Shier, D. R. (1996). An Improved Algorithm for Approximating the Performance of Stochastic Flow Networks. *INFORMS Journal on Computing*, *8*(4). doi:10.1287/ijoc.8.4.355

Jeffery, C., Mohamed, S., Parlett, R., & Pereda, R. (2008). *Programming with Unicon*. Unicon Project, http://unicon.org.

Karger, D. (1995). A Randomized Fully Polynomial Time Approximation Scheme for the All-terminal Network Reliability Problem. *Proceedings of the 27th ACM STOC*.

Keilson, J. (1979). *Markov Chain Models-Rarity and Exponentiality*. Springer-Verlag.

Levendovszky, J., Jereb, L., & Elek, Zs., & Vesztergombi, Gy. (2002). Adaptive Statistical Algorithms in Network Reliability Analysis. *Performance Evaluation*, 48.

Levy, Y., & Wirth, P. E. (1989). *A Unifying Approach to Performance and Reliability Objectives. Teletraffic Science for New Cost-Effective Systems, Networks and Services, ITC-12*. Elsevier Science.

Li, V. K., & Silvester, J. A. (1984). Performance Analysis of Networks with Unreliable Components. *IEEE Transactions on Communications*, 32.

Lisnianski, A., & Levitin, G. (2003). *Multi-State System Reliability: Assessment, Optimization and Applications*. World Scientific.

Mendiratta, V. B. (2001). A Hierarchical Modelling Approach for Analyzing the Performability of a Telecommunications System. *PMCCS-5: The Fifth Int'l Workshop on Performability Modelling of Computer and Communication Systems*.

Messac, A., Ismail-Yahaya, A., & Mattson, C. (2003). The Normalized Normal Constraint Method for Generating the Pareto Frontier. *Structural and Multidisciplinary Optimization*, 25.

Meyer, J. F. (1995). Performability evaluation: where it is and what lies ahead. *1st IEEE Computer Performance and Dependability Symposium (IPDS)* (pp. 334-343).

Oikonomou, K. N. (2006). Unreliable Components with Fast Repair and Dynamic Network Restoration. In S. Raghavan & G. Anandalingam (Eds.), *Telecommunications Planning in Proceedings of the 8th INFORMS Conference on Telecommunications*. Kluwer Academic.

Oikonomou, K. N., Ramakrishnan, K. K., Doverspike, R. D., Chiu, A., Martinez-Heath, M., & Sinha, R. K. (2007). Performability Analysis of Multi-Layer Restoration in a Satellite Network. In Mason, L., Drwiega, T., & Yan, J. (Eds.), *Managing Traffic Performance in Converged Networks, ITC 20 (LNCS 4516)*. Springer-Verlag. doi:10.1007/978-3-540-72990-7_11

Oikonomou, K. N., & Sinha, R. K. (2006). A New State Generation Algorithm for Evaluating the Performability of Networks with Multimode Components. *8th INFORMS Telecommunications Conference*.

Oikonomou, K. N., & Sinha, R. K. (2008). Techniques for Probabilistic Multi-layer Network Analysis. *Proceedings of the IEEE Globecomm*.

Oikonomou, K.N., & Sinha, R.K. (2009). Improved Bounds for Performability Evaluation Algorithms using State Generation. *Performance Evaluation*.

Padhye, J., Firoiu, V., Towsley, D., & Kurose, J. (2000). Modeling TCP Reno Performance: A Simple Model and Its Empirical Validation. *IEEE/ACM Transactions on Networking*, 8.

Phillips, S., Reingold, N., & Doverspike, R. (2002). Network Studies in IP/Optical Layer Restoration. *Optical Fiber Communications Conference*.

Phillips, S., & Westbrook, J. (1999). Approximation Algorithms for Restoration Capacity Planning. *Proceedings of Algorithms-ESA'99: 7tt Annual European Symposium,* Prague.

Probability, C. (2000). Kluwer Academic.

Ramirez-Marquez, J. E., & Gebre, B. A. (2007). A Classification Tree Based Approach for the Development of Minimal Cut and Path Vectors of a Capacitated Network. *IEEE Transactions on Reliability*, *56*, 2007. doi:10.1109/TR.2007.902788

Rauzy, A. (2005). An *mlogm* Algorithm to Compute the Most Probable Configurations of a System with Multi-Mode Independent Components. *IEEE Transactions on Reliability*, *54*(1), 156–158. doi:10.1109/TR.2004.842084

Rocco, C., & Muselli, M. (2005). Approximate Multi-state Reliability Expressions Using a New Machine Learning Technique. *Reliability Engineering & System Safety*, 89.

Sahner, R., Trivedi, K., & Puliafito, A. (1996). *Performance and Reliability Analysis of Computer Systems*. Kluwer Academic.

Shier, D. R. (1991). *Network Reliability and Algebraic Structures*. Oxford: Clarendon Press.

Shier, D.R., Bibelnieks, E., Jarvis, J.P., & Lakin, R.J. (1990). Algorithms for Approximating the Performance of Multimode Systems. *Proceedings IEEE Infocom*.

Yang, C. L., & Kubat, P. (1989). Efficient computation of most probable states for communication networks with multimode components. *IEEE Transactions on Communications*, 37.

Yang, C.L., & Kubat, P. (1990). An Algorithm for Network Reliability Bounds. *ORSA Journal on Computing, 2*.

ENDNOTES

1. In the U.S. there are even some legal requirements on businesses to be aware of their network's "risk profile", e.g. the US Sarbanes-Oxley act of 2002.
2. The reader who finds our usage of the terms reliability, performability, etc, unfamiliar may wish at this point to look at the note on terminology in *The Theoretical Problem*.
3. There are more complex DWDM systems with various optically-transparent "add/drop" capabilities which, for simplicity, we do not discuss here.

APPENDIX: PARETO FILTERING PROBLEM

The Pareto filtering problem is: if V is a set of d-dimensional vectors in R^d, we want to eliminate from V all points v that are *dominated* by another point $u \in v$: u dominates v, denoted $u \phi v$, iff $u_i \geq v_i$ for all i with strict inequality for at least one i. Suppose that determining whether $x \phi y$ takes d units of time. Then given n points, a straightforward algorithm for eliminating all dominated points takes time $O(dn^2)$; see e.g. Messac, Ismail-Yahaya, and Mattson (2003). A much better algorithm for $d = 2$, requiring only $O(n\log n)$ time, is given as follows. Denote a point by (x_i, y_i) and sort the points in order of increasing x, with increasing y for equal values of x_i. Without loss of generality, let the result be $(x_1, y_1), (x_2, y_2),..., (x_n, y_n)$. Then it is clear that any point can be dominated only by a point to its right. Starting from the end, (x_{n-1}, y_{n-1}) is dominated by a point on its right iff $y_{n-1} < y_n$. (x_{n-2}, y_{n-2}) is dominated by a point on its right iff $y_{n-2} < \max(y_{n-1}, y_n)$, etc. Finally, (x_1, y_1) is dominated by some point to its right iff $y_1 < \max(y_2, y_n)$. Denoting the maxima by m_i, the algorithm finds them in linear time by making a backward pass through the list to determine $m_{n-1} = y_n$, $m_{n-2} = \max(y_n, y_{n-1})$, $m_{n-3} = \max(y_{n-1}, y_{n-2}),..., m_1 = \max(y_n,..., y_2)$.

The filtering problem may also be solved in $O(n\log n)$ time with the more general bi-criterion shortest path algorithm of Climaco and Martins (1982), but the procedure we described here is much simpler.

This work was previously published in International Journal of Interdisciplinary Telecommunications and Networking, Volume 1, Issue 3, edited by Michael R. Bartolacci, and Steven R. Powell, pp. 1-30, copyright 2009 by IGI Publishing (an imprint of IGI Global).

Chapter 9
Next-Generation Strategic Business Model for the U.S. Internet Service Providers:
Rate-Based Internet Subscription

Charles C. Willow
Monmouth University, USA

ABSTRACT

The (information digital) network bandwidth and its usage through Internet subscription, by far, is perhaps the only uni-modal commodity provided to today's consumers at a flat rate. Regardless of his/her actual bandwidth usage, a consumer is charged a uniform subscription fee by the Internet Service Providers (ISP). Meanwhile, the ISPs have a considerable stake in the overall infrastructure which supports the local perimeters of the Internet. This article presents a rigorous statistical analysis with the objective of determining the optimal billing or pricing policy for the ISPs of the U.S. on the basis of the proposed 'variable subscription-rate' business model. Two leading global consumer profiles were selected for data comparison on a cardinal scale; NYC, NY, U.S.A. and Seoul, South Korea.

INTRODUCTION

The (information digital) network bandwidth and its usage through Internet subscription, by far, is perhaps the only uni-modal commodity provided to today's consumers at a *flat* rate. Regardless of his/her actual bandwidth usage, a consumer is charged a *uniform* subscription fee by the Internet Service Providers (ISP). Meanwhile, the ISPs have a considerable stake in the overall infrastructure which supports the local perimeters of the Internet.

DOI: 10.4018/978-1-60960-505-6.ch009

Constant upgrades encompassing but not restricted to network infrastructure, servers, proxies, security, middleware, applications software, fault tolerance, redundancy, contingency plans, and even customer service are expected on a daily basis, as the Internet usage increases geometrically if not exponentially. Moreover, statistics indicate that approximately 10% of the Internet consumers are roughly occupying 90% of the bandwidth at all times (FCC, 2007). In addition, *digital convergence* of household electronic products comprising the voice-phones, video-phones, Internet Protocol Television (IPTV), digital Hi-Definition

TV, home networks and automation, and even IP-based refrigerators or washing machines is rapidly reshaping the Internet usage among consumers across the globe. As a consequence, the network bandwidth shortage and even depletion problem is further aggravated.

Early research on 'rate-based' variable Internet subscription fees was performed by the British Telecom (BT), a state-owned organization centrally located in London, U.K. This new business model for the ISP has been actively penetrating most areas of Britain, other than Wales and Scotland (KT, 2006). The success of the business model has brought a number of imitators, with such leaders as South Koreans. The U.K and Korea models have close ties in that most ISPs are State-funded and managed and that they are single-state nation, relative to the U.S. Their differences, however, are yet to be explored.

The purpose of this article is to investigate the *pros* and *cons* of the 'variable-rate' business model by conducting an empirical analysis before its introduction to the U. S. The estimated market – after adoption – is approximately $3.0B annually in the U.S. alone, covering 50 continental States other than Hawaii, Puerto Rico, and Virgin Islands (FCC, 2007).

Unlike their South Korean counterparts, the U.S. Internet Service Providers (ISP) carry major disadvantages:

- Extension of the information network infrastructure which far outsizes that of Korea due to its geographical coverage area.
- Network bandwidth bottleneck problem, often characterized as the 'last-mile problem'. Whereas the population density and congested residential apartments benefit the South Korean ISPs, the reverse is true for the U.S. counterparts. By way of an example, a resolution to a single last-mile problem by using say, fiber optic cable, which may deliver data transfer rate of up to 1.3 Giga bits per second (Gbps) for an apartment complex may enable a Korean ISP to acquire thousands of additional customers. In contrast, as for the U. S. consumers, however, their majority reside in private houses, apart from one another, which eventually may require the ISP to tackle the 'last-mile' problem individually on a per household basis.

The outlined advantages may justify the world's highest Internet penetration among consumers in nations such as Korea and Japan (Park, 2001). Thus, the adoption of partly successful 'variable rate-based' Internet subscription pricing or utility business model of South Korea may not be adequate for the U. S. ISPs. The purpose of this article, however, is to investigate the effectiveness of such adoption (of business model) to the U. S. through empirical analysis. In order to perform a sound comparative statistical analysis, a sound design of experiments is the prerequisite. Therefore, the cardinal scale of comparative data is sought by restricting the sample to be selected in highly dense metropolitans in both nations; Manhattan, or formally NY, NY against Seoul, Korea.

Determination of the optimal pricing policy for such service-oriented commodity as *knowledge* from the Internet may confound both the consumers and ISP, the business entity. Possible ramifications are:

- The consumer does not have sufficiently accurate *a priori* information as to how s/he will be charges or billed for the service s/he seeks to receive from the ISP via the Internet, as opposed to other utility services such as power, energy, traditional entertainment, or even commodities such as food.
- Associated with the aforementioned, the units of billing are unclear to the consumer, and may in fact bring their reluctance if not complete resistance to continued subscrip-

tion. This clearly indicates that the pricing may be *elastic*, depending on external variables, which motivated the initiation of this research.

This article is comprised of five sections, and is organized as follows. Studies of relevant literature are presented in the following section. The third section focuses statistical input data analysis to validate the samples prior to performing statistical analysis in the fourth section for the proposed price discrimination model. Summary, conclusions, and advice on extensions to this research are contained in the final section.

REVIEW OF LITERATURE

In his confines of proposition, Odlyzko (2004) delivers the rationale behind the ultility- or demand-based business model (Rappa, 2008) for Internet subscription through price discrimination. Indeed, he further suggests in his remarks for future research the possibility of such qualitative incumbents (*i.e.* variables in the business model) such as price fluctuations, elasticity of price itself for a serviced utility, and consumer behavioral characteristics.

Han (2005) explored the correlation between the price and product searches on the Web through a series of statistical hypotheses testing and Analysis of Variance (ANOVA). His conclusion was that there was a marked difference in consumer behavioral patterns between the traditional and electronic markets. The rational consumer in traditional market did not necessarily continue his/her behavior in the e-Market.

Porter & Donthu (2006) studied consumer acceptance for technology and in particular, the Internet. Some clear patterns were obtained relative to the demographics data by incorporating the enhanced Technology Acceptance Model (TAM). Factors such as age, income level, education, and race were associated differentially with beliefs about the Internet. In effect, the authors (Porter & Donthu, 2006) conjecture that these beliefs influence a consumer's attitude toward and the use of the Internet, based on their Chi-Square (χ^2) goodness-of-fit tests.

Roycroft & Anantho (2003) report on the demographics of Internet subscription in the African continent. Proposed models of this article may be extended to include future global consumers of the Internet.

A reliable and tolerant, yet secure digital information network is critical, should ISPs seek the proposed 'variable-rate subscription' policy of this article strategically. A middleware dedicated to providing a seamless connection with the application-layer network protocol was suggested by Sousa and Freitas (1998). Network node tolerance to better locate the faulty connections caused primarily by inoperable devices is delineated in Horneffer (1998). Simulation studies of this location detecting techniques and its extension with Genetic Algorithm were sought by Luo *et al.* (2004). Wills *et al.* (1998) seek to tackle this tolerance problem by a preventative means. By far, according to them, the design of a network at the outset of its construction outperforms any known mechanism or method for managing a fault-tolerant network. A comprehensive design method, such as the Integrated Systems Design Methodology (ISDM), developed by Willow (2007*b*) may be employed. Further, the ISDM may be applied to a wide array of problems associated with Information Systems and Technology (IST), as described in Willow (2007*a*).

Critical mass provides the ISPs with maximized profit, and is reached only when the demands of consumers are met continuously. Internet subscribers demand a variety of digital contents through an integrated, reliable, and ubiquitous access to the ISP network. Target market segments encompass but are never limited to medical imaging, its compression, and even telemedicine (Choong *et al.*, 2007), multimedia (Hanaro Telecom, 2007), legal boundaries of the Peer-to-Peer networking

(P2P) (Banerjee *et al.*, 2008; Viswanathan, 2007), and/or live broadcasts (Sorte *et al.*, 2008).

INPUT DATA ANALYSIS

Thus far, a common practice related to statistical data analyses has been associated with over-emphasis on output data exclusively (*i.e.* verification tests). To the contrary, however, the key to ensure success for empirical studies lies in accurate *input* data analysis. That is, the procedures governing the sample data collected as to why and how they fit into the scope of the research, ensued by rigorous statistical analysis must be completed prior to making any immature and at times incomplete conclusions. Unfortunately, the majority of empirical research to date still relies on simple deliberations provided by automated software such as SAS and/or SPSS/GPSS without a clear illustration why such a sample was selected and how. This indispensable procedure in statistical analysis, often referred to as the 'Input Analysis' or 'Validation Tests' should be sought to increase the cogency of the research. In essence, whereas the verification test alone may be regarded as the necessary condition for a successful empirical study, it is the validation test which provides the sufficient condition, its integrity.

Units of price discrimination for Internet subscription may be inconsistent and at times, dubious, as emphasized in section I. Arguably, usage time, similar to 'air time' for cell phones, may be a possibility. However, digital contents vary widely in data size measured in bits, and consumer utility time may well distort the pricing policy. For example, a single two-minute High Definition (HD) digital movie, which averages 200 Mega Bytes (MB), is equivalent to 1,600 Mega bits (Mb) or alternatively 1.6 Giga bits (Gb), assuming a 8-bit Byte. This translates to a transfer speed of approximately one second over a fiber-optic network, whereas for the Asymmetrical Digital Subscriber Line (ADSL) on Plain Old Telephone System (POTS) network, up to 400 seconds or as much as 7 minutes of download. To this end, the ISPs may not introduce price discrimination on utility time alone, unless comparable performance of their network infrastructure (*viz.* bandwidth) is available to every subscriber.

Another conflicting yet compelling factor, which may be unpredictable if not completely random, is the resource capacity of the ISP. That is, the information as to which customer may occupy how much bandwidth as a result of being serviced with which type of digital content at a certain time instant is completely stochastic in nature. In essence, utility time of the Internet produced by each subscriber, coupled with his/her desired digital content, as well as the total number and the variety of the content heavily influence both the capacity and performance of the ISP network resource. When the maximum capacity of the (ISP) network bandwidth is reached, other customers may be denied service, similar to the traffic congestion on major motor vehicle highways. However, the financial budget limitation on information infrastructural improvement may not always allow the ISP to increase its network bandwidth constantly to meet the subscriber demands. These resource constraints precisely dictate the problems of most ISPs of the U. S., which support the rationale for the proposed 'variable-rate price discrimination' model of this article. Figure 1 follows to illustrate these correlations based on the data provided by the FCC (2007). Note the depiction is not calibrated to scale.

A series of hypothesis tests were completed to *validate* the design of experiment, and thus the input data prior to (experiment) sampling. In essence, the input data analysis is analogous to a pilot run of statistical sampling for the entire empirical experiment. A section follows to illustrate.

Figure 1. Correlation of resource constraining variables

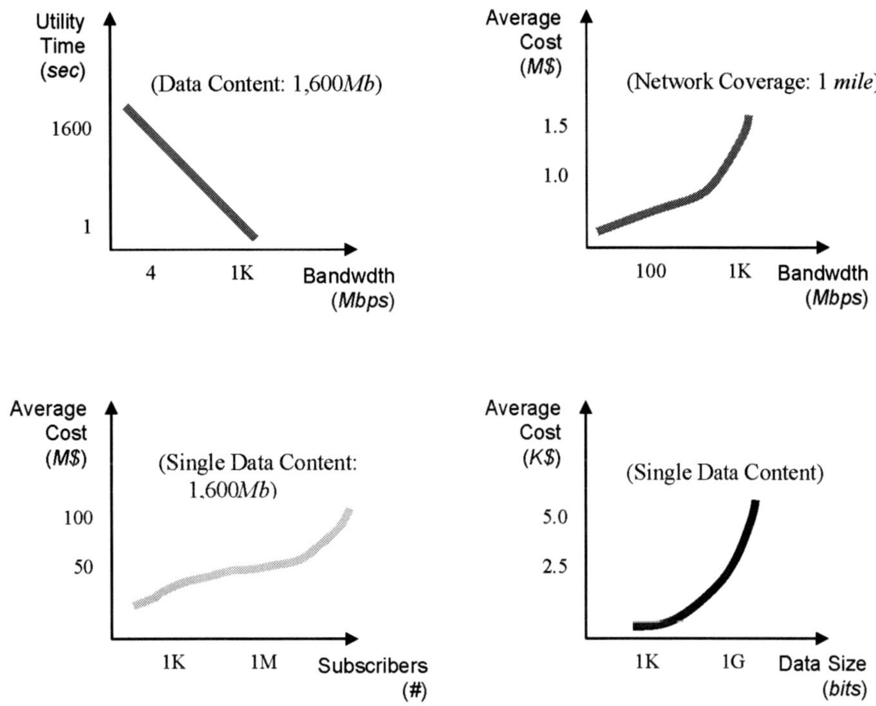

Hypotheses

Both Korea Telecomm (2006) and Hanaro (2007) of South Korea report a clear and consistent pattern of acceptance for price discrimination among Information Technology (IT) professionals. Thus, an assumption is made in this article to restrict the demographics samples to be chosen to IT professionals. IT professionals were defined as those who engage more than 4 hours a day in networked applications software, middleware, and/or hardware (Korea Telecomm, 2006).

A total of three consumer behavioral attributes, derived from Figure 1 above, were selected for the tests:

H1: *IT-savvy professionals are willing to pay more for the digital content they desire, regardless of their physical location for the ISP service. That is, given the flexibility in logistics (viz. roaming), they are not reluctant to the idea of subscribing to digital contents with price discrimination.*

H2: *IT professionals are willing to accept variable rate pricing for different digital content of the same type, regardless of the utility time. For instance, they are willing to pay more for a digital movie A (60-minute running time), compared with movie B (120-minute running time), provided that they could watch it online without interruptions. Notice the online running time depends on the technical parameter, bits of the data content, as opposed to its analog counterpart, which is the actual movie running time designed by the movie producers and/or studios.*

H3: *IT professionals are willing to accept variable rate pricing for digital content of different types, regardless of the utility time. For example, they opt to assign higher priority to a digital movie A (120-minute running time) than a piece of digital audio such as a major hit song β (5-minute running time), provided that they could watch or listen to it online without interruptions. Note again the online running time depends on the technical*

Table 1. Input data analysis for hypotheses [unit: frequency]

	NYC, NY, USA			Seoul, Korea		
H-1	**Software**	**Middleware**	**Hardware**	**Software**	**Middleware**	**Hardware**
Managerial	16	12	9	30	20	14
Development	32	28	5	35	35	10
Support	30	2	20	27	0	9
	NYC, NY, USA			**Seoul, Korea**		
H-2	**Software**	**Middleware**	**Hardware**	**Software**	**Middleware**	**Hardware**
Managerial	33	28	18	2	7	3
Development	28	28	11	4	1	1
Support	19	18	6	0	0	0
	NYC, NY, USA			**Seoul, Korea**		
H-3	**Software**	**Middleware**	**Hardware**	**Software**	**Middleware**	**Hardware**
Managerial	16	27	30	30	29	35
Development	32	28	34	35	35	26
Support	30	21	22	27	30	30

parameter, bits of the data content. It is highly affected by the attributes of the original data content (i.e. knowledge) such as (analog) resolution, 3-D images, Virtual Reality (VR), stereo, dolby systems, or super woofer, among others, as well as technical attributes encompassing analog-to-digital conversion, digital resolution, etc.

Pilot Sampling

IT professionals, as discussed in the previous section, were classified into three categories relative to their duties; managerial, development, and support. Measurement was then carried out based on the IT architecture; software, middleware, and hardware. To achieve the cardinal scale for data comparability, not less than 20 but no more than 35 individuals from each IT category and job-duty class were selected for the sample. In effect, a randomized *stratified cluster sampling* was completed to maximize the statistical soundness for the design of experiment. Clusters of the sample were the IT category, whereas the strata were the job duties of the IT professionals. Demographics were limited to Korea Telecom (KT), Korea and Cisco, Inc. of NY. Table 1 summarizes the pilot sampling for NYC, NY, USA and Seoul, Korea for the three hypotheses.

Table 1 clearly indicates data distortion even at the pilot-experiment level, and indeed serves the purpose of this input data analysis. Both hypotheses, H-1 and H-2, are therefore, rejected without further statistical analysis. With the acceptance of H-3 exclusively, χ^2tests, followed by its Analysis of Variance (ANOVA) are sought in the next section.

STATISTICAL ANALYSIS

Once the design of empirical experiment is validated with the input data analysis, successful pilot experiment is extended to incorporate the entire data necessary to complete the experiment. Therefore, the stratified cluster sampling to test the hypothesis H-3, delineated in the previous section, is extended to its full scale. The sample size is now increased to 50 (for each stratified cluster), with data listed in Table 2.

Table 2. Sampled data for hypothesis H-3 (unit: %)

	NYC, NY, USA			Seoul, Korea		
H-3	**Software**	**Middleware**	**Hardware**	**Software**	**Middleware**	**Hardware**
Managerial	82.8	84.1	90.7	94.6	92.1	95.8
Development	91.0	97.5	98.2	98.9	99.0	99.2
Support	89.8	68.6	52.4	93.4	90.3	99.2

Table 3. Chi-square statistic values for strata and clusters

χ^2	NYC, NY, USA			Seoul, Korea		
Cluster: *IT arch*	4.43	14.46	24.57	**2.89**	4.59	**1.96**
Strata: *Job Duty*	4.24	3.97	18.76	**1.89**	**0.15**	4.52

The Chi-Square (χ^2) goodness-of-fit tests are particularly effective to cases in which the demographic characteristics are suspected to be contaminating each other (Law, 2007; Han, 2005). Note the major statistical demographics data for H-3 are the IT architectural tier and the job-duty category for the IT professionals. In order to ensure their fit to the Normal Distribution at both the 95% and 99% confidence levels, the following parameters are observed:

$n = 150$;
Degree of Freedom = (# clusters) –
(# estimated parameters of the Normal Distribution)
= 3 – 2
= 1
χ^2 0.95, 1= 3.841; χ^2 0.9, 1= 6.635

Test statistic, X^2 values follow, and are summarized in Table 3.

A considerable lack of fit to the Normal distribution at both the 95% and 99% confidence levels were observed, to be statistically conservative. In fact, H-3 may be accepted only for those IT professionals of South Korea for this particular empirical study. Table 3 indicates that H-3 may hold for the following groups: IT software and hardware persons, management, and the development professionals.

The following statistical conclusion may be drawn from the results:

1. The U.S. ISPs should seek alternative price discrimination strategies to their South Korean counterparts', based on their own customer demographics. Further, they are advised to research the differences in their market with the premise that *localization* strategy is the key to adopting 'variable-rate' subscription policy in the U. S. market.
2. Future research should be pursued to identify the possible correlation between the *data-content* consumer groups and *utility-time* groups. As the U. S. population is educated to become more IT-savvy in general, subscriber behavioral patterns may alternate drastically, and the possibility to accept an optimal discriminant pricing policy offered by the ISPs may become the reality for the management in the near future.

CONCLUSION

A complete iteration of a statistically sound empirical study was conducted in this article, with the objective of performing a comparative analysis. The 'variable-rate' subscription for the Internet with price discrimination was tested and compared between the data of the USA and South Korea.

An input data analysis to complete a pilot run of the statistical analysis was presented to achieve the cogency of this article, followed by a full-scale data analysis. Stratified cluster sampling was utilized to maintain the statistical soundness of the study, while comparable markets were selected as the sample space (*i.e.* NY, NY, USA *vs* Seoul, Korea). The Chi-Square (χ^2) goodness-of-fit tests were applied to the hypotheses constructed in this article. Results indicate it may yet be too early for the U. S. ISPs to introduce price discrimination to the market.

Future research may be pursued in two major directions:

- Optimal price discrimination strategy should be sought for each market segment that is of interest to the management of ISPs.
- Accurate correlation study between the transfer rate of the network, size of the digital data contents, types of the data contents, and consumer behaviors on the Internet should be the impetus for future extensions to this article.

ACKNOWLEDGMENT

This article was funded by the 2007 Business Council Research Grant from the School of Business Administration, Monmouth University, West Long Branch, NJ 07764-1898, USA.

The author is indebted to his colleagues, Drs. Robert Scott Jr. and Edward Christensen for their time and thoughtful insights during *ad hoc* research gatherings. In addition, the author wishes to extend his gratification to the anonymous reviewers for their time, assistance, and patience.

As always, this article is dedicated to his little precocious son, Ocean D. Willow.

REFERENCES

Banerjee, A., Faloutsos, M., & Bhuyan, L. (2008). The P2P War: Someone is Monitoring your Activities. *Computer Networks*, *52*(6), 1272–1280. doi:10.1016/j.comnet.2008.01.011

Choong, M. K., Logeswaran, R., & Bister, M. (2007). Cost-effective Handling of Digital Medical Images in the Telemedicine Environment. *International Journal of Medical Informatics*, *76*(9), 646–654. doi:10.1016/j.ijmedinf.2006.05.007

Federal Communications Commission. (2007). *Statistical Reports and FCC State-Link*. Retrieved May 1, 2007, from http://www.fcc.gov/wcb/iatd/stats.html

Han, T. (2005). Exploring Price and Product Information Search in e-Market. In *Proceedings of the International Conference on Information Technology* (pp. 397-402). Las Vegas, NV, U.S.A.: IEEE Computer Society Press.

Hanaro Telecom. (2007). *Multi-bell and Various Value Added Services*. Retrieved July 7, 2007, from http://www.hanaro.com/eng/iproduct/home.asp

Horneffer, M. (1998). Methods for performance-analysis of internet access points. *Computer Networks and ISDN Systems*, *30*(16), 1607–1615. doi:10.1016/S0169-7552(98)00169-X

Korea Telecomm. (2006). *Consumer Demographics White Pages*. Retrieved July 20, 2006, from http://english.ktf.com/eng/

Law, A. (2007). *Simulation Modeling and Analysis* (4 ed.). McGraw-Hill Publishing.

Luo, S., Kezunovic, M., & Sevick, D. R. (2004). Locating Faults in the Transmission Network using Sparse Field Measurements, Simulation Data, and Genetic Algorithm. *Electric Power Systems Research*, *71*(2), 169–177. doi:10.1016/j.epsr.2004.01.011

Odlyzko, A. (2004). The Evolution of Price Discrimination in the Transportation and its Implications for the Internet. *Review of Network Economics*, *3*(3), 323–346. doi:10.2202/1446-9022.1055

Park, K. A. (2001). Development of ICT Indicators of Korea. *2001 Annual Government Statistics Report*, Tokyo, Japan. Retrieved June 1, 2007, from http://www.stat.go.jp/English/info/meetings/iaos/pdf/park.pdf.

Porter, C. E., & Donthu, N. (2006). Using the Technology Acceptance Model to explain how Attitudes determine Internet Usage: The role of Perceived Access Barriers and Demographics. *Journal of Business Research*, *59*(9), 999–1007. doi:10.1016/j.jbusres.2006.06.003

Rappa, M. (2008). Business Models on the Web. *Managing the Digital Enterprise, Lesson 5*. Retrieved January 9, 2008, from http://digitalenterprise.org/models/models.html

Roycroft, T. R., & Anantho, S. (2003). Internet subscription in Africa: Policy for a Dual Digital Divide. *Telecommunications Policy*, *27*(1), 61–74. doi:10.1016/S0308-5961(02)00091-5

Sorte, D. D., Femminella, M., & Reali, G. (2008). QoS-enabled Multicast for delivering Live Events in a Digital Cinema scenario. *Journal of Network and Computer Applications*.

Sousa, P., & Freitas, V. (1998). A Framework for the Development of Tolerant Real-Time Applications. *Computer Networks and ISDN Systems*, *30*(16), 1531–1541. doi:10.1016/S0169-7552(98)00188-3

Viswanathan, V. (2007). Information Hiding in Wave Files through Frequency Domain. *Applied Mathematics and Computation*.

Willow, C. C. (2007a). Comprehensive Design for Information Systems with Integrated Systems Design Methodology. *International Journal of Taiwan Academy of Business Management Review*, *3*(1), 49–72.

Willow, C. C. (2007b). Qualitative Decision Making with Integrated Systems Design Methodology. *Journal of Engineering Management and Technology*, *24*(3), 262–282. doi:10.1016/j.jengtecman.2007.05.003

Wills, C. E., Brown, D. C., Dunskus, B. V., & Kemble, J. (1998). Evaluating Network Serviceability. *Computer Networks and ISDN Systems*, *30*(24), 2283–2291. doi:10.1016/S0169-7552(98)00154-8

This work was previously published in International Journal of Interdisciplinary Telecommunications and Networking, Volume 1, Issue 3, edited by Michael R. Bartolacci, and Steven R. Powell, pp. 31-41, copyright 2009 by IGI Publishing (an imprint of IGI Global).

Chapter 10
Application of Parametric Cost Estimation Model to Telecommunication Networks

Swadesh Kumar Samanta
University of Essex, UK

John Woods
University of Essex, UK

Mohammed Ghanbari
University of Essex, UK

ABSTRACT

The parametric cost estimation approach has proved to be an efficient method for analyzing complex systems such as spacecraft, missiles, ships, buildings, etc where cost varies according to a number of parameters. The cost to provision a telecom network also depends on a number of parameters; but little research effort has been applied to estimate cost using this approach. In estimating the cost of a telecom network, most published research has considered two parameters; distance and bandwidth of a link and ignored the effects of other parameters. We have modelled the cost based on distance, bandwidth, geographical terrain and technology simultaneously using a parametric cost estimation methodology applied to real data obtained from the Indian Telecom Company, BSNL. Using the model, we show how a cost optimized network can be designed given the real world constraints. The applicability of our model to determine revenue sharing mechanism for an international call is also demonstrated.

INTRODUCTION

The origins of the parametric cost estimation model date back to the Second World War when US Department of Defense (DoD) wanted a quick and accurate estimate of the cost to build aircraft of different capacities (Brundick, 1995). The model helped DoD to re-engineer business processes and reduce overall cost. Since then the model has been extensively used for estimating the financial cost for civil infrastructure such as buildings ("Parametric Cost Estimating", 2008)

DOI: 10.4018/978-1-60960-505-6.ch010

and complex systems such as spacecraft, space transportation systems, aircraft, missiles, ships, and land vehicles (Mahr & Richardson, 2003; Meisl, 1993; "NASA Cost Estimating Handbook", 2008).

The model has proven to be effective in the practical world where cost varies according to a number of parameters and is now widely applied in equipment system cost estimation (Jiang, et. al., 2007). However the model's application in telecom networks is rare; Somani et. al (2004) have developed a model for WDM (Wavelength Division Multiplexing) networks to compare the cost between multi-fiber and single fiber links. The work has demonstrated a tradeoff between the cost of multi-fiber and the cost of a single fiber with a wavelength converter. This is not however applicable for a network with heterogeneous links such as SDH (Synchronous Digital Hierarchy) on Optical Fiber Cable (OFC), microwave and satellite. Cost Models for a heterogeneous network can potentially serve a number of purposes such as estimation and optimization of financial cost (i.e. capital and maintenance expenditure) to provision the network and price the services (e.g. interconnection service) which are provided through the network.

The published literature naturally focused on both the above aspects but from two distinct disciplines; engineering and economics. However none of the works have followed the approach of a parametric model and there appears to be no published work to estimate or optimize the cost of a telecom network (of heterogeneous link technologies) using parametric models either. Engineers tend to be more concerned about the network cost optimization and have focused on modeling the cost to provision a telecom network based on distance and bandwidth of the interconnected links (Bley, et. al., 2004; Cabral, et. al., 2007; Harmatos, et. al., 1999; Mudchanatongsuk, et. al., 2008; Pioro, et. al., 2001; Sigurdsson, et. al., 2004; Yoon & Current, 2008;). In contrast, economists estimate the cost of an existing network to decide the prices (e.g. interconnection price) having made a number of assumptions. The assumptions include: the same technology ("An Analytical Cost Model", 2000; Kennet & Sharkey, 2001) for a country wide network, which is clearly untrue in the real world situations.

The interconnection price is the money charged by operators between themselves when a service such as a voice call is provided by more than one operator. Ideally the revenue generated from a call should have been shared according to the cost incurred by the respective operators. There is no quick and easy mechanism to decide the interconnection cost when the call originates and terminates at different geographical regions (e.g. an international long distance call). A preliminary study from a regional tariff group within the ITU ("ITU Regional Tariff Group", 2000) has indicated that even within the same country the cost of terminating international calls may differ by about 35 percent depending upon the volume of traffic, transmission media and technology of a particular route. In a heterogeneous geographical terrain a combination of link technologies (shown in Figure 1) such as OFC, Microwave and satellite are employed depending upon distance and bandwidth requirements. We show that there are certain scenarios where terrain and the associated technology of the link are the main parameters influencing the cost rather than distance and bandwidth.

In this work we model the network cost based on four parameters: distance, bandwidth, geographical terrain, and technology. Using the model, we show how a cost optimized network can be designed considering the real world constraints. We show that capital cost to the extent of 25% can be reduced if cost variations on distance, capacity and link technology are considered simultaneously.

The model can also potentially address the issues of interconnection pricing for an international long distance call which is influenced more by the policy (Antonio, et. al., 2005; Galbi, 1998)

Figure 1. A network with heterogeneous link technologies

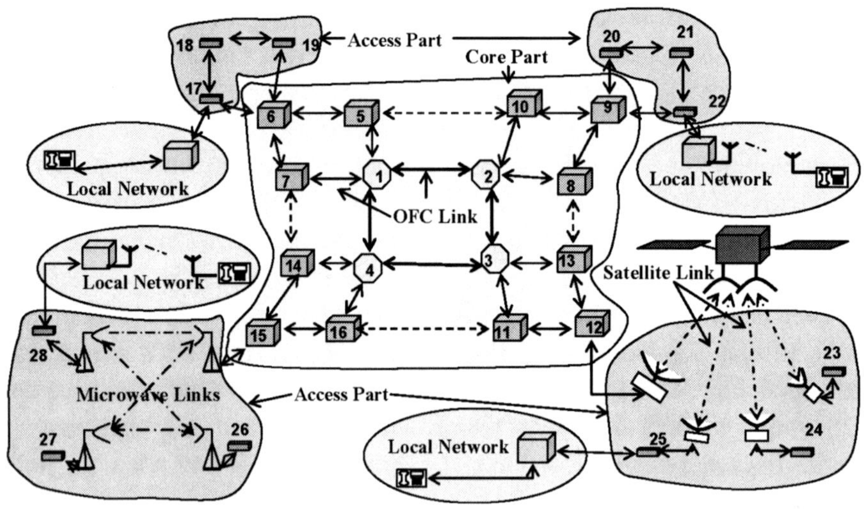

than by the incurred cost in the two countries. We demonstrate that the parametric cost estimation approach is more applicable for determining the interconnection pricing of an international long distance call.

The rest of the paper is organized as follows: In next section we present the benefit of parametric cost model. The subsequent two sections discuss the components of a parametric Link cost model and its development using the General Error Regression (GER) technique. The following section demonstrates the applicability of the model to design a cost optimized network topology resolving the *Capacity Distance Conflict* (CDC). The model's applicability for determining interconnection pricing for an international long distance call is demonstrated and some concluding remarks with direction for future research is presented.

BENEFIT OF A PARAMETRIC COST MODEL

A telecom network consists of nodes and their interconnected links. While locations of nodes (access or core) are generally fixed and dictated by population density, the link path mostly depends on both the population density and the geographical terrain of the nodes locations. Even though the link bandwidth can be estimated according to the population around the nodes and inter node distances (Kowalski & Warfield, 1995), the selection of link technology is dependent on factors such as distance, bandwidth and geographical demography between the nodes. The time taken to deploy a network can also influence the technology selected e.g. a microwave link may be used to serve a semi urban area or a satellite link to connect islands with small populations (Jhunjhunwala, et. al., 1998). This will reduce the installation time and improve the time to market. For microwave technology a tower has to be erected every 50KM which houses a repeater to regenerate and relay the microwave signal. Therefore for connecting islands which are separated by more than 50KM, microwave technology cannot be employed as placing towers in the ocean is neither practical nor cost effective.

The current research ("Cost-efficient network architecture", 2007) from a leading telecom equipment manufacturer shows that cost of links can constitute 15% to 80% of the total network cost depending upon the location of the network; core or access part (Figure 1). In the core part, bandwidth

Figure 2. Cost variation on technology for a 2Mbps link

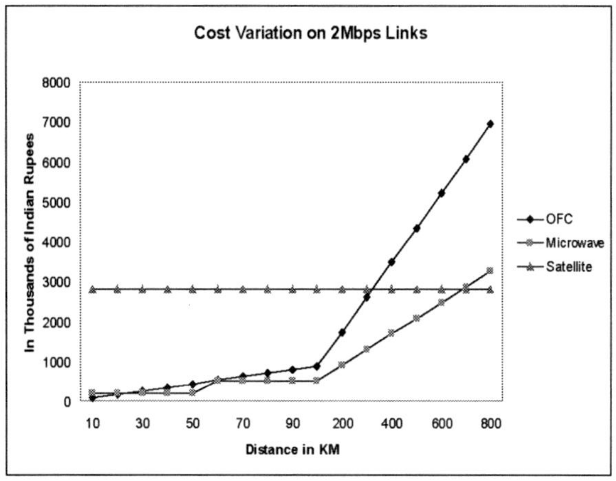

requirement is of the order of Gbps and therefore SDH and WDM technology on OFC provides a cost effective solution due to scale economies. In the access part, distances between nodes are longer and bandwidth requirement is smaller than the core part and therefore the cost of link constitutes a higher proportion of the total cost. A 10% reduction of link cost will result in 8% overall cost reduction in the access part compared to 1.5% in the core part. This shows that a cost estimation/optimization model which is based upon single technology such as OFC may not produce an optimum solution for a network where different link technologies are required to be deployed due to the geographical terrain constraints. Based on real data from BSNL, an Indian telecom company, we show (in Figure 2) the variation of cost on technology for a 2Mbps (millions of bits per second) link in a semi urban terrain.

Figure 2 shows that although the microwave link is cost effective for a distance up to 700Km, it can not be deployed for oceanic terrain. The same is also true for the OFC technology (even for higher bandwidth) since laying cable under the ocean is expensive and takes a longer time. Therefore there is no option other than to deploy a satellite link between the islands even though the cost is higher than microwave and OFC. However the research efforts have tended to ignore this constraint and have focused on single link technology such as OFC ("An Analytical Cost Model", 2000; Bley, et. al., 2004; Cabral, et. al., 2007Harmatos, et. al., 1999; Kennet & Sharkey, 2001; Mudchanatongsuk, et. al., 2008; Pioro, et. al., 2001; Sigurdsson, et. al., 2004; Yoon & Current, 2008) in designing a cost optimized network. To capture the engineering constraints the model needs to consider all the cost variable parameters. Using a parametric approach the cost can be modelled with parameters such as distance, bandwidth, geographical terrain and technology of interconnected links. The model will help to estimate and optimize the cost considering the variation on the above parameters simultaneously. Network operators are now realising the benefits of considering the link technology in network cost estimation and optimization (Hanna & Ramarao, 2006) but more research work is needed to concentrate in this direction.

The cost estimation based on parametric approach also provides additional tool to the operators to negotiate the revenue sharing mechanism (in the form of interconnection price) in a fair and transparent manner at the national and international level. In the case of international long distance call the interconnection price is generally

Figure 3. Cost components and their interdependencies

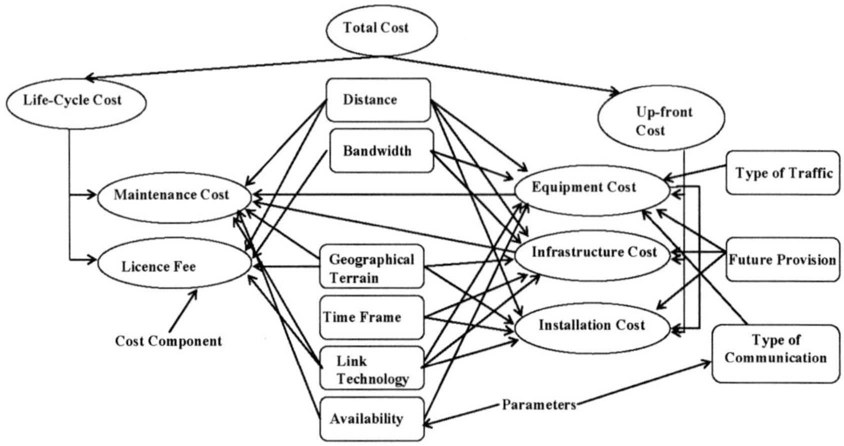

decided as per the US Federal Communications Commission (FCC) benchmarking order August 1997. The Order sets "benchmark" settlement rates for the termination of international traffic, based on a calculation of the "average" per minute incremental cost according to the teledensity and economic development of a country (Wettemann & Kelly, 1998). The average incremental cost may have little relationship with the actual cost of provisioning the service at the respective countries. To address this problem ITU (International Telecommunications Union) have undertaken substantive research to estimate the cost of provisioning an international long distance call ("ITU-T Study Group 3", 2000). Although, ITU Study Group 3 (representing members of 80 countries) has agreed for a cost-orientated asymmetric settlement rates, no appropriate costing methodologies have yet been approved and accepted (as US is opposed to asymmetric settlement rate). We show that our methodology can potentially address some of the issues.

PARAMETRIC LINK COST MODEL

In the previous section inclusion of parameters such as distance, bandwidth, geographical terrain and type of link technology in the cost model were discussed. Although these are the main parameters, some additional ones can be identified, including:

1. Type of Traffic to be carried (voice, video or data).
2. Type of Communication (simplex or duplex).
3. Link availability.
4. Time frame (time to deploy the network).
5. Future Provision (spare bandwidth).
6. Political environment in a country.

We have considered units of distance and bandwidth in kilometres (km) and millions of bits per second (Mbps) respectively and quantify the terrain in a scale of 1 to 9, with increasing cost. Quantifying terrain is a difficult task as it depends on several parameters including population and geographical demography. We consider demography as: plain land, hilly areas and oceanic island, and consider them in the context of population density when quantifying the terrain. We use a terrain index of 1 to reflect the lowest cost and 9 as the highest whilst other parameters (distance, bandwidth and link technology) are fixed.

In our current implementation parameters such as type of traffic is only voice and to simplify the model and make it comprehensible only the fundamental parameters; distance, bandwidth, terrain and link technology have been considered. The costs of the parameters mentioned in c), d) and e) are taken as a percentage of the total cost. Figure 3 shows the component costs of a link and their interdependencies. There are two main types of cost component: the up-front or capital cost and the life-cycle or maintenance cost for 10 years.

We regard a link as two end terminals with intermediate repeaters spaced at equal distances between them. For satellite links only a single repeater (satellite) is considered for distances up to 5000 km and the leasing cost of the transponder bandwidth has been considered as the cost of the repeater. For OFC and microwave, a repeater spacing of 50km is assumed. In this model the following cost components are considered:

1. Equipment cost includes the cost of the terminal, the required repeaters and is dependent on bandwidth and distance.
2. Infrastructure cost includes land, buildings, microwave towers, OFC cable etc and depends on distance and terrain.
3. Installation and maintenance cost is estimated as a percentage of the equipment and infrastructure cost.
4. License fee costs are related to spectrum charges or Right of Way to be paid to the local authority for cabling and both depend on distance and bandwidth.

With the above considerations a technology dependent link cost model can be defined as:

Cost*Link*(x) = $f(x)$ (1)

Where, Cost*Link*(x), represents the financial cost to provision a link and vector x represents the cost variable parameters; distance, bandwidth and terrain. We name (1) as the Cost Estimating Relationship (CER), and its form and coefficients are derived using a General Error Regression (GER) technique (Book & Young, 1997).

MODEL DEVELOPMENT USING GER

Most cost models develop a CER based on the classical least-squares linear regression equation

$$Y = a + b.X + E. \qquad (2)$$

where, Y is the cost, X is the numerical value of a cost variable, E is a Gaussian error term, and a and b are the coefficients. This relationship is only applicable if cost varies linearly with X, but if the relationship is non linear, then the following equation is usually assumed

$$Y = a.X^{b}.E. \qquad (3)$$

where coefficients a and b are derived by taking logarithms of both sides to convert in to the form

$$\log(Y) = \log(a) + b \log(X) + \log(E). \qquad (4)$$

There are theoretical and practical difficulties in this approach that have been demonstrated by Bradu & Mundlak (1970). The main weakness of this form is that it is fixed and other nonlinear forms such as $Y = a + b.\ X^{\wedge c}$, which are more appropriate can not be applied because the logarithmic transformation does not condition it to the classical linear-regression format. The problems associated with the classical least-squares linear regression can be solved by using a GER, which allows to chose any form of equation and determine the optimized coefficients for any non-linear form.

When cost varies with several parameters, a general form of the cost function is:

$$y_i = f(x_i, \hat{a})\ \varepsilon_i. \qquad (5)$$

where x is a vector representing different parameters, ε_i is a random error related to the actual value y_i corresponding to one sample x_i and $\hat{a}$ is the vector of coefficients. The term ε_i can be expressed as a ratio of y_i to $f(x_i,\hat{a})$.

Thus,

$$\varepsilon_i = \frac{y_i}{f(x_i, \hat{a})} \tag{6}$$

where, $\varepsilon_i = 1$ for all i indicates no estimation error. The aim is to find the vector parameter â that minimizes the sum of squared error. In our case y represents the cost of a link; vector x represents parameters distance, bandwidth and terrain. The important issue which decides the accuracy of the CER is the functional form $f(x, \hat{a})$, which depends on the nature of the data.

To understand this we have calculated cost values for different combinations of distance, bandwidth and terrain for different link technologies, from the data gathered from BSNL. They clearly indicate that there are certain combinations where link technology is the main parameter influencing cost rather than the distance and bandwidth. For a link bandwidth of 2 Mbps, microwave will be cheaper than OFC if the distance is longer than 20km, and satellite will be the cheapest if distance is longer than 700km. Further analysis indicates that for lower bandwidths (< 155 Mbps) satellite will be cheaper than microwave and OFC if the distance is longer than 1000km. We find that for different combinations of distance, bandwidth and terrain, the optimal solution will result from a combination of different link technologies.

With the exception of satellite where cost versus distance is fixed, the cost is non-linear with distance, bandwidth and terrain. Accordingly we have considered the following functional form of the CER with three parameters x, y and z representing distance, bandwidth and terrain respectively as

$$\text{Cost}Link(x, y, z) = a_x.x^{b_x} + x.a_y.y^{b_y} + x.a_z.z^{b_z} \tag{7}$$

Our aim is to find optimum values for the coefficients a_x, b_x, a_y, b_y, a_z, and b_z so that the estimation error is minimized. To find the optimum value in a limited time we have used a stochastic iterative method (Corne, et. al., 1999) in our optimization process.

To validate the model with the actual data gathered from BSNL, we tested each type of link technology against the parameters of distance, bandwidth and terrain and assessed the quality based on the standard deviation of the estimated error. We have considered 621 link combinations for distances 1-10,000km, bandwidths of 2-320,000 Mbps, and three terrains, semi-urban, hilly and metro city. Empirically the nature of data is such that higher error will result if a single set of coefficients is used to estimate all link combinations. Therefore two separate sets are used for bandwidth of up to 10,000Mbps and from 10,000-320,000Mbps. Our simulations show that the error is below 2% for over 80% of links, below 5% for 95% of links and below 10% for 99.7%.

NETWORK COST OPTIMIZATION; RESOLVING CDC

The link cost model from the previous section can be used to design a cost optimum heterogeneous network (as shown in Figure 1), where different link technologies are involved because of geographical constraints. We assume that the capital cost of a node is proportional to the traffic to be switched by it, and the total number of nodes and the capital cost for all nodes are fixed. Therefore we can exclude the cost of nodes for the cost optimization.

The model is applied to a hypothetical core network with a total of n nodes with the following assumptions.

1. Two physically and geographically separate paths are available between any two nodes in order to ensure full availability in case of single physical path failure (Bley, et. al., 2004).

2. Additional bandwidth is reserved for point to point links so that in the event of a link failure traffic can be re-routed via these links.
3. The number of nodes per ring is not predetermined and there are no boundary conditions.
4. The terrain among the links can be different.

We assume nodes are interconnected by *k* rings subject to the constraint of two physically and geographically separate paths between them. Distance, bandwidth and terrain among the nodes are generalized between nodes *i* and *j* as x_{ij}, y_{ij} and z_{ij}.

By replacing $x = x_{ij}$, $y = y_{ij}$ and $z = z_{ij}$ in (7), the cost of link between nodes *i* and *j* becomes

$$\mathrm{Cost}Link(i,j)(x_{ij}, y_{ij}, z_{ij}) =$$
$$a_x.x_{ij}^{b_x} + x_{ij}.a_y.y_{ij}^{b_y} + x_{ij}.a_z.z_{ij}^{b_z} \qquad (8)$$

We compare the cost based on different link technologies and the lowest cost is selected to be used in (9) or (10).

In general the total cost for a network of *k* interconnected rings with a total of *n* nodes connected as SDH rings can be estimated as,

$$\mathrm{Cost}_{KRings} \sum_{i=1}^{n}\sum_{j=1}^{n} = \mathrm{Cost}_{Link(i,j)}.l_{ij} - \sum_{k=1}^{k} n_k .\mathrm{Cost}_{KTer} \qquad (9)$$

Where Cost_{KRings} is the cost with *k* rings, *n* is total number of nodes and n_k refers to the number of nodes in the kth ring. Cost_{KTer} is the cost of terminal equipment of the kth ring and depends on the bandwidth of the kth ring and

$$l_{ij} \begin{cases} =1 & \textit{for direct physical link between nodes i and j} \\ =0 & \textit{otherwise} \end{cases}$$

For networks that use point to point links in cost analysis, the following cost is used.

$$\mathrm{Cost}_{PLinks} \sum_{i=1}^{n}\sum_{j=1}^{n} = \mathrm{Cost}_{Link(i,j)}.l_{ij} \qquad (10)$$

The cost obtained from (9) or (10) depends on the topology; there can be a variety of topologies for a network of n nodes. The cost for each topology can be compared and the optimum one is ascertained.

In seeking an optimum network topology for a large numbers of nodes, most published work ("An Analytical Cost Model", 2000; Bley, et. al., 2004; Harmatos, et. al., 1999; Kennet & Sharkey, 2001; Pioro, et. al., 2001; Sigurdsson, et. al., 2004) optimize cost related to distance and bandwidth in two separate stages. In addition most researches assume a fixed number of nodes for an interconnected ring and attempt to minimize the overall link distance. It may be possible that minimizing overall link distance may not always provide optimum solution. There may a situation (i.e. distance, bandwidth and terrain between nodes) where overall network cost for a topology can be lower even when overall link distance is larger than the minimum. This we termed as *Capacity Distance Conflict* (CDC). We show that by modeling cost variations for capacity and distance simultaneously and constructing rings of variable numbers of nodes, better solutions to the overall cost can be determined. Consider for example a five node network of Figure 4.

Figure 4. A network of five nodes

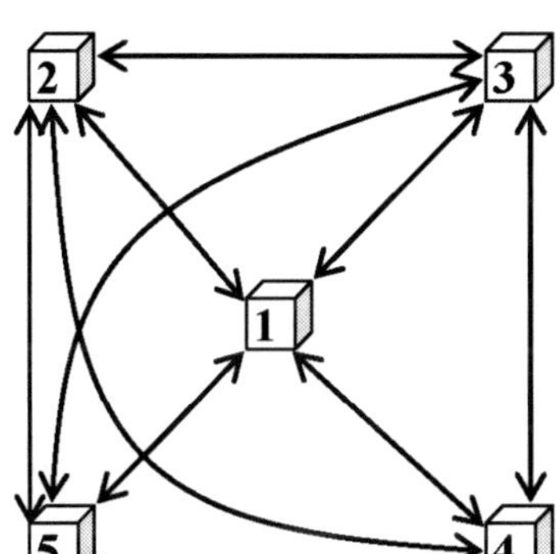

Figure 5. a) Five nodes connected in a single ring path b) Five nodes connected in two rings path

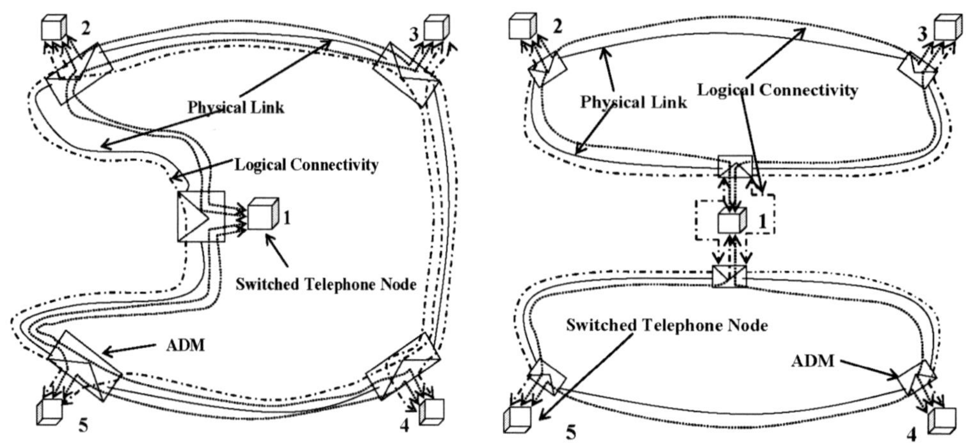

The inter node bandwidth requirements can be met by either a single or a two ring topology (Figures. 5a&b). The paths can be achieved by an SDH ring or alternatively by point to point links. In the case of a single link failure, the SDH technology provides the connectivity by routing the traffic in the opposite direction around the ring. For higher link bandwidth (>10Gbps), point-to point DWDM (Dense WDM) links are normally used and we have considered additional bandwidth to provide full redundancy in case the single physical path fails. This aspect has not been considered in the earlier research and therefore they do not provide full availability in case of a single path failure.

Consider also the traffic in a single ring topology (Figure 5a). If the amount of traffic from node 1 to other nodes increases then the required bandwidth of the ring will be affected, resulting in an increased cost for the entire ring. Breaking the single ring into two rings as shown in Figure 5b will increase the total physical link distance, but it will in turn reduce the required bandwidth of the individual rings. There are situations where the cost of two rings will be less than that of one.

To show this we have generated different combinations of inter-node link distance, bandwidth and terrain. Using (8) and (9) the capital costs of the network, for 32896 combinations are calculated for the single and two ring topologies. It is found that in 4063 out of 32896 combinations the cost of two ring topology is less than that of a single ring. Figure 6 plots the relative cost of a single ring versus two ring topology for the above 5 nodes for a limited number of combinations. It shows very clearly that there are cases where the single ring is not cost effective and the two ring solution should be chosen. In some cases a cost saving in the order of 25% is possible even for this simple five node network.

We have applied a similar analyses to the point-to-point links for the above five nodes. Figures. 7a and 7b show how the inter node bandwidth requirements can be met using single ring and two rings topology for point-to-point links. Similar to SDH technology we considered 32896 different combinations and found that in 16481 combinations the two ring topology had a lower cost. This clearly reinforces our *Capacity Distance Conflict* (CDC) statement.

To show that this phenomenon is generic rather than being specific to simple topologies, we have taken an 11 node network shown in Figure 8. The 11 nodes can be connected forming from a single ring path up to five interconnected rings as shown in Figures 9a, b, c, d & e.

A total of 2.45 million combinations of link characteristics have been considered. For each

Figure 6. Relative cost for different combinations of inter-node link capacity

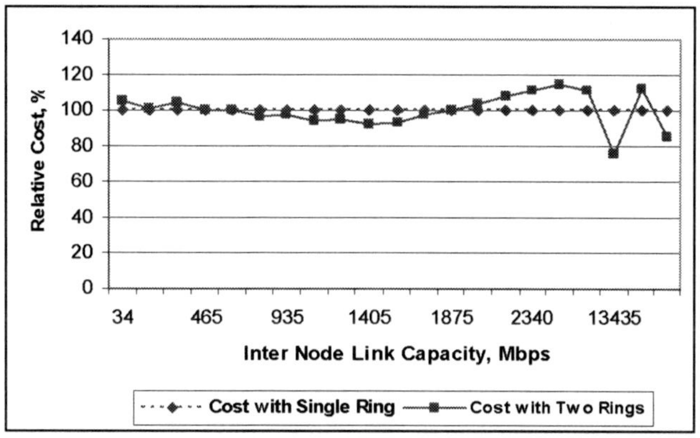

Figure 7. a) Point-to-Point links in a single ring path b) Point-to-Point links in two rings path

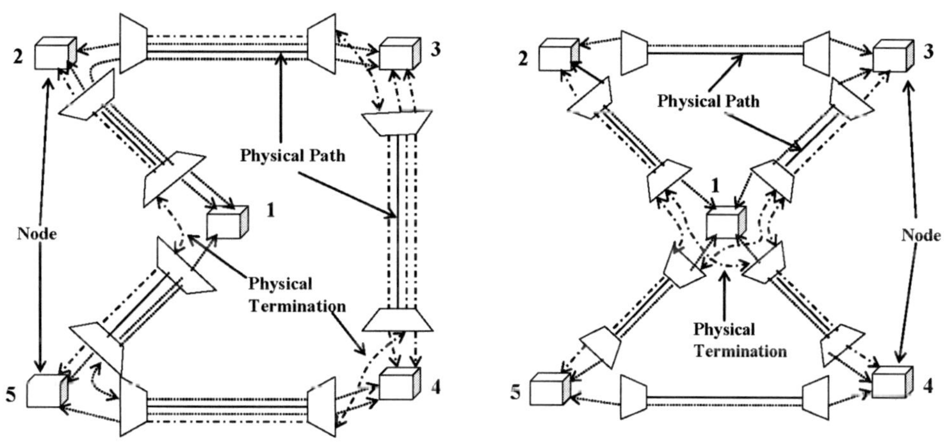

Figure 8. A network of eleven nodes

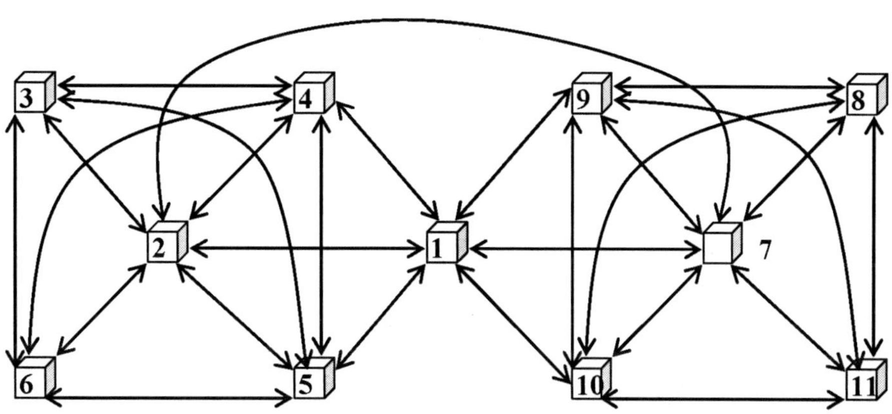

Figure 9. Five different topologies for a network of eleven nodes

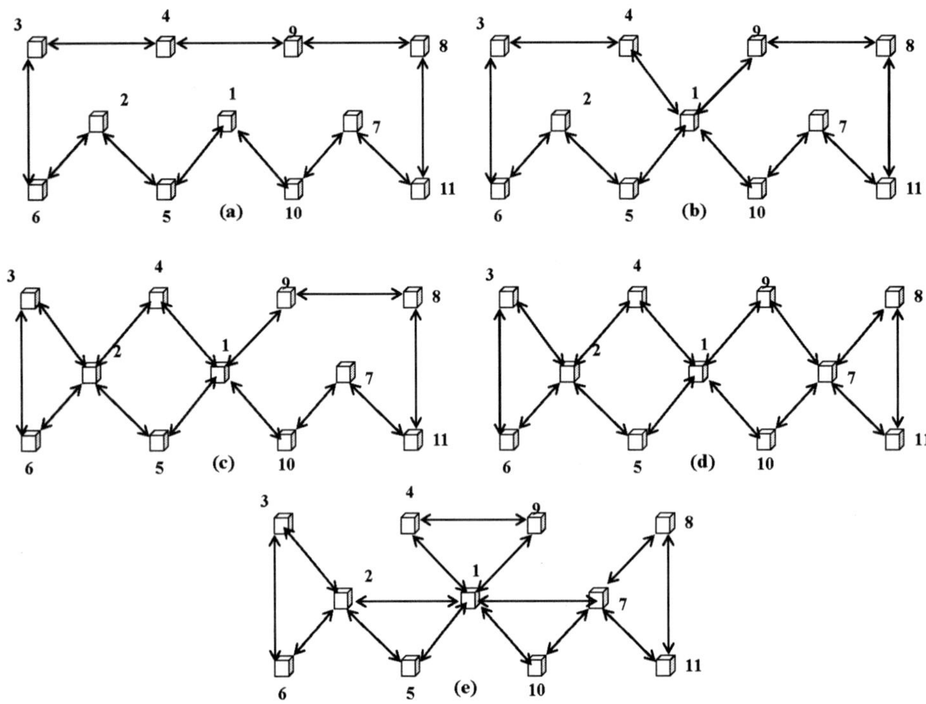

combination, the total cost of the network has been calculated for the single up to the five ring topologies. We found the least costs for the multiple ring topologies: 0.52, 0.68, 0.6, and 0.5 million combinations for 2, 3, 4 and 5 rings respectively. Further analysis show for 0.27 million combinations the cost of three rings is the lowest of all, in 0.29 million combinations four rings provide less cost than the three rings and in 0.10 million cases where four rings provide the least overall cost. These all validate our hypothesis on CDC and illustrate the need for a fundamentally different approach in network cost optimization.

To show that this approach scales with increase in number of nodes we have taken a 20 node network. Twenty nodes can be a representative of the real world networks which are generally hierarchical and have similar number of nodes in each level. As an example, the British Telecom (BT) has decided to use 20 nodes in its core part of the proposed Next Generation Network for the UK ("BT's 21CN Technology", 2008). In India, BSNL, the largest fixed and mobile operator has 23 nodes in the core part and their interconnection topology is designed independent of other nodes of the country. Using our methodology a cost optimized topology for 20 node network is obtained in less than two minutes on a Pentium PC Intel® Core™2CPU 6400@2.13GHz and 1.99Gb of RAM.

The model can also be used to optimize the capital cost of the access network where ordinarily star, bus and tree (Bley, et. al., 2004) topologies are encountered according to operator strategy. The operator must strike a compromise between cost and lack of service due to physical path failure. For the access part, a star topology is a good choice since only one node becomes unavailable in the event of a single path failure. However terrain can prohibit a star topology such as where a hill obstructs a microwave link. With the help of our model the cost for complicated hybrid topologies for an access network can also be optimized.

APPLICATION FOR INTERCONNECTION PRICE DETERMINATION

The model can also potentially address the issues of interconnection pricing for an international long distance call which is influenced more by the policy (Antonio, et. al., 2005; Galbi, 1998) rather than by the incurred cost in two countries. Historically the termination of an international call is controlled by government companies in the respective countries and revenue settlement is based on the accounting rate system (Galbi, 1998; "ITU-T Study Group 3", 2000). In an accounting rate system the net payment outgoing from some countries (especially developed countries) is significantly higher when there is asymmetric flow of traffic between the countries. As an example, in 1995, 70 percent of the (US$6 billion in net settlement) payment from USA (for international telephone interconnection service) was made to developing countries where the top four net receiving countries (i.e. Mexico, China, India, and the Philippines) accounted for 30 percent of the total payment (Braga, et. al., 1999).

Therefore developing countries were under constant pressure from USA and other international bodies such as the World Trade Organisation (WTO) to introduce competition for both domestic and international long distance services (Jain, 2006). In this regard the US Federal Communications Commission (FCC) had taken an aggressive strategy and issued a benchmarking order in August 1997. The Order sets "benchmark" settlement rates for the termination of international traffic, based on a calculation of the "average" per minute incremental cost according to the teledensity and economic development of a country (Wettemann & Kelly, 1998). Even though operators in some countries have argued that the order violates US WTO market entry commitments, most of the regulatory authorities of developing countries where virtually forced to take steps to substantially reduce the termination charge for an international call. But the result is that the operators in those countries are now receiving less revenue compared to their foreign counter parts. As an example, since 1999, the Indian regulatory authority reduced termination charges for international calls by a factor of five and operators now have lower revenue.

An international call from the UK to India currently (As of April 2008) generates an on average revenue of 16 cents per minute. The revenue is shared between the operators in the UK, the international long distance carrier and operators in India according to 5:2:1 ratio. For an average International call, the distance carried by the Indian operator is greater than that of the UK. The average link bandwidth in rural parts of India is of the order of 34Mbps which is much less than that in the UK. Moreover certain parts e.g. the north east part of India are very difficult to reach involving higher cost to deploy and maintain the network. Therefore the sharing should have been reversed if the operator's provisioning cost for an average call is considered. To address this problem researcher in academic institutions and international organisations such as the ITU (International Telecommunications Union) have undertaken substantive research to estimate the cost of provisioning an international long distance call ("ITU-T Study Group 3", 2000). Although, ITU Study Group 3 (representing members of 80 countries) has agreed for a cost-orientated asymmetric settlement rates, no appropriate costing methodologies have yet been approved and accepted (as US is opposed to asymmetric settlement rate). The Study Group has formulated the basic principles to be observed when developing a costing methodology. However, it has been accepted that the development of a cost estimation model which can be applied to all countries is arguably complex.

A preliminary study from a regional tariff group within the ITU ("ITU Regional Tariff Group ITU", 2000) has indicated that even within the same country the cost of terminating international calls may differ by about 35 percent depending upon the volume of traffic, transmission media and technology of a particular route. This indicates that a parametric cost estimation approach (which can consider these parameters) is much more applicable for determining the interconnection pricing for an international long distance call. The parametric estimation approach will also adhere to the ITU recommendation ("ITU-Recommendation D.140", 1998). The recommendation considers the cost of two parts; international transport (transmission & switching) and national transport of terminating countries. To achieve fairness, ideally a third part should be considered, that of the national transport in the originating country. The relative costs incurred in the different parts can then be calculated.

The cost is usually estimated using the accounting data of the dominant operators. When there is a limitation on the availability of data for national transport, the price of a national long distance call is generally taken as a proxy. The FCC in the calculation of the benchmark rate has taken this approach for developing countries for which adequate cost data are not available. However this poses a number of problems (Wettemann & Kelly, 1998); the major one being operators provide free local and national calls in lieu of a fixed charge included in the monthly subscription fee. This is also true where operators cross subsidize local and national calls with monthly subscription fees. In most cases these adjustments are not visible and therefore create a problem in determining the actual price of a national long distance call.

We have examined this issue in a case study within the current (as of April 2008) international call pricing policy between the UK and India. In Figure 10 we show the different routes through which an international long distance call from the UK to India is terminated. A user from the UK can make a call to India in two ways; 1) by directly dialing the international access code (00) + country code (91) + national access code followed by local number of India, and 2) through a calling card where the user pays in advance for making the calls. Using a card an international call can be made either from a fixed line or mobile phone by dialing a UK national access number (0800 or 02xx) where the call is first terminated to a server/switch located in the UK. A user can dial the international number after the server verifies

Figure 10. Revenue sharing for international long distance call

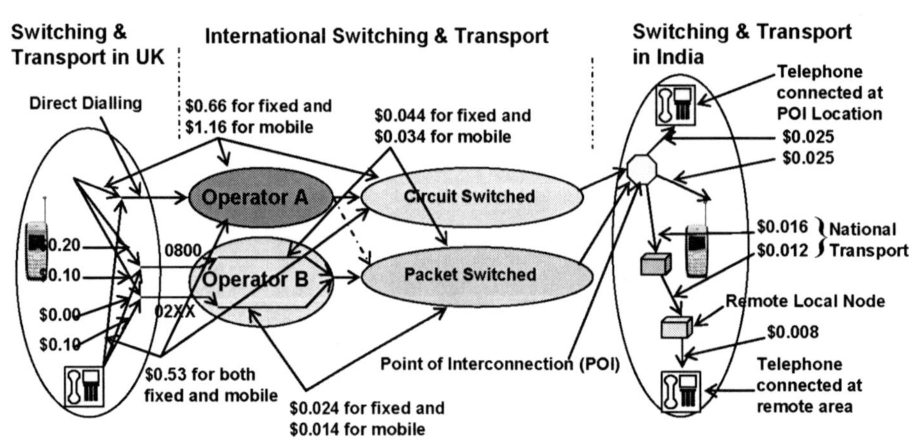

and authenticates the details of the card (e.g. type of card and credit amount).

Operators (e.g. Operator B of Figure 10) other than the local and national access providers also provide international call services to India through the calling card. For this specific case study the price data for different parts of the call both for direct dialing and using a calling card have been taken from data contained on the websites of BT-UK (UK fixed telephone operator), O2-UK (UK mobile operator) and LycaTalk, an operator who provides international call service using calling card ("Rates for International Call", 2008). For the Indian part of the call, the prices are taken from BSNL, Indian fixed and Mobile operator.

The operators BT and O2 in the UK provide local, national and international access using fixed and wireless technologies. The local and national access in a fixed line telephone is highly subsidized and the operator compensates the revenue from the fixed monthly subscription fees. As a result of the higher subscription fee the fixed line operator even subsidizes the international long distance calls which are evident if compared with charges for mobile phones (see Figure 10). Therefore, in a situation like this the price cannot be taken as a proxy to determine the operators cost of provisioning the call. In Table 1 we show the percentage share of revenue for the operator for international call flow through different routes. It is observed that the percentage share for Indian operators varies from 3.8% to 31.3%, which is less compared to their UK counterparts. Moreover the percentage share varies according to the technology used for international carriage and type of phone from which calls originate. However in absolute terms the operators in India receive a fixed amount $0.025 per minute of call irrespective of technology and type of originating phone.

The percentage share of revenue for UK operators and International operators has been derived from the price information mentioned earlier for UK national and International calls to India. If the international operator and UK operator are part of the same business house then it may be possible for the UK operator to subsidies the local transport and charge heavily for international carriage in order to justify the local transport costs.

In this specific case study we can find that when a mobile user in the UK terminates a call to an

Table 1. Percentage share of revenue for international call flow from UK to India

International call flow from UK to India			Percentage share of revenue by operators			Remarks
Type of phone which originate call	Method of Dialing	Technology for International carriage	At UK	International operator	AT India	
Fixed	Direct Dialing	Ckt Switched	18.9	76.4	4.7	
Fixed	Direct Dialing	Pkt Switched	50.0	29.2	20.8	
Fixed	Calling Card	Ckt Switched	N/A	N/A	N/A	
Fixed	Calling Card	Pkt Switched	25	43.7	31.3	
Mobile	Direct Dialing	Ckt Switched	15.2	81.0	3.8	% of Indian share is less for call terminated to mobile phones.
Mobile	Direct Dialing	Pkt Switched	N/A	N/A	N/A	
Mobile	Calling Card	Ckt Switched	N/A	N/A	N/A	
Mobile	Calling Card	Pkt Switched	76.9	13.5	9.6	

Indian fixed phone (by direct dialing), the Indian operator gets only 3.8% of the total revenue. This is four times less than that of the mobile operator in the UK. In general the provisioning cost for a one minute call in a fixed line phone is much higher compared to a mobile phone. The sharing should have been reversed if the technology of the originating and terminating phone is considered. The need to incorporate technology in the cost estimation is evident if we consider the international carriage through both circuit switched (ckt switched) and packet switched (pkt switched) routes. The provisioning cost for an international carriage in a packet switched route is 4-6 times less than the circuit switched equivalent due to efficiency gains from statistical multiplexing and voice coding. Consider a packet switched route (international transport) is four times more cost effective, and assume that the cost of the UK and Indian operators remains the same as for a circuit switched route. According to our calculation the mobile operator in the UK should have retained four times the revenue of the Indian operator compared to the currently observed factor of eight (see Table 1).

The case study highlights that for estimating the cost to provision an international long distance call the parametric approach is very useful even if four basic parameters (i.e. distance, bandwidth, technology and geographical terrain) are considered. The geographical terrain parameter can effectively capture the cost variation between two countries for a network of similar volume. The model we proposed will facilitate the operators to estimate the cost incurred by their counterparts and help to negotiate the sharing mechanism in a fair and transparent manner. The model will also allow international bodies (e.g. ITU, WTO) to mediate and resolve any dispute which may arise in deciding interconnection price for telephone calls or internet communication (Faratin, et. al., 2007) between two countries.

CONCLUSION

A parametric cost estimation approach is very useful for system optimization where cost varies with the number of parameters. Even though the approach is widely employed for complex systems such as spacecraft, space transportation systems, aircraft, missiles, ships, and land vehicles, little research has focused on its application to telecommunication networks. This work highlights the potential benefits of a parametric cost estimation approach for a telecom network where a combination of link technologies such as OFC, microwave and satellite are employed due to the technological constraints associated with geographical demography.

We have developed a model to estimate and optimize the cost of a telecom network based on parameters such as distance, bandwidth, geographical terrain and technology of the interconnected links. Using a General Error Regression and stochastic iterative optimization the model equation is obtained in less than two minutes on a Pentium PC Intel® Core™2CPU 6400@2.13GHz and 1.99Gb of RAM. Our model has addressed the shortfalls of the approaches currently used by researchers who optimize cost on distance and bandwidth separately.

The validity of the *Capacity Distance Conflict* (CDC) hypothesis has been demonstrated, indicating that cost optimization on distance and bandwidth separately can overlook better minima. Application of the model in determining the interconnection price for an international long distance call between two countries has also been demonstrated. Using a case study for an international call between the UK and India, we highlighted the anomalies of the current interconnection pricing mechanism where the developing countries are now placed in a disadvantageous position compared to their developed counterparts. We showed that the proposed cost estimation approach can effectively address the current anomalies and foster network development in developing countries. The model can also be used to decide the interconnection price for an internet network based on distance, total bandwidth, technology and location of the interconnected link for the two countries. We are undertaking further research in this direction.

ACKNOWLEDGMENT

The authors deeply acknowledge the hard efforts made by the officers of Department of Telecommunications, and BSNL of India for compiling and providing the detailed data which has been used in this model. The authors specially acknowledge the tireless efforts of Shri J. Kole, Deputy General Manager and Shri K.S. Mondal, Assistant General Manager of BSNL, in providing commercial and technical clarifications on international long distance call termination at Indian networks. The authors also thank both Govt. of India and British Govt. for providing the financial support for this work under UK-India Education and Research Initiative scheme.

REFERENCES

An Analytical Cost Model-National Core Network Version 2.0. Consultative Document 2.0. (2000). Retrieved January 20, 2007, from, http://www.regtp.de/en/reg_tele/start/in_05-07-00-00-00_m/fs.html

Antonio, O. R. L. N., & Ricardo, R. (2005). Mexico's international telecommunications policy: Origins, the WTO dispute, and future challenges. *Telecommunications Policy*, *29*(5-6), 429–448. doi:10.1016/j.telpol.2005.03.005

Bley, A., Koch, T., & Wessäly, R. (2004). Large-scale Hierarchical Networks: How to Compute an Optimal Architecture? *11th International Telecommunications Network Strategy and Planning Symposium. Networks*, *2004*, 429–434.

Book, S. A., & Young, P. H. (1997). General-Error Regression for Deriving Cost-Estimating Relationships. *The Journal of Cost Analysis* (pp. 1-28).

Bradu, D., & Mundlak, Y. (1970, March). Estimating in Lognormal Linear Models. *Journal of the American Statistical Association*, *65*, 198–211. doi:10.2307/2283587

Braga, C.-A. P., Forestier, E., & Stern, P. A. (1999). Developing Countries and Accounting Rates Reform--A Technological and Regulatory El Niño? *Public Policy Journal*, (173). Retrieved May 14, 2008, from http://rru.worldbank.org/PublicPolicyJournal/Summary.aspx?id=173

Brundick, B. (1995). *Parametric Cost Estimating Handbook*. Retrieved April 20, 2008, from http://cost.jsc.nasa.gov/PCEHHTML/pceh.htm

BT's 21CN Technology -21CN domain descriptions. (2008). Retrieved June 16, 2008, from http://www.btplc.com/21CN/Thetechnologyofthenetwork/21CNdomaindescriptions/21CNdomaindescriptions.htm

Cabral, E. A., Erkut, E., & Laporte, G. (2007). The network design problem with relays. *European Journal of Operational Research*, *180*(2), 834–844. doi:10.1016/j.ejor.2006.04.030

Corne, D., Dorigo, M., & Glover, F. (1999). *New Ideas in Optimisation*. Berkshire, England: McGraw-Hill Publishing Company.

Cost-efficient network architecture for addressing low-income segments. (2007). Retrieved April 20, 2008, from http://www.ericsson.com/technology/whitepapers/wp_nw_arch_%20for_low_arpu_rev_a.pdf

Cost Estimating Handbook, N. A. S. A. 2008. (2008). Retrieved July 1, 2008, from http://www.nasa.gov/offices/pae/organization/cost_analysis_division.html

Faratin, P., Clark, D., Gilmore, P., Bauer, S., Berger, A., & Lehr, W. (2007). Complexity of Internet Interconnections: Technology, Incentives and Implications for Policy. *TPRC 2007: The 35th Annual Research Conference on Communication, Information and Internet Policy*. Retrieved April 20, 2008 from, http://people.csail.mit.edu/wlehr/Lehr-Papers_files/Clark%20Lehr%20Faratin%20Complexity%20Interconnection%20TPRC%202007.pdf

Galbi, A. G. (1998). Distinctive arrangements for international interconnection. *Telecommunications Policy*, *22*(11), 945–951. doi:10.1016/S0308-5961(98)00058-5

Hanna, M., & Ramarao, B. (2006). *Cost optimization for transmission and backhaul technologies*. Retrieved April 18, 2008, from, http://www.accenture.com/Global/Services/By_Industry/Communications/R_and_I/acs_cost_optimization.htm

Harmatos, J., Juttner, A., & Szentesi, A. (1999, September). Cost-based UMTS Transport Network Topology Optimisation. *International Conference on Computer Communication*, Tokyo, Japan. Retrieved April 18, 2008, from, http://www.cs.elte.hu/~alpar/publications/ proc/ICCC99.pdf

ITU-Recommendation D.140; Annex A: Guidelines for the cost elements to be taken into account when determining accounting rates and accounting rate shares for the international telephone service. (1998). Retrieved April 16, 2008, from, http://www.itu.int/osg/csd/intset/itu-t/d140/d140_e_rev.html

ITU Regional Tariff Group on Asia and Oceania (TAS). TAS Cost Model. (2000). Retrieved April 16, 2008, from, http://www.itu.int/ITU-T/othergroups/tas/index.asp

ITU-T Study Group 3; Special Projects and Issues: Accounting Rate Reform undertaken by ITU-T Study Group 3 (2000). Retrieved April 16, 2008, from, http://www.itu.int/ITU-T/studygroups/com03/accounting-rate/index.html

Jain, R. (2006). Interconnection regulation in India: Lessons for developing countries. *Telecommunications Policy*, *30*(3-4), 183–200. doi:10.1016/j.telpol.2005.11.005

Jhunjhunwala, A., Ramamurthi, B., & Gonsalves, T. A. (1998). The role of technology in telecom expansion in India. *IEEE Communications Magazine*, *36*(11), 88–94. doi:10.1109/35.733480

Jiang, T., Guo, S., Wang, S., & Wei, R. (2007, September 21-25). Universal Study on Parametric Models in Equipment System Cost Estimation. W*iCom 2007: International Conference on Wireless Communications, Networking and Mobile Computing*, 2007 (pp. 5228–5231).

Kennet, D. M., & Sharkey, W. W. (2001). Calculating the Long Run Incremental Cost of Interconnection using a Network Cost Simulation Model. *Journal of Computing in Economics and Finance, 9*.

Kowalski, J. P., & Warfield, B. (1995, December). Modelling traffic demand between nodes in a telecommunications network. *Australian Telecommunications and Networks Conference*. Retrieved March 12, 2008, from, http://citeseerx.ist.psu.edu/viewdoc/summary?doi=10.1.1.56.9702

Mahr, E., & Richardson, G. (2003). Development of the Small Satellite Cost Model (SSCM). *IEEE Aerospace Conference Proceedings 2003, 8*, 3831-3841.

Meisl, C. J. (1993, December). Parametric Cost Model for solar Space Power and DIPS Systems. *IEEE AES Systems Magazine* (pp. 24-29).

Mudchanatongsuk, S., Ordonez, F., & Liu, J. (2008). Robust solution for network design under transportation cost and demand uncertainty. *The Journal of the Operational Research Society*, *59*(5), 652–662. doi:10.1057/palgrave.jors.2602362

Parametric Cost Estimating Model for Buildings. (2008). Retrieved April 16, 2008, from http://www.aacei.org/technical/BuildingModel.shtml

Pioro, M., Juttner, A., Harmatos, J., Szentesi, A., Gajowniczek, P., & Myslek, A. (2001, September). Topological Design of Telecommunication Networks- Nodes and Links Localisation under Demand Constraints. *17th International Teletraffic Congress*, Salvador da Balia, Brazil. Retrieved March 12, 2008 from http://www.cs.elte.hu/~alpar/publications/proc/ITC-TopDesTelNel.pdf

Rates for International Call to India from UK. (2008). Retrieved April 18, 2008, from http://www.lycatalk.com/res_rates.htm

Sigurdsson, H. M., Thorsteinsson, S. E., & Stidsen, T. K. (2004). Cost optimization methods in the design of next generation networks. *IEEE Communications Magazine*, *42*(9), 118–122. doi:10.1109/MCOM.2004.1336727

Somani, A. K., Mani, M., & Li, L. (2004). On Trading Wavelengths With Fibers: A Cost-Performance Based Study. *IEEE/ACM Transactions on Networking*, *12*(5), 944–951. doi:10.1109/TNET.2004.836130

Wettemann, R., & Kelly, T. (1998, March). The Changing International Telecommunications Environment: Country Case Studies. *2nd World Telecommunication Policy Forum* (pp. 16-18), Geneva. Retrieved April 20, 2008, from, http://www.itu.int/osg/csd/wtpf/wtpf98/cases/index.html

Yoon, M. G., & Current, J. (2008). The hub location and network design problem with fixed and variable arc costs: formulation and dual- based solution heuristic. *The Journal of the Operational Research Society*, *59*(1), 80–89. doi:10.1057/palgrave.jors.2602307

This work was previously published in International Journal of Interdisciplinary Telecommunications and Networking, Volume 1, Issue 3, edited by Michael R. Bartolacci, and Steven R. Powell, pp. 42-61, copyright 2009 by IGI Publishing (an imprint of IGI Global).

Chapter 11
Performance and Complexity Evaluation of OTR–UWB Receiver

Hossein Gharaee
Tarbiat Modares University, Iran

Abdolreza Nabavi
Tarbiat Modares University, Iran

Jalil (Joe) Etminan
Rock Valley College, USA

ABSTRACT

This article presents a new transmitted reference UWB receiver, which utilizes the orthogonal property of even and odd order derivatives of Gaussian pulses in neighboring chips for synchronization. This system, referred to as orthogonal TR-UWB (OTR-UWB), employs only a single spreading code, which results in much lower mean detection time compared to DS-UWB systems. The hardware complexity for OTR-UWB receiver is significantly reduced against conventional TR-UWB systems. In addition, simulation results show that BER performance is improved, while the new system is capable of supporting higher data rates. Also, this article presents the FPGA implementation of OTR-UWB, with a bit-rate of 25Mb/s without using equalizer. In addition, we present the DSP algorithm of baseband. Hardware of this system is implemented on two different FPGAs from ALTERA and XILINX, CycloneII (EP2C35F672C6) and Spartan 3 (3s4000fg676-5). Gate estimation and power analysis are performed by Quartus II 7.2 (ALTERA) and ISE 8.1 (XILINX) softwares.

INTRODUCTION

Ultra-wide band (UWB) radio has attracted much more attention as an excellent candidate for various applications own to its appealing feature, such as less complex hardware, low power consumption and high bit rate, as well as high-precision ranging capability 0. UWB impulse radio operates by transmitting very short pulses with bandwidth about several gigahertzes, illustrating numerous advantages compared to narrowband systems including robustness to multipath fading and potential for achieving high capacity 0. It's very wide bandwidth, however, poses several design

DOI: 10.4018/978-1-60960-505-6.ch011

challenges, among which channel estimation and synchronization appear to be the two major ones. At this point, three merged technical approaches, referred to as multi-band orthogonal frequency division multiplexing (MB-OFDM), direct-sequence ultra-wideband (DS-UWB), and time-hopping ultra-wideband are considered as strong candidates for some applications such as ranging, positioning, sensor networks and WPAN standard. The direct-sequence (DS) spread spectrum technique is a well-known and powerful multiple-access technology. Antipodal signaling, such as binary phase shift keying (BPSK), can be employed with DS-UWB and is supported by the current technology. In DS-UWB, multiple pulses are transmitted during one data symbol duration using bipolar modulation based upon a certain spreading sequence. This method has many attractive properties, including low peak-to average power ratio and robustness to multiuser interference. On the other hand low complexity of DS-UWB is an important advantage of DS-UWB than MB-OFDM. For DS-UWB, which preserves the original UWB pulse nature, due to the short length of the impulse, in the order of nanosecond, one of the tough challenges existed in UWB system is high performance and low complexity timing acquisition. DS-UWB is used with one pseudo random sequence (PN) code for spreading each bit. Transmitted reference (TR) UWB is a DS-UWB system that used two PN codes which it explain in follow.

TR modulation schemes have been proposed because of the ease of channel estimation in a dense multipath environment and the low implementation complexity 0. In TR-UWB systems, the energy in the multipath can be collected by autocorrelation of the received signal and a delayed version of itself. TR-UWB system with DS signaling consists of an *outer* code and an inner code, where the former modulates both the reference and the data pulses, while the latter modulates only the data pulse. The polarity of every alternate received pulse in the product signal depends only on the modulating bit and the inner code. Therefore, given the phase of the inner code, the bit can be demodulated without knowledge of the phase of the outer code 0.

The authors have proposed a new TR-UWB system with BPSK modulation, which employs even and odd derivatives of Gaussian pulse in neighboring chips along with a single spreading code 0. Since the two neighboring pulses are orthogonal, autocorrelation between the received signals with a delayed version of itself approximately gives zero DC. From this, the receiver rapidly acquires chip's timing (coarse synchronization) because the search space is a function of sampling frequency. Then, the system, referred to as orthogonal TR-UWB (OTR-UWB), performs the code acquisition with a search space linearly proportional to code length. Lower search space reduces the complexity of the system by decreasing the number of multipliers, adders, and memories. Therefore, the OTR system illustrates better acquisition performance and higher data rate than that of conventional UWB counterparts.

The reminder of this article is organized as follows. In Section 2, we review some applications of UWB which used characteristics of UWB pulses or standards. The properties of Gaussian pulses are reviewed in Section 3. In Section 4, the OTR-UWB system is described. Section 5 explains and reviews acquisition approaches in UWB communication system. The results of performance analysis and simulation are presented in Section 6. Section 7 describes the DSP algorithm of baseband implementation. In Section 8, the results of analysis and simulation are presented. Finally, the article is concluded in Section 9.

UWB APPLICATIONS

In February 2002 the Federal Communications Commission (FCC) has given the permission for the marketing and operation of certain types of new products incorporating UWB technology

(FCC, 2002).The FCC decided for the UWB systems the allocation of a7.5 GHz unlicensed band in the range of the radiofrequency spectrum 3.1-10.6 GHz, through a modification of the 47 CRF Part 15 regulations (FCC, 2002). The FCC has provided a mask of the maximum power spectral density (PSD) allowed for UWB devices set to very low values (-41.3 dBm/MHz in the 3.1-10-6 GHz band), since UWB systems are intended to operate in regions of spectrum in which other services already operate. Three different approaches were in UWB: Multi-Band approach combining frequency hopping with Orthogonal Frequency Division Multiplexing (MB-OFDM UWB), Direct-Sequence UWB (DS-UWB) and Time-Hopping UWB(TH-UWB). It was also considered a potential candidate for alternate physical layer protocols for the high-rate and the low-rate wireless personal area network (WPAN) standards (FCC, 2002; Staderini, 2002). The IEEE 802.15.4a and SG3aTask Groups have developed an UWB based physical layer standard for short-range networks with a precision ranging capability and optimized for low data rate application. Among the possible approaches, Direct Sequence UWB (DS-UWB) was designed to exploit the available spectrum by the use of spreading codes, in a similar way to classical Direct Sequence Spread Spectrum systems (Staderini, 2002). Two bands (Low and High) were proposed for DS-UWB in IEEE 802.15.3 which they are between 3-5 and 6-10 GHz respectively.

UWB offers some unique and distinctive properties that make it more suitable for some applications and less suitable for other applications. UWB has the potential for very high data rates using very low power, however only at very limited range. This will lead to applications well suited for WPAN. UWB will not be suitable for applications such as WLAN (Wireless Local Area Networking) or WWAN (Wireless Wide Area Networking, cell phone radios), both of which having greater range expectations.

UWB will provide enhanced value for WPAN in three areas. First, peripheral connectivity through cable-less connections to applications like storage and I/O devices will improve the ease and value of using PC's (Personal Computers) and Laptops. Second, enabling high-speed content transfers between computers and consumer electronics like digital cameras, video cameras, MP3 players, projectors, televisions and automobile applications will present new experiences in home and personal entertainment. Third, as data becomes an integral part of daily communication, mobile users will grow to expect higher performance low power solutions from WPAN leading to new freedom and expectations.

Other interesting applications are low-rate systems which they need low power system: ground penetrating radar (GPR) systems, medical imaging systems, Wireless sensor network (WSN), wall and through wall imaging systems, surveillance systems and high data rate communication systems. Consequently, highly miniaturized, low-power and low-cost UWB radar sensor could be integrated as system-on-a-chip, realizing a new class of wireless sensors (Zito, 2002).

For instance, these sensors can be integrated in a wireless body area network (WBAN). WBAN consists of a network of sensors placed around the body, for the human health monitoring.

Another instance today, sensors are used for specific local applications. Sensor networks suggest the use of many low-cost low-powered sensors, on a wider more generalized scale, networked to provide ubiquitous access. Sensor networks such as this will provide information that can make life easier. Sensors in these types of networks will work together to provide information that could: maintain environmental conditions across large buildings or many buildings, identify empty conference rooms or help one find an empty parking place in a huge parking lot.

The success of these kinds of applications will depend on industry adoption. As with any

Figure 1. Correlation between Gaussian pulses

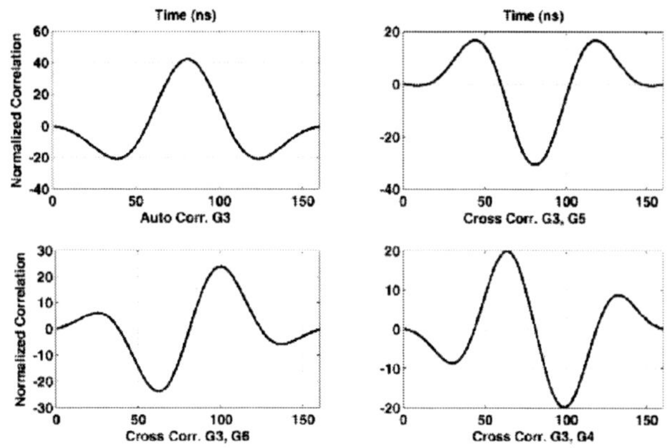

wireless technology targeted for consumer use, volume deployment will most likely require the cost advantages and interoperability enabled by open standards.

Position and location detection is another unique property of UWB. UWB is basically RADAR (Radio Direction and Ranging). Positioning and Locating were early focuses of wideband Research and Development. Short impulses (wideband signal) allow for very accurate delay estimates providing position and location capabilities within a few centimeters.

The RADAR like properties of UWB may also be applied to imaging applications. Some forms of imaging will be applicable in Surveillance Systems (i.e.Wall Imaging and Through-wall Imaging Systems or Intrusion Protection). Another example of imaging is a specialized Government application using UWB for GPR (Ground Penetrating Radar). This application was employed after the terrorist disaster on September 11th in New York to try and find survivors in the rubble. Other Military examples of UWB have been used to locate enemy objects behind walls and around corners in the battlefield. Imaging applications applied to commercial use find value in Medical Systems where X-ray Systems may be less.

BASIC OF UWB PULSES

A Gaussian pulse is a wide bandwidth signal in which the pulse duration sets the bandwidth and the center frequency of the pulse. The monocycle $G_n(t)$ is the nth derivative of Gaussian pulse $G(t)$.

$$G(t) = Ae - \frac{t^2}{2\sigma^2} \tag{1}$$

$$G_n(t) = \frac{d^n G(t)}{dt^n} \tag{2}$$

where A represents signal power adjustment, and σ^2 is the variance of Gaussian distribution which determines the temporal width of the pulse. The fourth and sixth derivatives (i.e., G_4 and G_6) of Gaussian pulse are suited for indoor and outdoor UWB transmission, respectively, as they efficiently meet the FCC masks. Interestingly, odd and even derivatives of Gaussian monocycles are orthogonal, and hence the cross-correlation between ideal pulses gives zero DC level. On the other hand, the cross-correlation between two different odd (even) monocycles results in non zero DC value. We use the above properties in the receiver for acquisition, as described below.

Figure 2. OTR-UWB transmitter

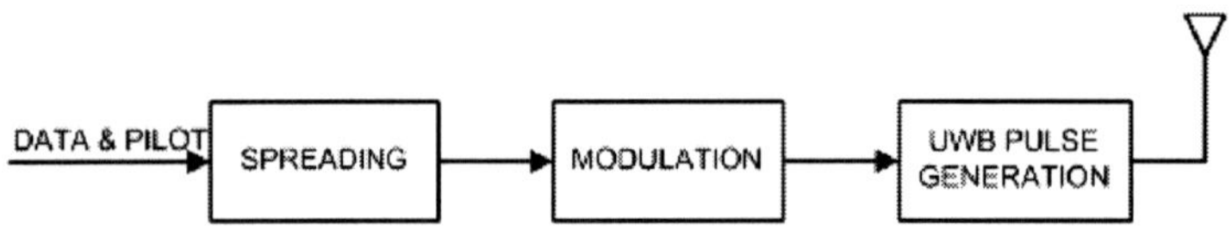

Box 1.

$$x(t) = \sqrt{P}\left(\sum_{k=-\infty}^{\infty} d(2k)G_{n-1}(t - kT_f) + d(2k+1)G_n(t - kT_f - T_d)\right) \quad (3)$$

OTR-UWB SYSTEM

Transmitter Structure

Figure 2 shows the structure of OTR-UWB transmitter, involving spreading, modulation, and UWB pulse generation. Using an even number of spreading code length, l_c, each bit of data and pilot is first spread to even number of symbols. The even code length will guarantee that the subsequent frames use similar monocycles for the start (end) bits. By using BPSK modulation, each pair of neighboring symbols is sent by orthogonal pulses of G_{n-1} and G_n.

The transmitted signal in OTR-UWB system is shown in Box 1 where P is the transmitted power, $T_f = N_f T_c$ is the frame duration containing a pulse pair, T_c denotes the chip duration, N_f represents the number of chips per frame, and $d_k \in \{-1, 1\}$. The delay between two symbols is $T_d = N_d T_c$, where N_d is the number of chips left between G_{n-1} and G_n.

Channel Model

The channel impulse response is modeled by the well-known tapped delay line expression for UWB indoor environment 0:

$$h(t) = \sum_{i=0}^{L-1} a_i \delta(t - \tau_i) \quad (4)$$

Where α_l represents the channel coefficients, L is the impulse response length, and τ_l denotes the delay of the path l.

Assume that the channel is quasi-static and the interference sources of the channel are zero-mean additive white Gaussian noise, $n(t)$. The received signal can be expressed as Equation 5 in Box 2.

Receiver Structure

Figure 3 illustrates the block diagram of OTR-UWB receiver, involving autocorrelation, code tracking, and matched filter.

Box 2.

$$r(t) = \sqrt{P}\left(\sum_{i=0}^{L-1}\sum_{k=-\infty}^{\infty} a_i d(2k)G_{n-1}(t - kT_f - \tau_i) + a_i d(2k+1)G_n(t - kT_f - T_d - \tau_i)\right) + n(t) \quad (5)$$

Figure 3. Baseband of OTR-UWB receiver

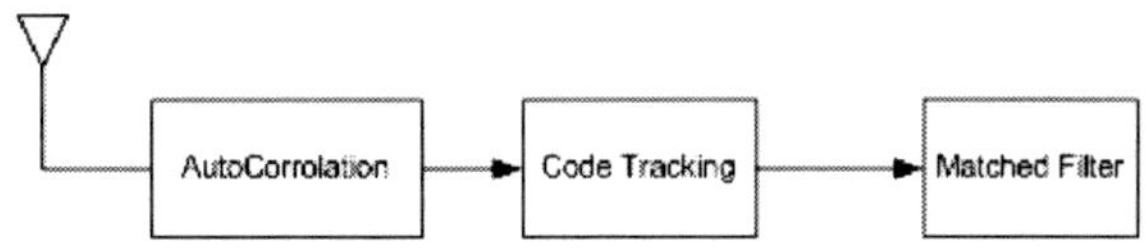

Figure 4. Block diagram of the autocorrelator

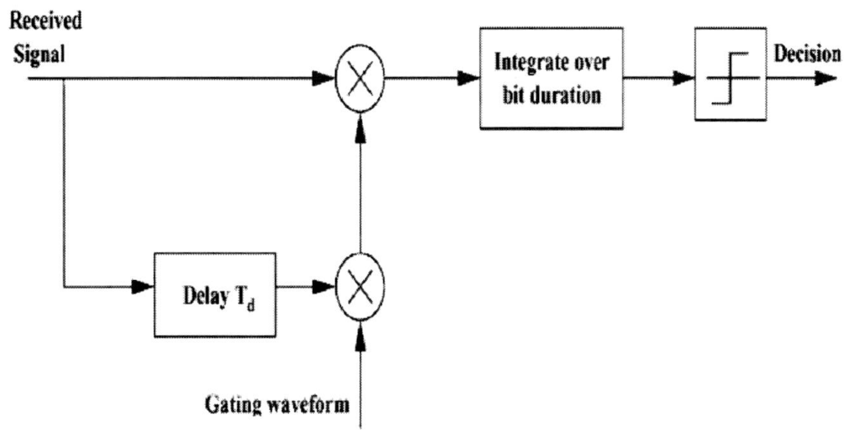

The synchronization is achieved in two steps: acquisition and tracking. First, the autocorrelation is performed between each pilot pulse with a delayed (T_d) version of itself (Figure 4). The result of autocorrelation is assumed zero when it is below a certain threshold level adjusted by simulating the channel effects. In this way, the chip's timing (coarse acquisition) is acquired. Note that in this step, no inner code is involved and the search space depends on the number of samples per chip. The spreading template is obtained by replacing each pair of neighboring symbols in spreading code by two orthogonal pulses of G_{n-1} and G_n. Then, the code tracking is performed by a correlator using a serial search scheme.

Finally, the data is obtained using a pilot-based matched filter 0. The samples of the entire spreading template are used to demodulate the data, not just a single pulse waveform. Thus, the results of pilot-based matched filter accounts for all multipath energies:

$$g[k] = \sum_{t=1}^{x} r[t], template[t] \tag{6}$$

Where $r[t]$ represents the received signal after synchronization, $template[t]$ is the template extracted from $r[t]$, and $g[k]$ is the resulting correlation value. The output of matched filter is positive for bit 1, and negative for bit 0.

ACQUISITION APPROACHES IN UWB COMMUNICATION SYSTEM

Acquisition schemes for UWB systems are divided into two approaches: detection-based approaches and estimation-theoretic strategies. The acquisition methods which employ a detection based approach typically evaluate a candidate phase of PN or signal by first obtaining a measure of correlation between the received signal and a locally generated template signal offset by the candidate

Figure 5. Block diagram of the acquisition scheme proposed in 0

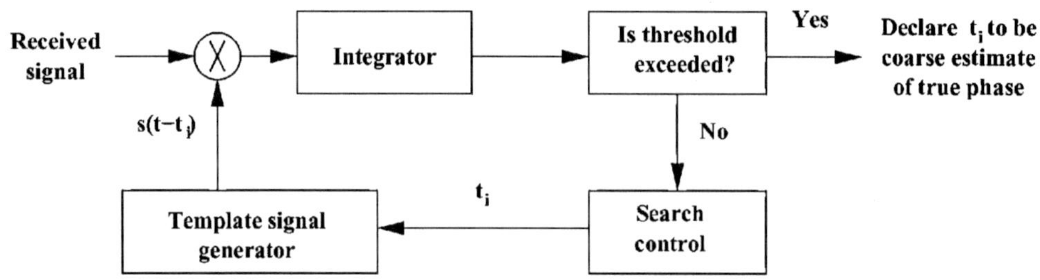

phase. This measure of correlation is then compared to a threshold in order to make a decision. These candidate phases could be evaluated in a serial, parallel, or hybrid manner. Among the detection based schemes for UWB acquisition some schemes focus exclusively on the development of efficient search strategies to quickly evaluate the candidate phases in the search space and certain other schemes propose two-stage acquisition methods that achieve a reduction in the search space itself. Conventional detection-based in serial search algorithm uses a fixed threshold to decide the occurrence of acquisition. However, in a practical communication environment, the channel is time-variant; the SNR in receiver is not a fix value. So, the threshold selection becomes a deliberate task. Many articles propose their approach to improve the performance of acquisition, but how to select an optimized threshold is a black box. Aims at this difficult, another articles propose the algorithm exploit the adaptive threshold to control the search process.

In the estimation based methods, an estimate of the timing is typically obtained by maximizing a statistic over a set of candidate phases. This statistic is usually obtained from correlation of the received signal with a template signal. These schemes thus do not involve a threshold comparison. Most of the estimation based schemes attempt to exploit the cyclostationarity inherent in UWB signaling due to pulse repetition.

Detection-Based

We review some important detection-based approaches for UWB acquisition in this article. In 0; Blazquez, 2003), the traditional coarse acquisition scheme where a particular phase ti in the search space is checked by correlating the received signal with a locally generated template signal with delay ti for the acquisition of TH-UWB signals in AWGN noise (Figure 5.).

In 0, there is a spreading code matched filter before the integrator. Then integrator output is sampled at multiples of mTc and compared to a threshold. The nonconsecutive search and a simpler version of the maximum selection and threshold crossing (MAX/TC) scheme called the global MAX/TC are applied to the acquisition of UWB signals in the presence of multipath fading and MAI 0. In the nonconsecutive search, only one phase in every D consecutive search space phases is tested by correlating the received signal with a template signal. The decimation factor D is chosen to be not larger than the delay spread N tap. In the global MAX/TC, a parallel bank of correlators is used to evaluate all the nonconsecutive phases, and the phase corresponding to the correlator output with maximum energy is chosen as the coarse estimate of the true phase.

More efficient nonconsecutive search strategies called the "look-and-jump-by- K-bins" search and bit reversal search are analyzed in the noiseless scenario with mean stopping time as the performance metric in 0. For the case when the acquisition phases are K consecutive phases in the uncertainty region, it has been claimed that the look-and-jump-by-K-bins search is the optimal serial search permutation when K is known, and the bit reversal is the optimal search permutation when K is unknown.

In order to use different threshold for acquisition, all known methods, in one form or another, must estimate the background power level or exploit the statistics of noise characteristics to control the preset threshold (Karaoguz, 2001). In 0, an adaptive threshold acquisition system with the constant false alarm rate (CFAR) was proposed. The correlator bank can search N chip positions at one time. Threshold changed depends on the correlator bank output except the maximum correlator output. If the maximum correlator output over the threshold TX, the chip position of maximum correlation occurs is the synchronization position. On the contrary, the PN sequence steps into next N chip positions for search.

Another scheme, named RASE (Rapid Acquisition by Sequential estimation) based on the property that the next combination of register states of PN sequence depends only on the present combination of the states. So, the acquisition system makes the best estimate of the first n (the number of stages in the PN code generator) received PN code chips, if this n code chips are correctly estimated, all the following states can be predicted based on the knowledge of only this initial condition. If the correct estimate is made, the PN code generation register is then closed within the PN tracking loop which is responsible for maintain code phase from that time on. Because the estimation process is performed on a chip-by-chip basis, it does not take advantage of the interference rejection capabilities of PN signals, the RASE technique has not the immunity of noise and ISI (Karaoguz, 2001).

In (Wang, 2006), they have presented the first method for achieving PN acquisition at low SNR as fast as full-parallel search, but with significantly lower complexity. This approach is based on the paradigm of message-passing on graphical models and more specifically, iterative message passing algorithms (iMPAs) and graphical models with cycles 0. This is a generalization of the "turbo" decoding algorithm. In this, no correlation to a single PN reference signal is computed explicitly. Rather, the global structure in the PN sequence is modeled as a set of coupled local constraints and correlations are formed against these incomplete, local structures. This may be viewed as an approximation to ML acquisition, much in the same way that a turbo decoder is an approximation of the ML decoder for a concatenated code.

Another detection-based approaches is called transmitted reference (TR) systems for impulse radio 0;0 due to their simple receiver structure because does not require a channel estimator and estimate the shapes of the received pulses. In typical TR signaling for UWB systems, each modulated pulse is preceded by an unmodulated (reference) pulse in the transmitted signal. At the receiver, the noisy reference pulse is demodulated the signal using a simple delay-and-correlate receiver 0;0. However, the BER performance of a TR receiver is usually worse compared to a Rake receiver due to energy wasted on the reference pulse 0; 0. However, the acquisition problem in TR-UWB systems is also less acute due to a relaxed timing requirement by the TR demodulator.

OTR-UWB receiver rapidly acquires chip's timing (coarse synchronization) because the search space is a function of sampling frequency. Then, it performs the code acquisition with a search space linearly proportional to code length. Lower search space reduces the complexity of the system by decreasing the number of multipliers, adders, and memories. Therefore, the OTR system illustrates better acquisition performance and higher data rate than that of conventional UWB counterparts.

Estimation-Based

Certain approaches toward acquisition in UWB systems have employed estimation theoretic methods to obtain timing information of the received signal. The nondata aided timing estimation approaches in 0; 0 exploit cyclostationarity, inherent in UWB signaling due to pulse repetition, to estimate timing information of the received signal. These schemes require frame rate sampling in the acquisition stage and pulse rate sampling during the tracking stage.

PERFORMANCE ANALYSIS AND SIMULATION

In this section, analytical methods are presented for computing the mean acquisition time and the hardware complexity for OTR-UWB system. In addition, the simulation results for bit error rate (BER) are given.

Mean Acquisition Time Analysis

Since an exact analysis of OTR-UWB system in a noisy and multipath channel is too complicated and requires many derivations, we analyze the system in a large-signal-noise-rate (SNR) for multipath channels without noise or false alarm for any search process.

Dividing the total mean acquisition time into two parts: code search time, $E[Tco]$, and chip search time, $E[Tch]$, the total mean acquisition time, $E[T]$, is (7):

$$E[T] = E[Tco] + E[Tch], \tag{7}$$

$$E[Tch] = \frac{1}{2}(K+1) \tag{8}$$

Where K denotes the number of search points. With the same assumption as that for (7) and using serial search algorithm, $E[Tco]$ is (9)

$$E[Tco] = \frac{(B-M)^2 + (3B-M)}{2B} \tag{9}$$

where $B = 2^{l_c}$ denotes the total number of codes and M is the number of true orthogonal codes. The total number of search points is $N = KBL_c$. Thus:

$$E[T] = \frac{\left(\frac{N}{Kl_c} - M\right)^2 + \left(3\frac{N}{Kl_c} - M\right)}{2\frac{N}{Kl_c}} + \frac{1}{2}(K+1) \tag{10}$$

To minimize $E[T]$, it is necessary to find the optimal search length, K_{opt}, by:

$$\frac{\partial E[T]}{\partial K} = \frac{-\frac{N^2}{K^2 l_c} + M^2 l_c - Ml_c + N}{N} \tag{11}$$

Let the left hand of (10) be equal to zero, we have:

$$K_{opt} = \sqrt{\frac{N^2}{M^2 l_c^2 - Ml_c^2 + Nl_c}} \tag{12}$$

Replacing K in (9) with K_{opt}, we have

$$\min\{E[T]\} = \sqrt{M^2 l_c^2 = Ml_c^2 + Nl_c} / l_c + 2 - M \tag{13}$$

The above formulation for OTR-UWB system can also be used for TR-UWB one. Note that chip search time in TR-UWB depends on the number of samples in a chip 0. Further, its code search time for inner code is similar to OTR-UWB receiver search for spreading code.

Figure 6 shows the normalized mean acquisition time versus consecutive search intervals (Peterson, 1995) during the serial search process for DS-UWB and OTR-UWB systems, where N = 8192 and l_c = 0. Clearly, the acquisition time

Figure 6. Normalized mean acquisition time, E(T)/N, versus different normalized intervals, M/N, for the search process in DS-UWB 0, and OTR-UWB

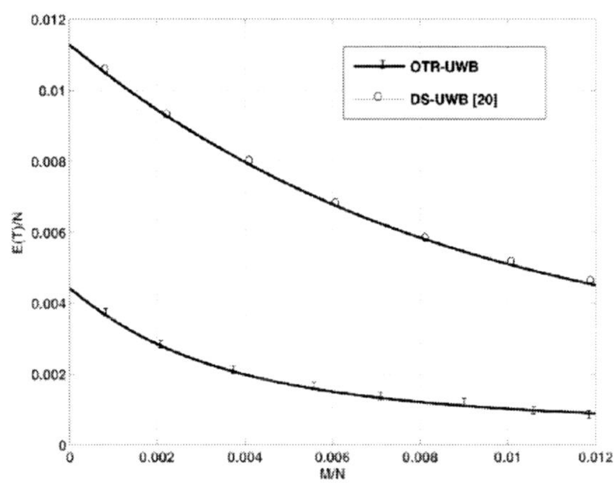

for OTR-UWB is much lower than that in 0 for DS-UWB.

Hardware Complexity

Assume N_s is the number of samples in chip, and l_i (l_o) are the length of inner (outer) code in TR-UWB.

In each correlation operation in TR-UWB system, two arrays of the received signal and its delayed version are multiplied to attain the polarity of inner code 0.The length of each array is $(l_i + l_o)N_S$, and the search space is N_S for coarse acquisition. Therefore:

$$g_{TR}[k] = \sum_{k=1}^{N_s} \left(\sum_{t=1}^{(l_i+l_o)N_s} r_{TR}[t-k],\ template_{TR}[t] \right) \tag{14}$$

In the next step for fine tuning (i.e., removing phase margin error), the search space equals l_i, and we have:

$$g_{TR}[k] = \sum_{k=1}^{N_s} \left(\sum_{t=1}^{(l_i+l_o)N_s} r_{TR}[t-k*N_s],\ template_{TR}[t] \right) \tag{15}$$

In OTR-UWB system, the received signal is correlated by a delay version of itself in each chip. Thus:

$$g_{OTR}[k] = \sum_{k=1}^{N_s} \left(\sum_{t=1}^{N_s} r_{OTR}[t-k],\ template_{OTR}[t] \right) \tag{16}$$

The length of each array for fine tracking (code search) is $(l_c)U_{cS}(l_c)N_S$ and the search space equals l_c.

$$g_{OTR}[k] = \sum_{k=1}^{l_c} \left(\sum_{t=1}^{l_c N_s} r_{OTR}[t-k*N_s],\ template_{OTR}[t] \right) \tag{17}$$

From (13)-(16) we can easily find the number of multiplications and adders for OTR-UWB and TR-UWB systems, as given in Table 1.

Simulation Results

Figure 7 depicts the BER of OTR-UWB receiver for UWB channel. The sampling frequency (f_s) is 8 GHz, chip duration is 1ns, bit rates are 100 and 125Mbps, pulse width equals 0.85ns, and spreading code lengths are 8 and 10. Obviously, the performance of OTR-UWB system is better when compared to TR-UWB.

DSP ALGORITHM OF BASEBAND IMPLEMENTATION

In this section, we introduce a DSP algorithm for implementing OTR-UWB baseband on FPGA. The baseband consists of two blocks: synchronization and matched filter. Autocorrelation and code tracking blocks are subset of synchronization.

The received waveform is sampled by a number of ADCs operating in parallel, where each ADC clock is slightly offset from the others (Figure 8). Thus, the effective sampling rate is the individual ADC sample rate multiplied by the total number

Table 1. Hardware complexity for OTR-UWB and TR-UWB

System	Synchronization		Matched filter	
	Multiplier	**Adder**	**Multiplier**	**Adder**
OTR-UWB	$(N_s+l)N_s$ (576)	$(N_s+l)N_s-(N_s+l)$ (504)	lN_s (512)	$L(N_s-1)$ (448)
TR-UWB	$(l_i+l_c)N_s(N_s+l_i)$ (2048)	$(l_i+l_c)(N_s-1)(N_s+l_i)$ (1120)	lN_s (512)	$L(N_s-1)$ (448)

Figure 7. BER performance of OTR-UWB versus SNR

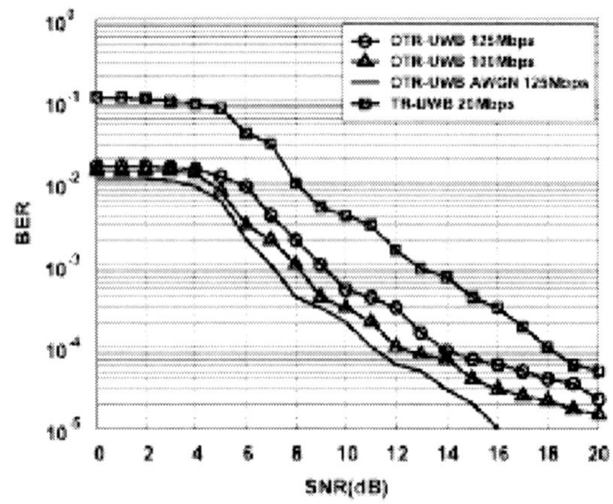

Figure 8. DSP algorithm of synchronization block

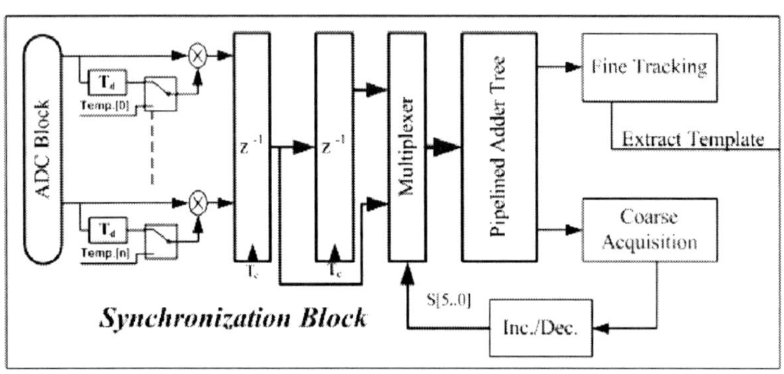

of ADCs0, which equals samples per chip (*num* = T_c/T_s).

Coarse acquisition is first performed, based on autocorrelation with delay version of received signal. It moves on the width of chip to find the start of chip as described above. We need the information of the two consecutive chips for autocorrelation of them. Therefore, we used two serial register blocks which the trigger clock is T_c. Multiplexer selects *num* samples from two chips to transfer adder block. The selector bits extract from coarse acquisition blocks, which compare consecutive autocorrelation energies for finding zero because two neighbor chips are orthogonal.

Table 2. Gate estimation on XILINX FPGA

XILINX	SPARTAN3: 3s4000fg676-5
Slices	2106 / 27648 (7%)
Slice Flip Flops	408 / 55296(1%)
4 input LUTs	3580/ 55296(6%)
bonded IOBs	337/489(68%)
MULT18X18	40/96(41%)

Table 3. Gate estimation on ALTERA FPGA

ALTERA	Cyclone II (EP2C35F672C6)
Total logic elements	7,268 / 33,216 (22%)
Total registers	640
Embedded Multiplier	70 / 70 (100%)

Table 4. Power analysis

Power	XILINX		ALTERA	
	I(mA)	P(mW)	I(mA)	P(mW)
Vccint 1.20V	120	144	88.55	123
VccIO 2.50V	20	50	13.42	33.55
Total Power	194		156.55	

To perform the autocorrelation, all multiplication results are accumulated using the Adder-Tree. However, the propagation delay of adder block is a main problem for digital implementation on FPGA. Pipelined-Adder-Tree is utilized here to overcome the propagation delay problem and to achieve the required throughput, despite increasing hardware complexity and critical path delay. Coarse acquisition block compares two consecutive autocorrelation's energies for acquiring the chip timing (coarse synchronization). Then, spreading code is searched by fine tracking block, which consists of two serial register blocks, a multiplexer, and a Pipelined-Adder-Tree. Samples of template are multiplied by multipliers which are shown in Figure 3 and control with fine tracking block. Length of each register block equals to code length. When spreading code was found, synchronization process is completed and template code is extracted from the receiving signal for data detection in matched filter. Data detection is performed using cross correlation between received signal and extracted template in synchronization block. If the result exceeds (smaller) zero, the bit is 1 (0).

FPGA IMPLEMENTATION

The sampling frequency (f_s) is 8 GHz, chip duration is 5ns, bit rate is 25Mbps, pulse width equals 1ns, and spreading code lengths are 8. We used 40 ADC with 200 MHz sampling frequency and 4 bit resolution. All VHDL codes are designed and implemented by XILIX and ALTERA FPGAs0,0. Power analysis and gate estimation have been done by ISE 8.1 and Quartus II 7.2. The results are presented in Table 1.

CONCLUSION

A new transmitted reference ultra-wideband receiver, namely OTR-UWB, is presented, which utilizes the orthogonal property of even and odd order derivatives of Gaussian pulses in neighboring chips for synchronization. This system employs only a single spreading code, and involves lower search space for coarse acquisition, which results in much lower mean detection time compared to DS-UWB systems. Analysis of hardware complexity shows significant reduction against conventional TR-UWB systems. Further, simulation results illustrate improvement for bit error rate (BER) performance, while the system potentially supports higher bit rates. In this research, we implemented OTR-UWB using FPGA, with a bit-rate of 25Mb/s without using equalizer. Detection-based technique is utilized for serial acquisition. Analog to digital conversion with 4-bit of resolution and 200MHz sampling rate is used. Hardware complexity on FPGA has been evaluated for implementation of OTR-UWB baseband algorithms using pipelined architecture. Hardware complexity on two different FPGAs from ALTERA and XILINX, CycloneII (EP2C35F672C6) and Spartan 3 (3s4000fg676-5), are investigated. Gate estimation and power analysis have been performed by Quartus II 7.1 (ALTERA) and ISE 8.1 (XILINX) softwares. The results show power performance of ALTERA FPGA is better than XILINX. Another result is implementing of UWB system in FPGA with low bit rate.

ACKNOWLEDGMENT

This work was supported in part by the Iran Telecommunication Research Center, ITRC.

REFERENCES

Aedudodla, S. R., Vijayakumaran, S., & Wong, T. F. (2005). Timing Acquisition in Ultra-wideband Communication Systems. *IEEE Transactions on Vehicular Technology*, (54): 1570–1583. doi:10.1109/TVT.2005.855495

Aedudodla, S. R., Vijayakumaran, S., & Wong, T. F. (2006, April). Acquisition of Direct Sequence Transmitted Reference Ultra-Wideband Signals. *IEEE Journal on Selected Areas in Communications*, *l*(24), 759–765. doi:10.1109/JSAC.2005.863825

Altera Inc. (2007). Data Books, http://www.altera.com.

Anderson, S. C. R. (2003). *Design of a transmitter for Ultra Wideband Radi.* Master dissertation, University of Linköpings, Linkoping.

Blazquez, R., Newaskar, P., & Chandrakasan, A. (2003, April). Coarse acquisition for ultrawideband digital receivers. In *Acoustics, Speech and Signal Processing*, 137–140, IEEE.

Callaway, E., Gorday, P., Hester, L., Gutierrez, J. A., Naeve, M., Heile, B., & Bah, V. (2002, August). Home networking with IEEE 802.15.4: A developing standard for low-rate wireless personal area network. *IEEE Communications Magazine*, (31): 70–77. doi:10.1109/MCOM.2002.1024418

Chang-Joo, K. (1998, May). Adaptive hybrid acquisition of PN sequencesfor DS/SS communications. *Electronics Letters*, (34): 939–940.

Chen, J., & Zhou, Z. (2005, March). Overview of Synchronization in DS-UWB. In *ISCIT 2005*, 951-954. IEEE.

Chugg, K. M., & Zhu, M. (2005, May). A New Approach to Rapid PN Code Acquisition Using Iterative Message Passing Techniques. *IEEE Journal on Selected Areas in Communications*, *l*(23), 884–897. doi:10.1109/JSAC.2005.845424

FCC. (2002). New public safety applications and broadband internet access among uses envisioned by fcc authorization of ultra-wideband technology. *Federal Communications Commission,* Retrieved, 2002 from http://www.fcc.gov/Bureaus/Engineering Technology/News Releases/2002/nret0203.html

Gharaee, H., & Nabavi, A. (2008, April). A New TR-UWB Transceiver with Orthogonal Pulses. In *ICEE (16th Iranian Conference on Electrical Engineering)*, 167–171, Tehran, Iran.

Gharaee, H., Nabavi, A., & Etminan, J. (2008, May). OTR-UWB System. In *WTS2008*, 325-329, California, USA, IEEE.

He, N., & Tepedelenlioglu, C. (2007, March). Fast and Low Complexity Frame-Level Synchronization for Transmitted Reference Receivers. *IEEE Transactions on Wireless Communications*, *l*(54), 1014–1023. doi:10.1109/TWC.2007.05266

Hoctor, R. T., & Tomlinson, H. W. (2002, January). An overview of delay-hopped, transmitted-reference RF communications. In Technical Information Series (pp. 1–29). GE Research and Development Center.

Homier, E. A., & Scholtz, (2002). Rapid acquisition of ultra-wideband signals in the dense multipath channel. In *Ultra-Wideband Systems Technology*, 105–109, IEEE.

Karaoguz, J. (2001). High-rate wireless personal area networks. *IEEE Communications Magazine*, (39): 96–102. doi:10.1109/35.968818

Ma, Y., Chin, F., Kannan, B., & Pasupathy, S. (2002). Acquisition performance of an UWB communications wideband system over a multiple access fading channel. In *Ultra-Wideband Systems Technology*, 99–104, IEEE.

Orndorff, A. M. (2004). *Transceiver Design for Ultra-Wideband Communication.* Master dissertation, University of Polytechnic Institute and State, Virginia.

Rushforth, C. K. (1964, January). Transmitted-reference techniques for random or unknown channels. *IEEE Transactions on Information Theory*, (40): 39–42. doi:10.1109/TIT.1964.1053641

Saleh, A., & Valenzuela, R. (1987, February). A statistical model for indoor multipath propagation. *IEEE Journal on Selected Areas in Communications*, *l*(5), 128–137. doi:10.1109/JSAC.1987.1146527

Soderi, S., Iinatti, J., & Hamalainen, M. (2003, June). CLPDI algorithm in UWB synchronization. In *Workshop on UWB Systems*, 759–763, IEEE.

Staderini, E. M. (2002). UWB radars in medicine. *IEEE Aerospace and Electronic Systems Magazine International*, *1*(17), 13–18. doi:10.1109/62.978359

Tian, Z., Yang, L., & Giannakis, G. B. (2002). *Symbol timing estimation in ultra wideband communications* (pp. 1924–1928). Signals, Systems and Computers.

Wang, H., Jiang, S., Wei, G., & Ji, F. (2006). A Rapid Synchronization Acquisition Scheme for UWB Communications. In *International Symposium on Systems and Control in Aerospace and Astronautics*, 844-848.

Win, M. Z., & Scholtz. (2000). Ultra-wide bandwidth time-hopping spread spectrum impulse radio for wireless multiple-access communications. *IEEE Transactions on Communications*, (48): 679–691. doi:10.1109/26.843135

Xilinx Inc. (2007). Distributed arithmetic filter v8.0, http://www.xilinx.com.

Yang, L., Tian, Z., & Giannakis, G. B. (2003). Non-data aided timing acquisition of ultra-wideband transmissions using cyclostationarity. In *Acoustics, Speech and Signal*, 121-124, IEEE.

Zhang, H., & Goeckel, D. L. (2003). Generalized transmitted-reference UWB systems. In *Ultra-wideband Systems and Technology Proc. 2003*, 147–151 IEEE.

Zito, D., Pepe, D., Neri, B., (2002). Feasibility Study of a Low-cost System-on-a-Chip UWB Pulse Radar on Silicon for the Heart Monitoring.

This work was previously published in International Journal of Interdisciplinary Telecommunications and Networking, Volume 1, Issue 3, edited by Michael R. Bartolacci, and Steven R. Powell, pp. 62-77, copyright 2009 by IGI Publishing (an imprint of IGI Global).

Chapter 12
Mobility and Enterprise 2.0

François J. N. Cosquer
Alcatel Lucent Enterprise, France

Annie Ohayon-Dekel
Alcatel Lucent Enterprise, France

ABSTRACT

The emergence of Web 2.0 and its rapid adoption by the corporate world, known as Enterprise 2.0, has radically modified access to tacit knowledge and significantly reduced business latency. Meanwhile, technology for wireless LAN and mobile communication, combined with advances in handsets, has allowed for superior quality of experience with acceptable productivity level. Combining mobility with Enterprise 2.0 is the next big step in evolution. This article presents the drivers for Enterprise 2.0 and the challenge of tapping tacit knowledge and, in parallel, the evolution of wireless and mobile technologies. The era of mobile broadband life is made possible, creating new ways of use and expectations for Millennials. The next business generation will be able to unleash the full potential of mobility and Enterprise 2.0. Three scenarios selected from different vertical domains—healthcare, education and emergency services—illustrate the benefits of mobility and Enterprise 2.0 in action. With the expected continued strong growth of wireless access, mobility support is one, if not the, key success factor for Enterprise 2.0.

INTRODUCTION

Many factors are aligning for significant change in communication and collaboration in the corporate world. At a time when competition is global and fierce, IT and communication tools are expected to be a real vector for differentiation. The objectives to thrive and survive include superior customer care, cost efficiency and ultimately innovation.

The constant and fast pace of change brings challenges to corporations which have to understand, deploy and massively adopt new technologies. At the same time organizations themselves are changing, moving from a strict and formal hierarchical model to ad-hoc virtual teams and networked communities. This task oriented, problem solving quest for knowledge approach is

DOI: 10.4018/978-1-60960-505-6.ch012

greatly facilitated by the emergence of the Web 2.0 in the corporate environment. For early adopters Enterprise 2.0 is already demonstrating its value.

On the infrastructure side, progress of wireless technology (WLAN, mobile) combined with advances in handsets has allowed for superior quality of experience with acceptable productivity level. The next business generation issued from the Millennials has always known Internet, cell phones and wireless connectivity. Mobility is an integral part of Web 2.0 and ultimately of Enterprise 2.0. The new era of "mobile broadband life" will impact most homes and corporate environments in ways we can only start to imagine.

This article is comprised of five sections and is organized as follows. Web 2.0 and Enterprise 2.0 is presented in the next section. The second section discusses the importance and the challenge of accessing tacit knowledge in corporate environments. The third section presents the technological trends and evolutions of wireless LAN and mobile communication. Section four motivates the role of mobile broadband life and the Millennials in the development and use of Enterprise 2.0 tools. Using selected examples from different domains, section five illustrates the potential of Mobility and Enterprise 2.0 in action. Summary and conclusions are contained in the final section.

In an economy where the only certainty is uncertainty, the one sure source of lasting competitive advantage is knowledge"

The Knowledge-Creating Company: How Japanese Companies Create the Dynamics of Innovation by Ikujiro Nonaka and Hirotaka Takeuchi (Oxford University Press, 1995)

WEB 2.0 AND ENTERPRISE 2.0

The *Financial Times* on June 17, 2009 stated "a request by the U.S. State Department asking Twitter networking Web site to delay a planned break in service has highlighted the vital role social media are playing in protest against a middle east country's election result." Not only Twitter, but also other social networking sites were blocked. Beyond the political meaning of the U.S. State Department request, it dramatically highlights that Web 2.0 tools have become a key information conduit.

Let's step back to the mid-1990s to the Web 1.0 time before it was called 1.0: connecting to what was simply called the "the Web," allowed users to surf through Web pages, sites, hyperlinks of which almost all the content was produced by the computer elite. It was a one way flow of information, read only material and only a few million people were on line—about 45 million—nevertheless it transformed the way users accessed the information. A new type of platform which was fundamentally communications oriented was created (Berners-Lee, 1989). Twenty years later, it is about participation and collaboration: publish, share, comment, and connect to networks. Blogs, wikis, social networking platform are flourishing and now used by a billion people online at home and at work. They have transformed the way of communicating and working together, while the core principle often cited is harnessing collective intelligence and network effect. "Web 2.0 is ultimately about understanding the rules of business in the network era … Web 2.0 is the design of systems that harness network effects to get better the more people use them, or more colloquially, as 'harnessing collective intelligence.' This includes explicit network-enabled collaboration … but it should encompass every way that people connected to a network create synergistic effects" (O'Reilly, 2008). There is a rich and diverse field of study and practice which concerns itself with "collective intelligence" and since the 1990s, a lot of definitions have flourished. One the earliest was given in 1994 by Pierre Levy, French philosopher and professor, "Within the space of Knowledge formed by cyberspace, a virtual world, individuals can think and discuss with each other freely. Once everyone is wired up, we will come together as the 'collective intelligence': an inclusive society borne out of the Net" (Levy, 1994). In 2004, one of the best known definition is given by James Surowiecki who argued that "the aggregation of information in groups, results in decisions that are often better than could have been made by any single member of the group" (2004). Still today, one should keep in mind that quantitative information, thoughts or discussions are not automatically generating qualitative or useful information and knowledge or even collective intelligence—as we all agree: the crowds are not always wise.

The pervasive presence of highly interactive social software is increasingly allowing us to experience a new way of life and consequently a new way of working. Today's pace of changes in business conditions are surely spectacular, but like most things in business, rapid change is a two edged sword—a threat but also an opportunity. The faster the enterprise adapts to change, the stronger the competitive advantage is. By ignoring rapidly changing circumstances, enterprises are expected to go the way of the dinosaurs. This is one of the main reasons why enterprises cannot stay out of this evolution and O'Reilly's definition on Web 2.0 effect can also be applied to Enterprise 2.0. Facing the actual business practices, organizations have to take new parameters into account:

- Boundaries between personal and professional lifestyles are blurred.
- Collaboration, mobility, consumerization, hyper-connectivity and sustainability are now part of employees' priorities.
- Organization hierarchy is starting to move from a strictly pyramidal structure to a team and networked communities. In addition the human resources departments realized that they need to recruit, develop and keep employees by providing them with a state-of-the-art workplace.
- Traditional tools (such as email for collaboration) considered revolutionary when they were invented, are becoming less efficient due to the volume of communications that users must manage every day.

Employees are bringing to the workplace new behaviors and processes whether that is communities, crowd sourcing, professional networks, micro blogging, etc. Some of the readers may have heard of the Innocentive marketplace: the company—launched in 2001 —built the first global community for open innovation, enabling scientists, professionals, entrepreneurs and business people to collaborate to deliver breakthrough solutions for organizations. Innocentive allows to the seekers to frame a "challenge problem" and open it up to anyone to solve it. Solvers who deliver the best solution—matching the criteria—receive financial award. Today, Innocentive counts more than 170,000 members. It becomes clear that organizations cannot afford the growing chasm between the world of business and the way their employees are doing things outside work. Despite their hierarchical nature, enterprises started to take these parameters in consideration, even though the principle of leaders and followers is still applicable, the bottom line is: "2.0 is here to stay"—Enterprise 2.0 tools have arrived. An intriguing report just released by Young, Burris, and Reiss (2009) states, "nearly one in two businesses will make use of Enterprise 2.0 software" (blogs, wikis, and social networking and messaging platforms), a staggering number given that the application of social software to employee collaboration and productivity goals is just three years old. One of the adoption drivers is to mitigate the risk of the introduction of untested consumer software by deploying enterprise class tools. However, the key driver is really to address the corporate "knowledge challenge."

CHALLENGES OF ACCESSING TACIT KNOWLEDGE

One of enterprises' biggest challenges is to face the new environment driven, among others, by the increasing speed and amount of data. According Gartner Group, 85% of business processes are slowed by human latency. Consequently, enterprises are challenged to accelerate "time to *x*"—time to decision, to revenue, to service, to support, to product, to manage the complexity of reducing business latency. This means enabling a mobile workforce, all in secure environment by running processes quickly and accurately, responding to customer preference in an agile manner and collaborating within and between organizations in real time. As a result, this is triggering new challenges, specially the one of managing corporate knowledge. What is it? Where is it? How to collect it and share it?

Knowledge is not just information Knowledge taps into the collective facts, concepts, experiences and know-how held by a group of people and requires an ongoing exchange between members of the group. It provides the context for how people think. When knowledge is accessible quickly and widely enough i.e. well managed, it becomes extremely useful and gives a competitive advantage that can be transformed into real opportunities. It should be noted that there are different types of knowledge in enterprises: structured knowledge and tacit knowledge. It is widely agreed among the analysts community that only 20% of any enterprise's knowledge is structured knowledge—the kind that business processes rely on to work properly, coming potentially from a variety of sources—databases, directories, files which is easily sharable. Unlike structured knowledge, tacit knowledge is the one that people carry in their minds or keep in their personal environments and is therefore more difficult to access. Its transfer requires several actions like search for the right expert that own the knowledge, find him and make personal contact with him. This tacit or unstructured knowledge is crucial to increase efficiency of processes that lack rigid pre-definition but where people work together in an adaptive way to innovate solutions. This is where the notion of Connected Knowledge becomes important (Ohayon-Dekel, 2009).

In the traditional enterprise communication landscape, the trend is to add intelligence on top of communications tapping part of the tacit knowledge. Thanks to Unified Communications Suites, people can now reach and be reached easily, whatever the device, whatever the time, and whatever the location. This is an efficient way to collect part of tacit knowledge. However, Unified Communications assume you know who you need to reach. On the other hand, introducing Enterprise 2.0 tools in the enterprise makes a significant part of the tacit knowledge more and more accessible. Blogs, wikis, micro blogging, social networking platforms makes the "search and find" easier. Remember the Google mantra "if you're not searchable, you won't be found." Searching by keywords, tags, comments, will lead you to the right expert and will allow you to collaborate. Adding these tools to the existing communication media like voice, emails, IM based on presence and SMS, increasingly feed and enrich the enterprise knowledge creation and flow. The impact of the multiple ways to communicate and to share, is especially measurable and tangible in scenarios of open innovation (peers creativity). Because enterprises' employees are more and more dispersed and mobiles—the ability to search, find and reach the right person, at the right time, on the right device —also increases productivity

According ABI Research, by 2010, about 400 million Internet-enabled mobile devices will ship; more than laptops and desktops combined. Today, mobile users are around 3 billion, and this number is expected to increase to more than 4 billion (GSMA, 2009) within 2 years. This evolution of the communication landscape will most likely lead the majority of Web use to be mobile within five years.

Figure 1. Technology trends WLAN

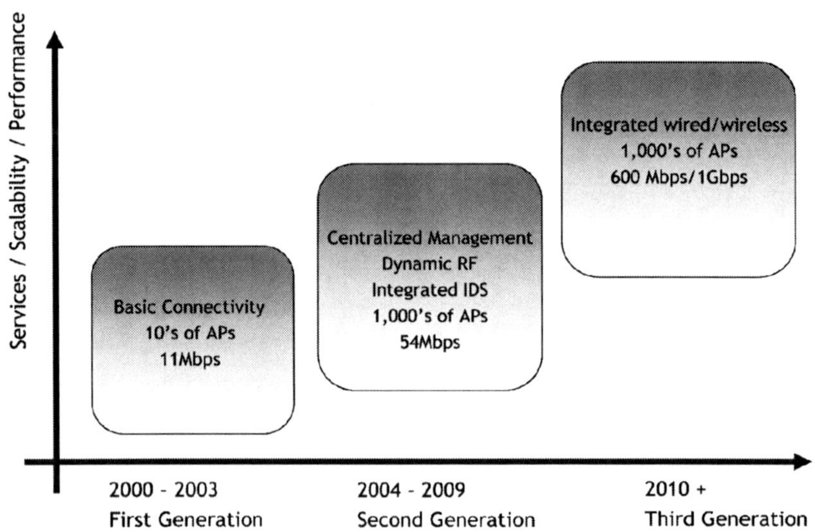

Bringing the benefits of communications enabled applications and business processes through mobile unified communications with the Enterprise 2.0 tools together, enables creation and retention of collective intelligence.

It is not the strongest species that survive, or the most intelligent, but the ones who are most responsive to change.

Darwin

If HP knew what HP knows, it would be three times more profitable.

Lewis Platt, former CEO HP

WIRELESS AND MOBILITY: TECHNOLOGY TRENDS

The evolution of technology is not the absolute driver for use and mass adoption. The success story of SMS (e.g., Short Messaging Service) in the mobile world is a prime example. Technology is often a true enabler, especially in the field of wireless and mobile communications. In the past 10 years, there has been a strong correlation between technology innovation which improves performance (speed, latency, batteries, displays, etc.) and superior quality of experience for the end user. When considering Mobility and Enterprise 2.0, two technology tracks are of importance:

- Wireless LAN which came as a natural extension of the corporate LAN, initially supporting data services, and
- Mobile communications which initially allowed for voice communication outside corporate boundaries.

Below an overview of their respective evolution.

Wireless LAN

Although initial experiments are dated back in the 1970s, wireless LAN technology really took off with the definition of IEEE 802.11 known as Wi-Fi. First generation deployed in the early 2000s had basic connectivity, low bandwidth and poor security. The second generation, currently being used, provides significant performance increase as well as stronger security required for corporate use. The next generation will provide even better performance and scale with wired/wireless integration (see Figure 1)

When considering the evolution and the role of WLAN technology in corporations there are clear signs that use and deployment is growing strong. Predictions from analysts also confirm the expected trends.

Demand for enterprise WLAN products continues to be strong; enterprises continue to see the benefits of mobility in the daily work environment and are extending the wireless network throughout the premises

IDC – December 2008 – Economic Crisis Response:

Western European WLAN Market 2008-2013

We expect that 5.8% of global enterprise LAN expenditure will shift from wired to wireless spending from 2007 to 2012. The number of laptop workers using wireless as the primary type of connection to the corporate LAN is expected to rise from approximately 4 million in 2007 to 54 million in 2012.

Gartner – October 2008 – Dataquest Insight:

Factors Driving the Worldwide Enterprise Wireless LAN Market, 2007-2012

As connectivity migrates from wired to wireless inside corporations as it is already the case in most public places (airports, hotels, shops, etc.) and home offices, the use of tools and employees expectations will also change. Highly interactive and collaborative tools and applications from the Web 2.0 are now reaching an acceptable use and productivity level over wireless network.

Mobile Communications

Mobile communications extend the reach of WLAN and thanks to wide coverage allow for connectivity on the move beyond corporate premises. The early 1990s marks the beginning of GSM (Global System for Mobile communications) and CDMA (Code Division Multiple Access) now referred to as 2G. The growth and use of voice service has been incredible to reach about 4 billion users today. Initially providing voice service, evolutions referred to as 2.5G started to provide data services using existing infrastructure (e.g., GPRS - General Packet Radio Service). It is only with introduction of 3G technology in the early 2000s and later improvement such as HSPA (High Speed Packet Access) protocols that downlink bandwidth made a significant jump (from a few Kbps to several Mbps). Further development in radio technology viewed as 4G like LTE (Long Term Evolution) is aiming at peak rate of several hundred Mbps. Besides increased bandwidth and associated spectral efficiency gain, it is the reduced latency that is also impacting use and acceptance of data services. (As noted earlier, the exception to this impressive technology roadmap and its impact is the unexpected success of the Short Messaging Service which has become a multi-billion dollars market.)

Looking closely at evolution depicted in Figure 2, it is clear that progress in battery, memory, processor, display, and mobile browsers has really been the additional factor in shaping the evolution of mobile communications. Started as a voice only service for a privileged few, mobile applications are now a mass market including messaging, video as well as real time context and content Web 2.0 tools. The resulting widespread changes in working practices and management styles enable true business in motion (Economist Intelligence Unit, 2007). The widgetization of most recent applications and adoption of platforms such as Iphone and Android is also accelerating the benefits of mobile communications.

Figure 2. Technology trends in mobile communications (source: ALU in collaboration with LGE)

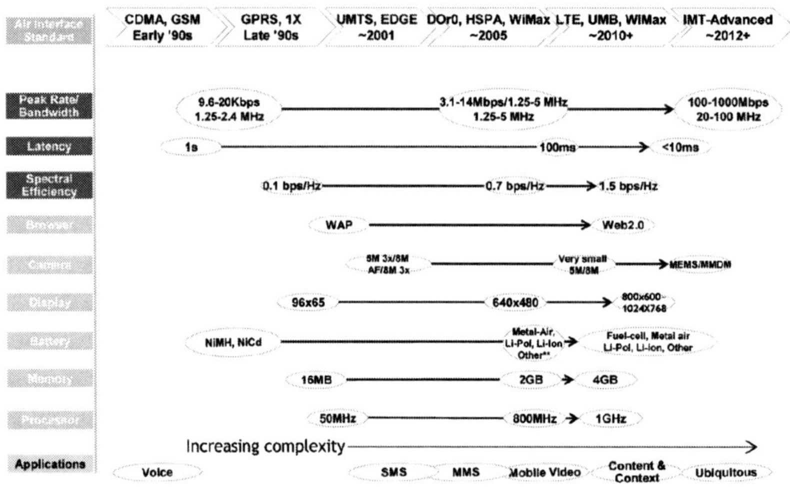

A NEW ERA: MOBILE BROADBAND LIFE AND THE MILLENIALS

The combined progress of both WLAN and Mobile Communications has opened a new era: the "mobile broadband life." In a few years, most new applications will have mobile users as a prime target, finally making themes such as "convergence" and "quad play" issues of the past. For the corporate world, that means that not only voice, video and data applications will be integrated but that mobility will just be considered native de facto Enterprise 2.0 use.

Mobile Broadband Life

In the same way that wireline broadband has impacted residential consumers with Internet connectivity, broadband wireless is becoming a basic expectation "everywhere". The fixed broadband life which is massively adopted (e.g., Facebook, Youtube, Googletalk, etc.) is now "exportable"—or already exported—to the mobile world. Furthermore, nomadic infotainement (e.g., Ipod, portable DVD, GPS, etc.) are also quickly being massively adopted pushing further and faster the evolution of mobile devices. End users do not think in terms of wireline versus wireless access and expectations are rising fast in terms of "everywhere, always on" connectivity.

The Millennials and the Next Business Generation

In the same way WLAN and Mobile technology is impacting corporations, a new population of users is shifting radically mobile services use, referred to as the "Millennials" or "Generation Y" which captures an age group of 11–30 year olds. It is a population which always had Internet and cell phones. The way they relate to technology will impact the corporations. For example, only a small percent of teen cell phone users are voice-only customers, a big difference with the adult population. Figure 3 below depicts the comparison across mobile service use.

Over the next five years Millennials will transform the way in which voice and data services are accessed and used. Adoption of peer to peer services, video and social networking give a preview to the services likely to dominate the mobile space in years to come. Their consumer behavior will translate into the enterprise. In fact,

Figure 3. Mobile service use (source: Pew Research Center)

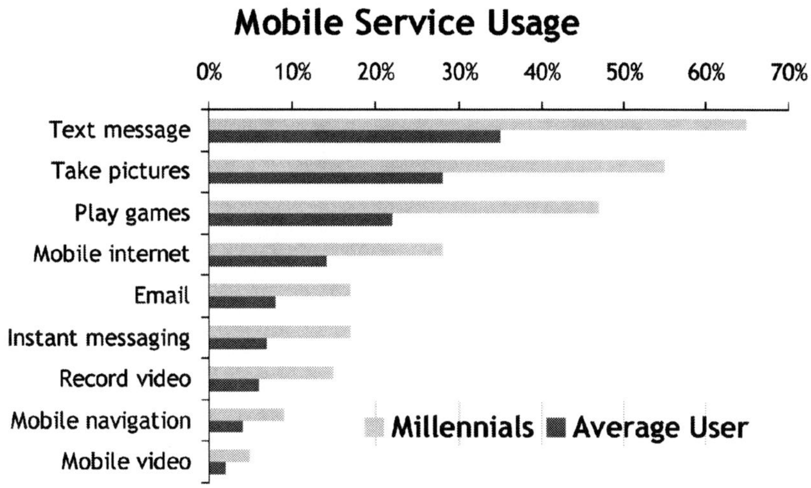

the consumer sector is now a strong driving force for open innovation in Enterprises (see Figure 4).

Furthermore, as carrier technologies and coverage improve, the full benefits (employee productivity, convenience) will rely on the seamless hand off between WLAN and mobile networks as an essential part of user experience.

MOBILITY AND ENTERPRISE 2.0 IN ACTION

Looking at both evolution of Enterprise 2.0 tools, mobility as well as the user population, it becomes clear that Enterprise 2.0 can be considered "native mobile." Mobility, in the wireless sense, brings the fundamentals for Enterprise 2.0 contextual communication:

- Available everywhere (much like wireline broadband)
- Always connected (solves the business latency issue)
- Location (beyond presence and availability, context, activity, surrounding)

Enterprise 2.0 allows context aware communication and collaboration. Mobility enriches context aware communication and collaboration. Below are three selected scenarios of mobility and Enterprise 2.0 in action, namely:

- Integrated Communication and Collaboration for Healthcare, part of Alcatel-Lucent (ALU) vertical market solutions for the healthcare sector (ALU, 2009).
- Participatory Learning in Higher Education, a collaboration with Abilene Christian University where Alcatel-Lucent is providing key technology components to the solution (Abilene Christian University, 2008a).
- Mobile Video Social Networking Application for Emergency Services, based on the mobile video social networking research at the Pennsylvania State University partially supported by Alcatel-Lucent (Gou et al., 2009).

Either fully deployed or in prototyping stage those scenarios illustrate the variety of applica-

Figure 4. Consumers applications' impact on enterprises

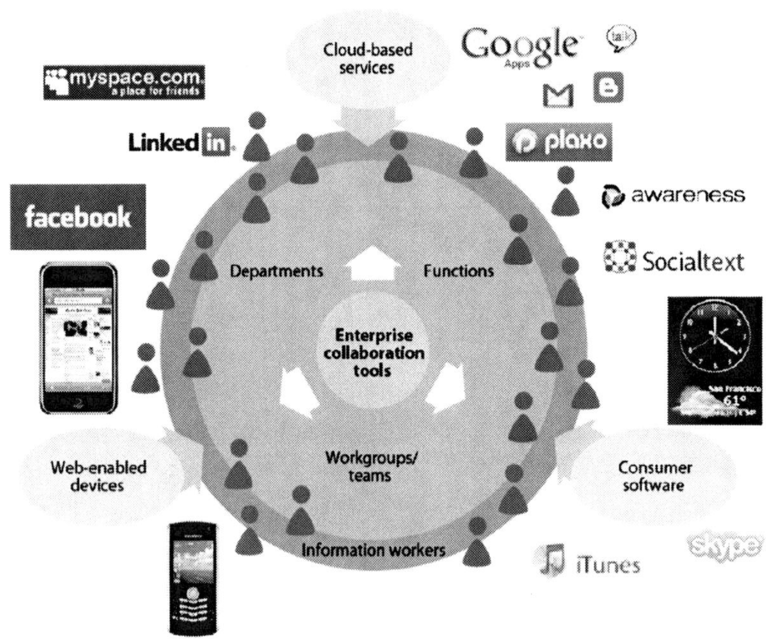

tion domains and potential improvements when mobility and Enterprise 2.0 join forces.

Integrated Communication and Collaboration for Healthcare

Healthcare is an easy first choice as it concerns all the population. In times of needs, health professionals and families expect the best for the patient. In many cases, situations are time critical and involve a high number of contributors in a suite of complex processes. Technology has already brought tremendous improvements to healthcare. However, in one critical area, progress has been slow: communications. Effective communications can have a huge impact on everyone's quality of life, including the patient, the patient's relatives and all the members of the healthcare team.

Another reason why healthcare is a good candidate for illustrating the benefits of Enterprise 2.0 and mobility is because it is by nature a very collaborative process. Usually several doctors, nurses, medical technicians and others work together to meet each patient's unique requirements. At any given time, you must be able to identify where each team member is, how best to reach them, and whether they are available for a consultation. However, there are a number of challenges to solve in order to provide effective communication and collaboration.

First, most healthcare team members are often on the move. Another important point is that of availability to communicate while busy with patient care. Adding critical situations, need to locate specialized skills, and emergencies in the mix and one can realize how quickly a situation can turn very complex.

The integrated communication and collaboration solution is based on presence, preferred media, data sharing and conferencing across all stakeholders to speed up collaborative patient care (ALU, 2008). Key features are:

- A client for collaboration and communication deployed on all capable devices (lap-

Figure 5. Simplified view of healthcare team in action

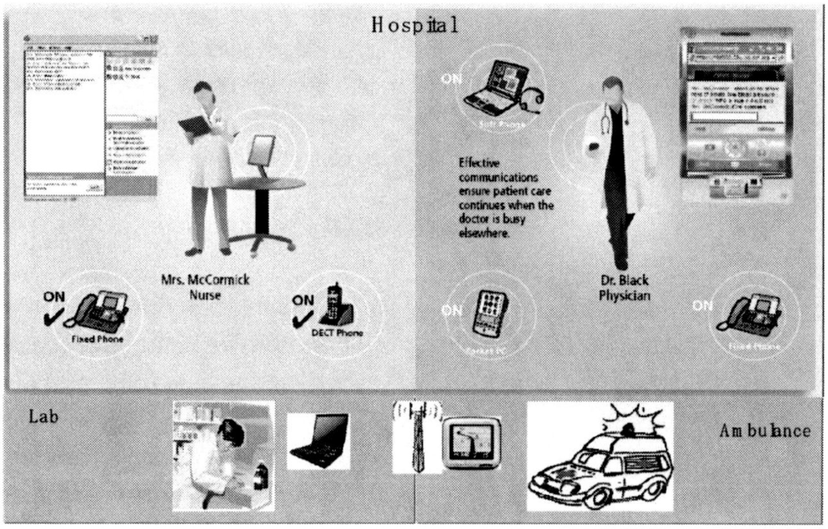

tops, pocket PCs, IP phones, Cart of Wheel computer, PDAs, etc.).

- One number rings through to whatever device is currently in use, integrated with healthcare systems' directories.
- Non-intrusive unique interface to manage all communications (voice, video, instant message, email, fax, etc.) according to sct prcfcrences and hospital established protocols.

One can check the contact's presence to determine which means of communications can be used, for example, emergency text for doctor in surgery. Contact can be initiated with the click of a button with appropriate channels (text, voice, video). Additional functions such as convert text to speech, so you can "listen" to your e-mail while on the move or momentarily unable to operate a mail client. The objective remains to be able to reach all team members quickly and share information efficiently.

The above deployment scenario can be extended to include external providers such as off site Laboratories. More advanced scenarios can include ambulance for remote diagnosis during emergencies and dynamic routing of patients to nearest hospital which can provide both specialized and available skills. In such highly collaborative and "moving" population, the potential of combined benefits and efficiency gains of using Mobility and Enterprise 2.0 are huge.

Participatory Learning in Higher Education

The field of Education is also a very good target for Mobility and Enterprise 2.0. Because this is a world more open to experiments and also because the user population (staff and students) are usually early adopters of technology. There are at least two key drivers for Universities: drive enrollment and retention and adapt to fast changing learning style and expectations (i.e., Millennials). (See Figure 6)

In 2008, part of a pilot mobile learning project, Abilene Christian University was the first to experiment large scale a participative learning experiment (Abilene Christian University, 2008b). The initial idea was to provide all incoming freshmen with converged media devices such as I-Phone or I-Pod Touch and a set of Web applications. Overview of the solution is depicted in Figure 7.

Figure 6.

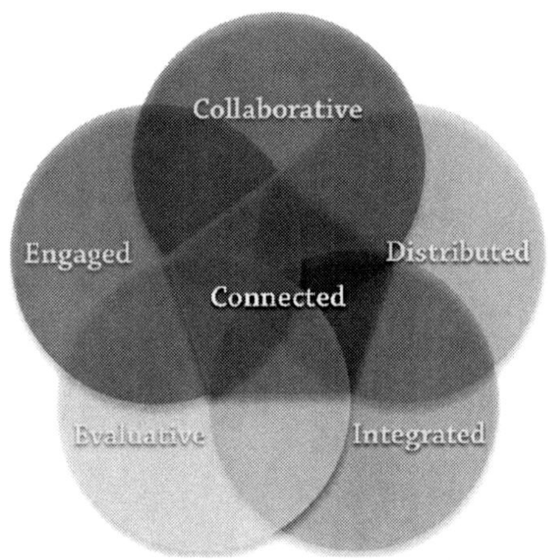

The type of applications provided are many types (Abilene Christian University, 2009):

- Academic such as providing location, direction and class details. Capability to perform dynamic quiz in classroom for immediate feedback.
- Utilities such as providing structured storage for class materials, student attendance, classroom control system for Web access but also an e911 client integrated with campus and local police and medical personnel for emergency situations.
- Social communitieswith access to multiple sites (Facebook, mySpace, etc.); presence function that provides dynamic feedback about a user's status.

The sharing of knowledge that is both transmitted and received via this connectedness is what is really amazing. In higher education, we are quickly finding that barriers such as geography, knowledge silos, and technical expertise are quickly eroding. As these barriers are removed, individuals now have access to a wealth of information, experiences, and observations that is literally at their finger tips. Add mobility to the mix, and now location dependencies are removed and this information is readily available whenever and where ever a person wants or needs to access the information. Beyond just accessing this information, now a person can be a contributor to this incredible resource

Arthur Brant, ACU

Although experiments are at the early stages, initial results for both higher education institutions, as well as students show many benefits:

- Increased students learning
- Increased students engagement
- Increased student enrolment and retention
- Foster a culture of innovation on campus

It is clear that the potential of Mobility and Enterprise 2.0 technologies is also very high in this environment. Furthermore, the fact that the model might further evolve from being IT driven to being enhanced by staff and students specific applications is another strong growth vector and additional adoption factor.

Mobile Video Social Networking Application for Emergency Services

Although hard to foresee initially, social networking tools (e.g., Facebook, Twitter, etc.) have now entered the corporation (LinkedIn, Yammer, etc.). Video social networking (e.g., YouTube, DailyMotion, etc.) are also entering the corporate world with video sharing sites to support employee/ management communication, foster collaborative innovation, etc. The corporate videos are usually more formal but this is changing fast as individual use their experience to add special effects and add humorous notes. Video chatting is also widely used as an additional channel or escalation from Instant Messaging. Inside corporation, video has been used for years with widely deployed video conferencing facilities.

Figure 7. Participative learning solution overview

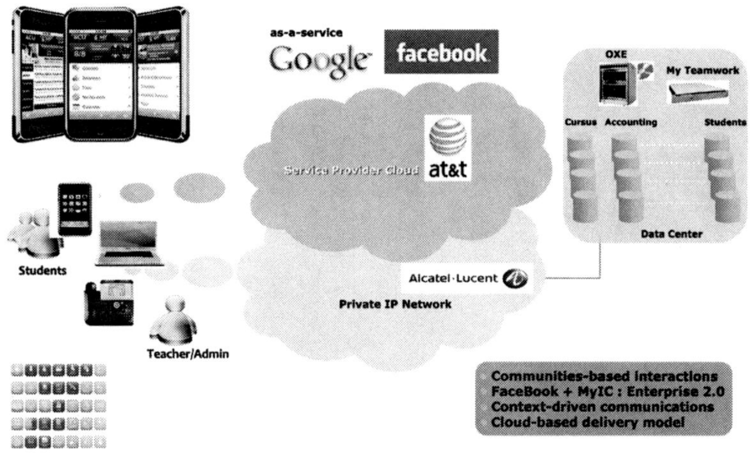

In parallel, another phenomena known as crowdcasting uses crowdsourcing and broadcasting techniques and tools for bringing an audience together in order to solve a (business) goal. The wide availability of high quality video enabled mobile devices now enables all sorts of video based collaboration applications. Not only mobile devices are always on and everywhere with their owners, they provide location and—when equipped with built-in camera—can see what their owners see. In case of natural disasters (fire, flood, earthquake, etc.) the challenge is very often for emergency services to prioritize rescue operations. This can only be achieved with precise status information on exact damages, number and location of injured victims, evolutions, and so forth. A great deal of effort and technology is constantly invested to further improve those lives saving mission where every support helps.

Using video of eye witnesses on the ground versus desperate unclear distress messages over the emergency phone lines can benefit operations and drastically reduce casualties. All witnesses can report the context around them by sending video feed from their mobile phone. This crowd-based real time video sharing allows for a better understanding of the ground in the control room. This enhanced content is then dispatched to emergency workers on the ground. The solution is depicted in Figure 8.

A recent example, dated January 2009, is the heroic landing of Flight 1549 in the Hudson River in New York City. All 155 passengers were saved but one can imagine the complexity of rescue operations had the weather been less cooperative or had it been night time. There were dozens of boats on site and one can imagine the value of bringing all video feeds back to control center in order to locate passengers in water. Beyond timeliness and precision of information, the additional benefit is that it removes all emotional factors of voice-only communication in face of such traumatizing events.

Few years ago, it was certainly hard to foresee that mobility and Enterprise 2.0 technologies and their use would potentially become critical so quickly. Today, the challenge is to imagine and build the next generation of applications that will enable more of such scenarios across many other vertical domains (transportation, energy, manufacturing, etc.).

CONCLUSION

The impact of Enterprise 2.0 on corporations will be similar to the impact of Web 2.0 on the Inter-

Figure 8. Crowdcasting and real time video sharing using mobile phone interface

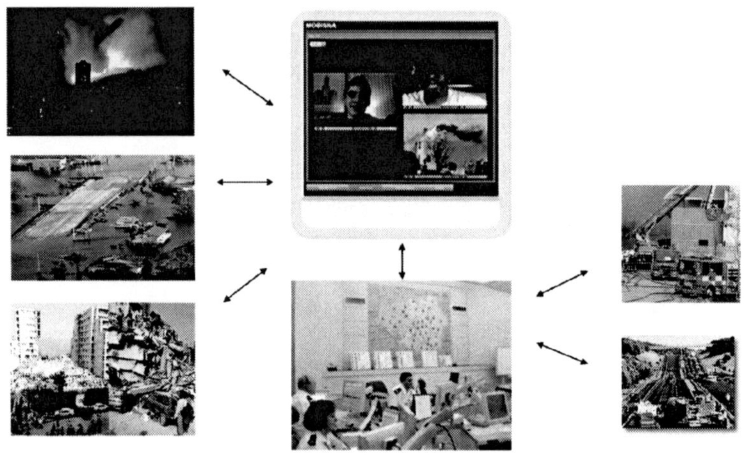

net. Enterprise 2.0 tools combined with Unified Communications help corporations access "tacit knowledge," one key asset which is spread and hidden across their organization. Mobility that initially provided convenience and increased productivity with voice only services is now ready to provide broadband connectivity and extend its reach. Technology is here to combine Mobility and Enterprise 2.0 in a way that significantly reduces business latency and increases productivity.

There is no restriction in terms of application domains as illustrated through selected examples in healthcare, higher education and emergency services. It becomes clear through initial scenarios that the power of combining Mobility and Enterprise 2.0 is certainly in several order of magnitude superior to the individual sum of their separate impact. This observation, combined with the continued strong growth of wireless access, leads clearly to believe that Mobility will be at the core of Enterprise 2.0 and one—if not the—key success factor. As always, challenges exist on the way with issues such as SLA (Service Level Agreement), security, mixed mode use and finding the appropriate business models for profitable growth.

ACKNOWLEDGMENT

The authors would like to thank all colleagues from Alcatel Lucent who provided their time, supporting material and insights during the preparation of this article: Xavier Martin, Chris Carfagnini, Jerry Power, Jean-Luc Ronarc'h, Jean-Marie Grospierre, Nicole Hill, Fernando Egea, Jessi Betjemann, Michael See and David Puglia. Arthur Brant from Abilene Christian University and C. Lee Rives from Penn State kindly shared views on ongoing experiments. Final thanks to Izabella Lokshina from SUNY Oneonta and Steven Powell from CalTech Pomona, for inviting us to submit a article based on the presentation made at the 2009 Wireless Telecommunications Symposium in Prague.

REFERENCES

Abilene Christian University. (2008a). *Mobile learning and the connected campus*. Retrieved from http://www.acu.edu/technology/mobilelearning/index.html

Abilene Christian University. (2008b). *ACU first university in nation to provide iPhone or iPod touch to all incoming freshmen.* Retrieved from http://www.acu.edu/news/2008/080225_iphone.html

Abilene Christian University. (2009). *Solutions for the 21st century university.* Retrieved from http://www.acu.edu/technology/mobilelearning/vision/solutions/index.html

ALU. (2008). *Creating more time for patient care with dynamic communication.* Retrieved from http://enterprise.alcatel-lucent.com/private/active_docs/EPG0550090601_Healthcare_EN_StraWhitePaper.pdf

ALU. (2009). *Healthcare solutions.* Retrieved from http://enterprise.alcatel-lucent.com/?solution=Healthcare&page=Homepage

Berners-Lee, T. (1989). *Tim Berners-Lee.* Retrieved from http://www.w3.org/People/Berners-Lee

Economist Intelligence Unit. (2007). *Business in motion: Managing the mobile workforce.* Retrieved from http://enterprise.alcatel-lucent.com/private/active_docs/WP-Business_in_motion-Managing_MobileWorkforce.pdf

Gou, L., Kim, J.-H., Chen, H.-H., Collins, J., Goodman, M., Zhang, X., et al. (2009). *MobiSNA: A mobile video social network application.* Retrieved from http://mobisna.ist.psu.edu/pods23-gou.pdf

GSMA. (2009). *Market data summary.* Retrieved from http://www.gsmworld.com/newsroom/market-data/market_data_summary.htm

Levy, P. (1994). *Collective intelligence.* New York: Basic Books.

O'Reilly, T. (2008). *Google, WalMart, and MyBarackObama.com: The power of the real time enterprise.* Retrieved from http://blogs.oreilly.com/cgi-bin/mt/mt-search.cgi?blog_id=57&tag=walmart&limit=20&IncludeBlogs=57

Ohayon-Dekel, A. (2009). Connected knowledge–connecting the brains of the enterprise in real time. *Enriching Communications, 3*(1). Retrieved from http://www.alcatel-lucent.com/enrich/v3i12009/article_c1a5.html?l=en

Surowiecki, J. (2004). *The wisdom of the crowds: Why the many are smarter than the few and how collective wisdom shapes business, economies, societies and nations.* New York: Doubleday.

Young, O., Burris, P., & Reiss, D. (2009). *The enterprise 2.0 buyer profile 2009–the top priorities and needs of firms looking at enterprise 2.0 technology.* Boston: Forrester.

KEY TERMS AND DEFINITIONS

Blogs: Web site in which entries are displayed in reverse chronological order. Blogs are made up of regularly published content, have a specific point of view or author and provide a feedback channel for comments

Wiki: Collaborative web site that can be directly edited by anyone with access to it. Wikis are most often used for knowledge management, project management, or customer contributed insight.

Social Networking: Use of self (or RH) generated user profiles to help facilitate and mediate social interactions. Social network users share details about themselves, search for others with similar interest, exchange messages and work together

RSS: Really Simple Syndication (RSS) is a content delivery mechanism by which users subscribe to content feeds like blogs, wikis or news sources. Users are automatically notified when new content is published.

Podcasting: Syndication of audio content to customers or employees. Closely associated with RSS, podcasting allows users to subscribe to an ongoing series of audio content that is delivered directly to their computer and digital music device.

Widgets: Stripped down web applications highly portable,. Make it easy for non technical users to add dynamic content or functionality such as search tools or maps, to different locations such as Web pages, portals, desktops, or mobile devices.

Mashups: Custom Web applications combining multiple, disparate data sources into something new and unique. Mashups platforms specifically allow nontechnical users to create and consume their own mashups with visual tools.

Microblogs: Allows blog posts with a short message visible to everyone or to a restricted group. Messages are published using any internet-connected device, phones, mobile devices, desktop software, or web sites (e.g. Twitter).

Social Bookmarks: Enable people to mark web pages (by selecting a button on the web page) of interest and to share those web locations with others. Users can add tags and textual description. Bookmarks are shared on a web site (e.g. Delicious) for others to see.

Chapter 13
On the Internationalization of the Wireless Telecommunications Industry:
A Market-Based Analysis of Six European Service Providers

Steven R. Powell
California State Polytechnic University - Pomona, USA

ABSTRACT

Companies can benefit from diversifying internationally. This paper analyzes the internationalization efforts of six major European wireless telecommunications service providers: Telefonica, Deutsche Telekom, France Telecom, Vodafone, Telenor and TeliaSonera. Although all six companies have invested heavily in foreign markets, their internationalization paths have not been the same, resulting in foreign market portfolios with different characteristics. Utilizing a proportionate customer weighting scheme based on the number of subscribers controlled by a company in each market, the paper examines how the companies compared in 2007 versus 2002 in their extent and scope of internationalization and in the attractiveness of their foreign market portfolios with respect to some key growth, profitability, and risk market characteristics.

INTRODUCTION

Internationalization can have benefits for companies. It can make companies more efficient in their current activities, as well as better at managing the risks associated with these activities. Internationalization can also develop the internal learning capabilities of companies to innovate and adapt to future changes. Companies can accomplish these goals by exploiting differences in input and output markets among countries and by benefiting from scale economies, synergies and economies of scope in their activities and organizations (Ghoshal, 1987). According to Ghemawat (2007) the particular countries in which a company invests matters a great deal. He has identified three strategies for global value creation based on their differences: *adaptation* (adjust to differences

DOI: 10.4018/978-1-60960-505-6.ch013

across countries), *aggregation* (overcome differences among countries by grouping them based on similarities), and *arbitrage* (exploit selected differences across countries instead of treating them all as constraints).

Dunning and Lundan (2008) have asserted that the extent and geographical pattern of a multinational's foreign investment will be determined by three factors: (1) the ownership-specific advantages (arising from access to income-producing assets and the ability to coordinate these assets across national boundaries) that a firm of a given nationality (or its affiliates) possesses compared with a firm of another nationality; (2) location-specific factors favoring the home or host countries; and (3) advantages gained from internalizing the markets to generate and use income-providing assets. The significance of these factors and their configuration is context-specific, depending on the industry, countries, and firms in question. In reality, Rugman and Verbeke (2004) have observed that most internationalization is not global, but regional in nature.

Strategy and scale have driven telecommunications service providers (TSPs) to internationalize (Sarkar, Cavusgil, & Aulakh, 1999). Strategic drivers for internationalization include the search for higher-growth markets abroad, the diversification of the companies' investment portfolios, enhanced leverage over regulators, the provision of seamless services for multinational customers, revenues from the transference of existing management and technical know-how from consulting operations abroad, and enhanced power in standard setting debates. Scale drivers for internationalization include enhanced negotiating power over equipment suppliers, cost reductions through volume accumulations across country locations and market segments, and economies derived from the optimization of network design. Many TSPs now have significant operations and market interests outside their home countries. The wireless communications industry is a good example.

According to Steinbock (2003) TSPs of developed countries are now in their third stage of wireless communications industry evolution: internationalization. This stage began at the end of the 2G era and has continued through the 3G era and into the beginning of the 4G era. It follows a first stage (from pre-cellular days through the 1G era) in which the TSP was a single state-owned or controlled monopoly and operated primarily in domestic markets, and a second stage (beginning at end of the 1G era and continuing through the 2G era) in which the TSP regionalized. Researchers have attempted to measure the degree of wireless communications industry internationalization and to determine if there is a link between internationalization and profitability.

Curwen and Whalley (2006) employed four measures to analyze the degree of internationalization of thirty-four companies representing all the mobile operators present in at least three countries at year-end 2005: (1) the companies' geographical dispersion, that is, the number of countries in which they were invested; (2) their psychic dispersion, that is, the number of zones with cultural affinity in which the companies operated; (3) the number of proportionate customers controlled by each company in each region; and (4) the percentage of proportionate customers outside their domestic markets. Echoing Rugman and Verbeke (2004) they found that most individual operators are not global in presence, but at best bi-regional. Their study did not examine how the degree of internationalization changed over time. Gerpott and Jakopin (2005) analyzed a sample of European mobile network operators to determine if their degree of internationalization and financial performance could be linked. The degree of internationalization ranged from offering connections to foreign destinations originated within the home country license area to majority control of a foreign business. Their research found no evidence over the 1997-2003 period of a link between higher degrees of internationalization and higher profitability levels.

This paper will investigate another aspect of wireless communications industry internationalization: how different internationalization strategies can affect the aggregate growth, profitability, and risk characteristics of a TSP's portfolio of foreign markets. It will also consider how the degree of internationalization of a TSP can change over time. The paper will focus on six prominent European TSPs that have internationalized their wireless operations: Telefonica, Deutsche Telekom, France Telecom, Vodafone, Telenor, and TeliaSonera. Telefonica, Deutsche Telekom, France Telecom, Telenor, and TeliaSonera are full-service TSPs in their domestic markets, and in addition to their wireless businesses, have expanded other lines of business abroad. Although it has some fixed line operations, Vodafone is primarily a wireless TSP. Some of these companies operate in the same markets. In particular, the paper will attempt to answer three questions. (1) How did the six TSPs rank with respect to their degree of internationalization in 2007 compared with 2002? (2) How did the six TSPs rank in the attractiveness (with respect to growth, profitability, and risk) of the foreign markets in which they operated in 2007 compared with 2002? (3) How did the six TSPs rank in terms of the difference (or distance) in growth, profitability, and risk between their foreign markets and their home markets in 2007 compared with 2002? The analysis will focus on the aggregate characteristics of the companies' foreign market portfolios and the characteristics of the home market.

The paper is structured in the following way: We first introduce the six TSPs and consider their internationalization efforts from 2002 to 2007; we then describe the method used to analyze the characteristics of the TSPs' market portfolios; and we also present the results of the analysis. The paper concludes with a summary of the results and some possible directions for future research.

OVERVIEW OF THE SIX TSPS

This section presents a brief overview of the six TSPs in the study: Telefonica, Deutsche Telekom, France Telecom, Vodafone, Telenor and TeliaSonera. Summary data on the state of the companies' internationalization efforts in 2002 and 2007 can be found in Tables 1 and 2. Detailed market data can be found in Tables 3 and 4.

Table 1. Comparison of the TSPs' Wireless Operations: 2002

	Telefonica		Deutsche Telekom		France Telecom		Vodafone		Telenor		TeliaSonera	
Characteristic	**Home**	**Abroad**	**Home**	**Abroad**	**Home**	**Abroad**	**Home**	**Abroad**	**Home**	**Abroad**	**Home**	**Abroad**
Locations	Spain	Latin America Africa	Germany	Europe US	France	Global	UK	Global	Norway	Europe Asia	Sweden Finland	Europe Asia
No. of Countries	1	10	1	5	1	18	1	26	1	10	2	11
Venture Customers (M)	18.4	25.5	24.6	33.8	19.2	40.2	13.3	280.1	2.4	23.5	6.1	23.7
Proportionate Customers (M)	17.0	13.1	24.6	27.8	19.2	26.5	13.3	103.9	2.4	10.1	6.1	10.4
Proportionate Customers (%)	56.5	43.5	46.9	53.1	42.0	58.0	11.3	88.7	19.2	80.8	37.0	63.0

Table 2. Comparison of the TSPs' Wireless Operations: 2007

	Telefonica		Deutsche Telekom		France Telecom		Vodafone		Telenor		TeliaSonera	
Characteristic	**Home**	**Abroad**	**Home**	**Abroad**	**Home**	**Abroad**	**Home**	**Abroad**	**Home**	**Abroad**	**Home**	**Abroad**
Locations	Spain	Latin America Europe Africa	Germany	Europe US	France	Europe Africa.	UK	Global	Norway	Europe Asia	Sweden Finland	Europe Asia
No. of Countries	1	19	1	11	1	24	1	24	1	16	2	15
Venture Subscribers (M)	22.8	146.3	36.0	83.8	24.2	95.0	18.5	742.6	2.9	135.0	7.3	97.4
Proportionate Subscribers (M)	22.8	123.7	36.0	75.8	24.2	87.1	18.5	242.0	2.9	77.0	7.3	43.9
Proportionate Subscribers (%)	15.6	84.4	32.2	67.8	21.7	78.3	7.1	92.9	3.6	96.4	14.3	85.7

Telefonica, formerly Spain's state-owned monopoly TSP, began implementing a Latin America regional strategy in the 1990s, which was carried out by its Telefonica Moviles subsidiary. At the end of 2002 Telefonica had wireless operations in Spain and ten countries abroad - nine in Latin America (Puerto Rico, Mexico, El Salvador, Guatemala, Argentina, Brazil, Chile, Peru, and Venezuela) and one in Africa (Morocco). The company had 30.1 million proportionate wireless subscribers (43.5% abroad) and overall revenues and a net loss of €28.4 billion and €5.6 billion, respectively. With the incorporation of O_2 and Cesky Telecom in Europe in 2005 Telefonica's internationalization strategy became bi-regional. By 2007 the company had expanded to nineteen countries abroad, having added wireless operations in five markets in Europe (UK, Ireland, Germany, Czech Republic and the Slovak Republic) and five in Latin America (Panama, Nicaragua, Colombia, Ecuador, and Uruguay). It no longer had an ownership presence in Puerto Rico. Telefonica had grown to 146.5 million proportionate wireless subscribers (84.4% abroad), with overall revenues and operating income of €56.4 billion and net income of €13.4 billion, respectively. It also had re-organized its fixed and wireless operations geographically.

Deutsche Telecom, formerly Germany's state-owned monopoly TSP, began internationalizing during the 1990s, and since 2002 its mobile business has primarily operated in all its markets under the T-Mobile brand. At the end of 2002 the company had wireless operations in Germany and five countries abroad - four in Europe (UK, Netherlands, Austria, and Czech Republic) and one in North America (United States). The company had 52.4 million proportionate wireless subscribers (53.1% abroad) and revenues and a net loss of €53.7 billion and €24.6 billion, respectively. During 2002-2007 Deutsche Telekom continued to expand in Europe adding six more markets (Hungary, Poland, Croatia, the Slovak Republic, Macedonia, Montenegro), two of which were minority-owned. In 2007 Deutsche Telekom had 111.8 million proportionate wireless subscribers (67.8% abroad) and revenues of €62.5 billion and net income of €600 million, respectively. Mobile communications revenues and profit from

Table 3. TSPs' Wireless Operations Market Characteristics: 2002

						Telefonica		Deutsche Telekom		France Telecom		Vodafone		Telenor		TeliaSonera	
Market	Risk	Population (M)	GDP/CAP (US $000)	Subs/ 100	CAG (%)	Prop. Subs. (M)	Equity %	Prop. Subs. (M)	Equity %	Prop. Subs. (M)	Equity %	Prop. Subs. (M)	Equity %	Prop. Subs. (M)	Equity %	Prop. Subs. (M)	Equity %
UK	92.6	59.1	26.7	83.3	41.0	0	0	12.4	100	13.3	100	13.3	100	0	0	0	0
Ireland	92.1	3.9	31.2	76.3	40.7	0	0	0	0	0	0	1.7	100	0	0	0	0
Germany	91.9	82.6	24.5	71.7	48.2	0	0	24.6	100	0	0	22.9	100	0	0	0	0
Hungary	71.0	10.2	6.5	67.7	57.7	0	0	0	0	0	0	0.8	83.8	2.45	100	0	0
Netherlands	93.5	16.2	27.3	75.3	47.8	0	0	1.4	100	1.0	100	3.2	97.2	0	0	0	0
Sweden	93.3	8.9	27.8	89.0	20.2	0	0	0	0	0	0	1.3	99.1	0	0	3.60	100
Norway	95.5	4.6	42.2	83.3	17.7	0	0	0	0	0	0	0	0	2.38	100	1.09	100
Finland	92.1	5.2	26.1	86.8	15.9	0	0	0	0	0	0	0	0	0	0	2.49	100
Denmark	94.3	5.4	32..3	83.2	25.4	0	0	0	0	0.3	67.2	0	0	0.59	53.5	0.47	100
Belgium	90.9	10.4	24.4	78.2	52.7	0	0	0	0	1.2	50.8	1.1	25.0	0	0	0	0
France	91.4	59.6	24.4	64.6	46.0	0	0	0	0	19.2	99.9	5.8	43.9	0	0	0	0
Poland	65.2	38.6	5.1	36.0	76.5	0	0	0	0	0	0	1.0	19.6	0	0	0	0
Switzerland	97.4	7.3	38.6	79.6	40.6	0	0	0	0	1.0	100	0.9	25.0	0	0	0	0
Italy	87.3	56.5	21.1	93.6	35.8	0	0	0	0	2.3	26.6	14.9	76.8	0	0	0	0
Albania	31.5	4.0	1.4	27.7	203.6	0	0	0	0	0	0	0.3	82.4	0	0	0	0
Greece	84.4	11.0	13.5	84.3	58.3	0	0	0	0	0	0	2.2	64.0	0.63	18.0	0	0
Malta	72.1	0.4	10.8	69.9	73.3	0	0	0	0	0	0	0.1	80.0	0	0	0	0
Portugal	86.3	10.4	12.4	83.9	41.9	0	0	0	0	0.4	20.2	2.9	94.4	0	0	0	0
Spain	88.5	40.7	16.5	80.6	50.5	17.0	92.4	0	0	0	0	9.1	100	0	0	0	0
Romania	46.3	22.3	2.1	23.3	91.0	0	0	0	0	1.5	67.8	0.5	20.1	0	0	0	0
Austria	92.8	8.2	25.6	82.9	42.2	0	0	2.0	100	0.2	17.5	0	0	0.23	17.45	0	0
Czech Republic	67.0	10.1	7.3	84.0	74.9	0	0	2.1	60.8	0	0	0	0	0	0	0	0
Slovak Republic	59.3	5.4	4.5	54.1	71.0	0	0	0	0	1.1	63.9	0	0	0	0	0	0
Montenegro	29.4	NA	NA	NA	NA	0	0	0	0	0	0	0	0	0.12	0.42	0	0
Moldova	30.1	2.3	0.4	8.0	173.8	0	0	0	0	0	0	0	0	0	0	0.10	74.0
Lithuania	57.2	3.5	4.1	47.5	58.3	0	0	0	0	0	0	0	0	0	0	0.5	55.0
Latvia	60.5	2.3	4.0	39.2	64.1	0	0	0	0	0	0	0	0	0	0	0.3	60.3

continued on following page

Table 3. continued

						Telefonica		Deutsche Telekom		France Telecom		Vodafone		Telenor		TeliaSonera	
Market	Risk	Population (M)	GDP/CAP (US $000)	Subs/ 100	CAG (%)	Prop. Subs. (M)	Equity %	Prop. Subs. (M)	Equity %	Prop. Subs. (M)	Equity %	Prop. Subs. (M)	Equity %	Prop. Subs. (M)	Equity %	Prop. Subs. (M)	Equity %
Estonia	63.9	1.4	5.4	65.3	43.6	0	0	0	0	0	0	0	0	0	0	0.2	49.0
Russia	46.0	145.3	2.4	12.1	91.0	0	0	0	0	0	0	0	0	1.5	29.0	1.3	43.8
Ukraine	33.9	48.0	0.9	7.7	130.1	0	0	0	0	0	0	0	0	1.0	54.2	0	0
Kazakhstan	50.6	14.9	1.5	6.9	146.9	0	0	0	0	0	0	0	0	0	0	0.2	37.7
Azerbaijan	40.2	8.3	0.8	9.6	81.8	0	0	0	0	0	0	0	0	0	0	0.3	37.9
Georgia	28.9	4.6	0.7	10.9	75.8	0	0	0	0	0	0	0	0	0	0	0.1	61.6
Turkey	43.6	70.3	3.3	33.2	70.7	0	0	0	0	0	0	1.4	60.0	0	0	5.9	37.3
Egypt	51.7	65.6	1.2	6.5	133.1	0	0	0	0	1.6	71.3	1.4	60.0	0	0	0	0
South Africa	57.9	45.5	2.4	29.4	49.5	0	0	0	0	0	0	2.8	35.0	0	0	0	0
Kenya	36.7	31.9	0.4	3.7	181.1	0	0	0	0	0	0	0.3	35.0	0	0	0	0
Morocco	55.2	29.6	1.3	20.6	142.1	0.5	29.0	0	0	0	0	0	0	0	0	0	0
Botswana	61.4	1.8	3.2	25.1	NA	0	0	0	0	0.1	51.0	0	0	0	0	0	0
Cameroon	31.1	15.5	0.7	4.5	178.3	0	0	0	0	0.2	70.0	0	0	0	0	0	0
Cote d'Ivoire	29.7	17.3	0.7	5.9	95.5	0	0	0	0	0.4	85.0	0	0	0	0	0	0
Madagascar	28.4	17.1	0.1	1.0	108.9	0	0	0	0	0.0	33.6	0	0	0	0	0	0
U.S.	95.6	288.4	36.1	48.9	20.7	0	0	9.9	100	0	0	14.8	44.4	0	0	0	0
Puerto Rico	NA	3.9	11.5	46.3	37.4	0.8	49.9	0	0	0	0	0	0	0	0	0	0
Dominican Republic	47.9	8.5	2.9	20.0	64.0	0	0	0	0	0.3	86.0	0	0	0	0	0	0
Mexico	61.4	101.9	6.3	25.2	71.6	2.1	85.0	0	0	0	0	0.7	34.5	0	0	0	0
El Salvador	50.6	6.5	0.2	13.6	85.8	0.2	83.5	0	0	0	0	0	0	0	0	0	0
Guatemala	45.1	12.0	1.8	13.5	89.7	0.1	92.4	0	0	0	0	0	0	0	0	0	0
Argentina	26.7	36.6	2.7	17.5	26.7	1.5	90.5	0	0	0	0	0	0	0	0	0	0
Brazil	44.6	173.9	2.8	19.5	50.3	6.3	46.2	0	0	0	0	0	0	0	0	0	0
Chile	65.1	15.1	4.3	39.6	72.4	0.8	43.6	0	0	0	0	0	0	0	0	0	0
Peru	46.3	26.8	2.1	8.6	40.5	1.1	90.6	0	0	0	0	0	0	0	0	0	0
Venezuela	39.9	25.4	3.7	25.8	43.6	0.2	6.9	0	0	0	0	0	0	0	0	0	0
Thailand	58.5	61.9	2.0	16.3	35.8	0	0	0	0	0.6	48.9	0	0	2.20	40.3	0	0
India	55.9	1041.9	0.5	1.2	71.3	0	0	0	0	0.2	26.0	0.0	20.8	0	0	0	0

Table 3. continued

						Telefonica		Deutsche Telekom		France Telecom		Vodafone		Telenor		TeliaSonera	
Market	Risk	Population (M)	GDP/CAP (US $000)	Subs/ 100	CAG (%)	Prop. Subs. (M)	Equity %	Prop. Subs. (M)	Equity %	Prop. Subs. (M)	Equity %	Prop. Subs. (M)	Equity %	Prop. Subs. (M)	Equity %	Prop. Subs. (M)	Equity %
Japan	88.3	127.3	30.7	63.6	16.2	0	0	0	0	0	0	9.7	69.7	0	0	0	0
China	57.9	1284.5	1.1	16.0	73.2	0	0	0	0	0	0	4.0	3.3	0	0	0	0
Bangladesh	38.7	134.0	0.4	0.8	110.5	0	0	0	0	0	0	0	0	0.4	46.4	0	0
Malaysia	61.0	24.0	4.0	37.8	35.3	0	0	0	0	0	0	0	0	1.0	61.0	0	0
Australia	90.4	19.7	21.1	64.0	22.6	0	0	0	0	0	0	2.6	100.0	0	0	0	0
New Zealand	86.5	3.9	15.5	64.9	34.0	0	0	0	0	0	0	1.3	100.0	0	0	0	0
Fiji	42.6	0.8	2.2	10.9	76.8	0	0	0	0	0	0	0.1	49.0	0	0	0	0

operations were €34.7 billion and €4.5 billion, respectively.

France Telecom, France's former state-owned monopoly TSP, began implementing an international strategy during 1999-2001 with the acquisition of Orange and other European interests. Its wireless operations now primarily operate under the Orange brand. At the end of its 2002 fiscal year France Telecom had wireless operations in France and eighteen countries around the world (UK, Netherlands, Denmark, Belgium, Switzerland, Italy, Portugal, Romania, Austria, Slovak Republic, Egypt, Botswana, Cameroon, Cote d'Ivoire, Madagascar, Dominican Republic, Thailand, and India). The company had 45.7 million proportionate wireless subscribers (58.0% abroad) and revenues and a net loss of €46.6 billion and €20.7 billion, respectively. The company continued adding and divesting foreign investments, primarily in Europe and Africa, during 2002-2007. By the end of its 2007 fiscal year the mobile business of France Telecom operated in 24 countries abroad, having expanded to Poland, Spain, Moldova, Jordan, Senegal, Mali, Mauritius, Equatorial Guinea, Guinea Bissau, Guinea, Central African Republic. It had left Denmark. The company had 111.3 million proportionate wireless subscribers (78.3% abroad) and revenues and net income of €53.0 billion and €6.3 billion, respectively.

Unlike the five other TSPs in this study, Vodafone was never a state-owned monopoly TSP and has always been primarily a wireless service provider. Initially a UK operator, Vodafone began an aggressive global expansion in the late 1990s. At the end of its 2002-2003 fiscal year Vodafone operated in the UK and 26 countries abroad including sixteen in Europe (Ireland, Germany, Hungary, Netherlands, Sweden, Belgium, France, Switzerland, Poland, Italy, Albania, Greece, Malta, Portugal, Spain, and Romania), three in Africa (Egypt, South Africa, and Kenya), two in the Americas (United States and Mexico), three in Asia (India, Japan, and China), and two in the South Pacific (Australia and New Zealand). The

Table 4. TSPs' Wireless Operations Market Characteristics: 2007

						Telefonica		Deutsche Telekom		France Telecom		Vodafone		Telenor		TeliaSonera	
Market	Risk	Population (M)	GDP/CAP (US $000)	Subs/ 100	CAG (%)	Prop. Subs. (M)	Equity %	Prop. Subs. (M)	Equity %	Prop. Subs. (M)	Equity %	Prop. Subs. (M)	Equity %	Prop. Subs. (M)	Equity %	Prop. Subs. (M)	Equity %
UK	91.0	60.8	44.9	118.5	7.9	18.4	100	17.3	100	15.6	100	18.5	100	0	0	0	0
Ireland	92.4	4.3	59.3	115.9	10.7	1.7	100	0	0	0	0	2.3	100	0	0	0	0
Germany	90.4	82.6	39.9	117.6	10.4	12.5	100	36.0	100	0	0	34.4	100	0	0	0	0
Hungary	69.0	10.0	13.8	110.0	9.9	0	0	2.9	59.3	0	0	2.3	100	3.4	100	0	0
Netherlands	92.0	16.4	45.9	105.9	9.3	0	0	4.9	100	0	0	4.3	100	0	0	0	0
Sweden	93.0	9.1	48.7	113.7	5.5	0	0	0	0	0	0	0	0	1.9	100	4.8	100
Norway	97.5	4.7	81.3	110.5	6.5	0	0	0	0	0	0	0	0	2.9	100	1.6	100
Finland	92.0	5.3	46.6	115.2	6.1	0	0	0	0	0	0	0	0	0	0	2.5	100
Denmark	93.4	5.4	56.6	114.5	6.8	0	0	0	0	0	0	0	0	1.7	100	1.5	100
Belgium	89.4	10.5	42.9	102.7	5.8	0	0	0	0	3.3	100	0	0	0	0	0	0
France	89.7	61.7	41.6	89.9	7.5	0	0	0	0	24.2	100	8.3	44.0	0	0	0	0
Poland	68.3	38.1	11.0	108.7	24.4	0	0	12.6	97.0	14.2	100	2.7	19.6	0	0	0	0
Switzerland	96.2	7.5	55.6	109.7	7.4	0	0	0	0	1.5	100	0	0	0	0	0	0
Italy	84.3	58.9	35.8	135.1	9.7	0	0	0	0	0	0	23.1	76.9	0	0	0	0
Albania	44.0	3.2	3.3	72.1	22.0	0	0	0	0	0	0	1.1	99.9	0	0	0	0
Serbia	43.5	9.9	4.2	85.7	NA	0	0	0	0	0	0	0	0	2.9	100	0	0
Greece	80.3	11.2	32.3	110.3	5.7	0	0	0	0	0	0	5..5	99.9	0	0	0	0
Malta	77.5	0.4	15.5	91.4	6.1	0	0	0	0	0	0	0.2	100	0	0	0	0
Portugal	82.4	10.6	20.7	126.6	9.2	0	0	0	0	0.4	19.2	5.2	100	0	0	0	0
Spain	86.4	44.3	32.3	110.2	7.8	22.8	100	0	0	11.1	100	16.0	100	0	0	0.3	76.6
Romania	57.4	21.4	7.7	106.7	35.0	0	0	0	0	9.8	100	8.9	100	0	0	0	0
Austria	92.3	8.4	45.1	118.6	8.0	0	0	3.3	100	2.0	35.0	0	0	0	0	0	0
Czech Republic	70.7	10.2	16.5	124.9	8.1	3.6	69.4	3.2	60.8	0	0	2.7	100	0	0	0	0
Croatia	59.6	4.6	11.2	110.5	16.6	0	0	1.2	51.0	0	0	0	0	0	0	0	0
Slovak Republic	67.4	5.4	13.9	112.6	15.7	0.6	100	1.2	51.0	2.9	100	0	0	0	0	0	0
Macedonia	47.2	2.0	3.7	69.6	44.4	0	0	1.2	33.6	0	0	0	0	0	0	0	0
Montenegro	44.7	0.6	5.9	107.3	50.0	0	0	0..2	45.4	0	0	0	0	0.4	100	0	0

continued on following page

Table 4. continued

						Telefonica		Deutsche Telekom		France Telecom		Vodafone		Telenor		TeliaSonera	
Market	Risk	Population (M)	GDP/CAP (US $000)	Subs/ 100	CAG (%)	Prop. Subs. (M)	Equity %	Prop. Subs. (M)	Equity %	Prop. Subs. (M)	Equity %	Prop. Subs. (M)	Equity %	Prop. Subs. (M)	Equity %	Prop. Subs. (M)	Equity %
Moldova	34.6	3.8	1.2	49.6	41.0	0	0	0	0	1.10	100	0	0	0	0	0.5	100
Lithuania	65.3	3.4	11.3	144.9	24.4	0	0	0	0	0	0	0	0	0	0	2.0	100
Latvia	62.8	2.3	11.9	97.4	19.3	0	0	0	0	0	0	0	0	0	0	0.6	60.3
Estonia	63.6	1.3	15.9	148.4	17.6	0	0	0	0	0	0	0	0	0	0	0.5	43.8
Russia	56.6	142.5	9.1	114.6	35.0	0	0	0	0	0	0	0	0	14.2	33.6	15.6	43.8
Ukraine	47.6	46.2	3.0	119.6	71.8	0	0	0	0	0	0	0	0	15.3	59.8	1.6	20.5
Kazakhstan	54.6	15.4	6.7	79.9	64.4	0	0	0	0	0	0	0	0	0.8	16.8	3.1	51.0
Azerbaijan	48.1	8.5	3.7	53.4	41.6	0	0	0	0	0	0	0	0	0	0	1.6	51.3
Uzbekistan	34.4	27.4	0.8	9.3	91.2	0	0	0	0	0	0	0	0	0.4	18.1	0.5	74.0
Tajikistan	28.6	6.7	0.6	4.1	171.8	0	0	0	0	0	0	0	0	0.1	33.6	0.4	60.0
Georgia	37.3	4.4	2.3	59.2	38.9	0	0	0	0	0	0	0	0	0.0	17.1	1.3	97.5
Armenia	42.3	3.0	3.1	10.5	64.6	0	0	0	0	0	0	0	0	0.2	33.6	0	0
Turkey	52.0	74.9	8.8	82.8	21.6	0	0	0	0	0	0	16.9	100	0	0	13.0	37.3
Jordan	49.3	5.9	2.7	80.5	31.4	0	0	0	0	1.5	100	0	0	0	0	0	0
Egypt	52.1	75.5	1.7	39.8	46.2	0	0	0	0	10.8	71.3	7.7	54.9	0	0	0	0
South Africa	60.5	48.6	5.7	87.1	25.3	0	0	0	0	0	0	15.4	50.0	0	0	0	0
Kenya	38.4	37.5	0.8	30.2	57.1	0	0	0	0	0	0	4.1	35.0	0	0	0	0
Morocco	54.3	31.2	2.3	64.2	26.4	2.0	32.2	0	0	0	0	0	0	0	0	0	0
Botswana	65.2	1.9	6.3	75.8	26.2	0	0	0	0	0.5	100	0	0	0	0	0	0
Cameroon	38.4	18.6	1.1	24.5	45.3	0	0	0	0	2.0	100	0	0	0	0	0	0
Cote d'Ivoire	31.9	19.3	1.0	36.6	47.0	0	0	0	0	2.5	100	0	0	0	0	0	0
Madagascar	35.3	19.7	0.4	11.3	68.6	0	0	0	0	1.3	100	0	0	0	0	0	0
Senegal	39.4	12.4	0.9	29.3	45.7	0	0	0	0	2.5	100	0	0	0	0	0	0
Mali	36.1	12.3	0.6	20.5	122.9	0	0	0	0	2.0	100	0	0	0	0	0	0
Mauritius	56.3	1.3	5.0	73.6	21.7	0	0	0	0	0.6	40.0	0	0	0	0	0	0
Equatorial Guinea	39.6	0.5	19.5	43.4	47.0	0	0	0	0	0.2	40.0	0	0	0	0	0	0
Guinea Bissau	25.6	1.7	0.2	17.5	NA	0	0	0	0	0.1	100	0	0	0	0	0	0
Guinea	27.3	9.4	0.5	21.3	85.6	0	0	0	0	0.2	100	0	0	0	0	0	0

continued on following page

Table 4. continued

						Telefonica		Deutsche Telekom		France Telecom		Vodafone		Telenor		TeliaSonera	
Market	Risk	Population (M)	GDP/CAP (US $000)	Subs/ 100	CAG (%)	Prop. Subs. (M)	Equity %	Prop. Subs. (M)	Equity %	Prop. Subs. (M)	Equity %	Prop. Subs. (M)	Equity %	Prop. Subs. (M)	Equity %	Prop. Subs. (M)	Equity %
Central African Rep.	30.2	4.3	0.4	3.0	59.5	0	0	0	0	0.1	100	0	0	0	0	0	0
U.S.	91.3	305.8	45.2	83.5	12.5	0	0	28.7	100	0	0	30.2	45.0	0	0	0	0
Dominican Republic	42.7	9.8	3.8	56.5	26.5	0	0	0	0	2	100	0	0	0	0	0	0
Jamaica	40.7	2.7	4.0	93.7	19	0	0	0	0	0	0	0	0	0	0	0	0
Mexico	63.5	106.5	8.4	64.1	21.4	12.5	100	0	0	0	0	0	0	0	0	0	0
El Salvador	46.5	6.9	2.9	89.5	47.2	1.2*	99.1	0	0	0	0	0	0	0	0	0	0
Guatemala	46.6	13.4	2.5	76.0	45.1	2.1*	100	0	0	0	0	0	0	0	0	0	0
Nicaragua	31.9	5.6	1.0	37.9	55.0	0.8*	100	0	0	0	0	0	0	0	0	0	0
Argentina	42.6	39.5	6.6	102.2	43.8	13.7	100	0	0	0	0	0	0	0	0	0	0
Brazil	56.3	191.8	6.9	63.1	28.2	0	0	0	0	0	0	0	0	0	0	0	0
Colombia	50.9	46.2	3.7	73.5	49.2	0	0	0	0	0	0	0	0	0	0	0	0
Ecuador	33.7	13.3	3.3	75.6	45.2	0	0	0	0	0	0	0	0	0	0	0	0
Chile	67.9	16.7	9.8	83.7	17.4	0	0	0	0	0	0	0	0	0	0	0	0
Peru	50.4	26.8	3.9	55.3	46.2	1.1	90.6	0	0	0	0	0	0	0	0	0	0
Panama	51.9	3.3	5.9	90.1	41.8	0	0	0	0	0	0	0	0	0	0	0	0
Uruguay	44.0	3.3	6.9	90.0	42.4	0	0	0	0	0	0	0	0	0	0	0	0
Venezuela	39.3	27.7	8.2	86.1	29.5	0.2	6.9	0	0	0	0	0	0	0	0	0	0
Thailand	58.2	63.9	3.8	123.8	50.7	0	0	0	0	0.6	48.9	0	0	7.8	65.5	0	0
Bangladesh	36.1	158.7	0.4	21.7	100.0	0	0	0	0	0	0	0	0	10.2	62.0	0	0
Pakistan	41.1	163.9	0.9	38.4	106.0	0	0	0	0	0	0	0	0	14.6	100	0	0
Malaysia	63.8	26.6	6.8	87.9	20.9	0	0	0	0	0	0	0	0	3.3	50.8	0	0
India	57.8	1169.0	1.0	20.0	78.2	0	0	0	0	0.2	26.0	0.0	20.8	0	0	0	0
Japan	87.0	128.0	34.2	83.9	5.8	0	0	0	0	0	0	9.7	69.7	0	0	0	0
China	60.5	1328.6	2.5	41.2	21.6	0	0	0	0	0	0	4.0	3.3	0	0	0	0
Australia	89.6	20.7	39.6	102.5	10.9	0	0	0	0	0	0	2.6	100.0	0	0	0	0
New Zealand	87.1	4.2	30.9	101.7	11.7	0	0	0	0	0	0	1.3	100.0	0	0	0	0
Fiji	38.2	0.8	4.1	63.2	42.6	0	0	0	0	0	0	0.1	49.0	0	0	0	0

company had 117.2 million proportionate customers (88.7%) abroad. and revenues and a net loss of £30.4 billion and £9.8 billion, respectively. The company then scaled back geographically. By the end of its 2007-2008 fiscal year Vodafone operated in 24 countries abroad having departed from Sweden, Belgium, Mexico, and Japan, but entered into the Czech Republic and Fiji. The company had 260.5 million proportionate customers (92.9% abroad) and revenues and net income of £35.5 billion and £6.8 billion, respectively.

Formerly Norwegian Telecommunications, Norway's state-owned monopoly TSP, Telenor began its international expansion into Europe and Asia in 1996. At the end of 2002 Telenor had wireless operations in Norway and 10 countries abroad (Hungary, Denmark, Greece, Austria, Montenegro, Russia, Ukraine, Thailand, Bangladesh, and Malaysia). The company had 12.5 million proportionate wireless customers (80.8% abroad) and revenues of NOK 48.8 billion (NOK 20.4 billion for Telenor Mobile) and a net operating loss of NOK 320.0 million (an operating profit of NOK 1.4 billion for Telenor Mobile). During 2003-2007 Telenor continued to increase its European and Asian presence, largely through its interest in Vimpelcom, which expanded throughout the former Soviet Union. By the end of 2007, Telenor's mobile business operated in Norway and sixteen countries abroad having departed from Greece and Austria, and entered Sweden, Serbia, Kazakhstan, Uzbekistan, Tajikistan, Georgia, Armenia, and Pakistan. At the end of 2007 the company had 79.9 million proportionate wireless subscribers (96.4% abroad), and revenues and operating profits of NOK 92.48 billion and NOK 15.0 billion, respectively.

TeliaSonera, a public limited company operating under Swedish law, is a result of the merger of Telia and Sonera in December 2002. Telia was formed in 1993 when Televerket, the only provider of public voice telecommunications services in Sweden, was reorganized as a limited company and renamed. The company was privatized in June 2000. Sonera was formed in December 1997 when the Finnish Parliament approved a gradual and partial privatization of Telecom Finland, formerly the monopoly long distance and international telephone service provider operating within the Post and Telecommunications of Finland, and renamed the company. At the end of 2002 TeliaSonera operated in its home markets of Sweden and Finland as well as eleven countries in Europe and Asia (Norway, Denmark, Moldova, Lithuania, Latvia, Estonia, Russia, Kazakhstan, Azerbaijan, Georgia, and Turkey). The company had 16.5 million proportionate wireless subscribers (63.0% abroad), and revenues of SEK 59.5 billion (SEK 21.6 billion for mobile communications) and an operating loss of SEK 10.9 billion (with operating income of SEK 1.7 billion for mobile communications). During 2003-2007 TeliaSonera continued to expand its presence in Europe and Asia. By the end of 2007 the company operated in its home markets of Sweden and Finland as well as fifteen countries abroad, having also entered Spain, Ukraine, Uzbekistan and Tajikistan. TeliaSonera had 51.2 proportionate wireless subscribers (85.8% abroad), and net sales and operating income of SEK 96.3 billion (SEK 44.5 billion in mobility services) and SEK 27.5 billion (SEK 9.0 billion in mobility services), respectively.

ANALYSIS METHOD

A four-step analysis method was used in the study: (1) identify the growth, profitability, and risk characteristic variables of interest and collect the requisite 2002-2007 market and company data to measure them; (2) divide each company's market portfolio into two parts, one consisting of the home market alone and the other consisting of an aggregate of all the company's foreign markets, and calculate the yearly scores of the characteristic variables for each part; (3) calculate the difference (or distance) of the aggregate foreign markets from the home market with respect to each characteristic variable by subtracting the variable's home market score from its aggregate foreign market score; and (4) compare the results from the beginning to the end of the time period ranking the companies on their extent of internationalization, the attractiveness of their foreign market portfolios, and the distance of their foreign market portfolios from their home markets.

Five market portfolio characteristic variables were identified: two growth variables - number of cellular subscribers per 100 (market penetration) and the five-year compound annual growth rate of cellular subscribers (market maturity); two market profitability variables - GDP per 100 (wealth) and population, that is, the number of inhabitants (scale); and one risk variable – country risk, measured by Euromoney Institutional Investor's country risk indicator. This indicator, which is published in March and September each year, incorporates the following criteria: a country's political risk, economic performance, debt indicators, debt in default or rescheduled, credit ratings, access to bank finance, access to short-term finance, access to capital, and discount on forfeiting. The highest possible score is 100 (Euromoney International Investor, 2003-2008). Data sources for the growth and profitability variables included the ITU statistical indicator database (International Telecommunications Union, 2002-2008) and the World Bank (2008). (Because ITU growth and profitability data for Montenegro in 2002 was not available, Montenegro's contribution to Telenor's attractiveness with respect to those two market characteristics was neglected in the analysis of the company for 2002. The effect of this assumption on the final results is minimal, since Telenor had only 0.12 million proprietary subscribers in Montenegro in 2002 out of a total of 10.1 million proprietary subscribers in foreign markets.) Sources for the companies' financial, market, ownership, and subscriber data included the companies' annual reports and SEC 20-F filings. No attempt was made to ascertain the validity of the data reported by the companies. End of fiscal year 2002-2007 operational and financial data was used. Since Vodafone's fiscal year ends March 31, for consistency March risk data at the end of the fiscal year was used for all the companies in the study. In order to determine variable scores across the aggregate of markets in which the company operated each company's market portfolio was separated into two mutually exclusive parts: a degenerate portfolio consisting of the home country only and a second portfolio consisting of all the company's foreign markets. To not distort the size of the revenue stream and profitability from any particular foreign market, proportionate weighting based on the company's equity stake in the foreign operator (i.e., number of subscribers the company controlled in each market) was used when determining aggregate market scores (Curwen & Whalley, 2006). Since TeliaSonera has two home countries, Sweden and Finland, the company's home market characteristics were given by an average of the two countries' market characteristics weighted by the number of subscribers in each market. The difference or distance with respect to each variable between the aggregate foreign market portfolio and the home market was found by subtracting the home market's score from the aggregate score across the foreign market portfolio.

The results of the extent of internationalization analysis are presented in Table 5 below. Curwen and Whalley (2006) ranked Vodafone as the most internationalized mobile operator in the world at year-end 2005, the company having scored very high on all four measures of internationalization (number of countries, proportionate subscribers, percentage of foreign proportionate subscribers, and psychic dispersion). Telefonica also scored high and was in the top ten on all four measures. France Telecom scored very well on three measures, but along with Deutsche Telekom did not rank in the top ten with respect to percentage of foreign proportionate subscribers. Although Deutsche Telekom ranked in the top ten with re-

Table 5. TSP Rankings: 2002-2007

Extent of Internationalization												
	Number of Foreign Markets				Foreign Proportionate Subscribers (Millions)				% of Foreign Proportionate Subscribers			
Rank	2002	2007	Ch.	% Ch.	2002	2007	Ch.	% Ch.	2002	2007	Ch.	% Ch.
1	VOD 26	VOD 24	TEF 9	DT 120.0	VOD 103.9	VOD 242.0	VOD 138.1	TEF 844.3	VOD 88.7	TN 96.4	TEF 40.9	TEF 94.0
2	FT 18	FT 24	FT 6	TEF 90.0	DT 27.8	TEF 123.7	TEF 110.6	TN 662.4	TN 80.8	VOD 92.9	TS 22.7	TS 36.0
3	TS 11	TEF 19	TN 6	TN 60.0	FT 26.5	FT 87.1	TN 66.9	TS 322.1	TS 63.0	TS 85.7	FT 20.3	FT 35.0
4	TN 10	TN 16	DT 6	TS 36.4	TEF 13.1	DT 75.8	FT 60.6	FT 228.7	FT 58.0	TEF 84.4	TN 15.6	DT 27.7
5	TEF 10	TS 15	TS 4	FT 33.3	TS 10.4	TN 77.0	DT 48.0	DT 172.7	DT 53.1	FT 78.3	DT 14.7	TN 19.3
6	DT 5	DT 11	VOD -2	VOD -7.7	TN 10.1	TS 43.9	TS 33.5	VOD 132.9	TEF 43.5	DT 67.8	VOD 4.2	VOD 4.7

Attractiveness of the Markets in the Foreign Market Portfolio																				
	Macro-Risk (High to Low)				Population (M) (High to Low)				GDP/CAP (US $) (High to Low)				Subs/100 (Low to High)				CAG (%) (High to Low)			
Rank	2002	2007	Ch.	% Ch.	2002	2007	Ch.	% Ch.	2002	2007	Ch.	% Ch.	2002	2007	Ch.	% Ch.	2002	2007	Ch.	% Ch.
1	DT 91.8	DT 84.8	TEF 16.6	TEF 35.8	VOD 142.3	VOD 286.5	VOD 144.2	VOD 101.3	DT 28.5	DT 35.8	TEF 13.2	TEF 400.0	TEF 20.2	TN 86.0	VOD 20.3	VOD 30.5	TS 67.4	TN 62.5	TN 0.6	TN 1.0
2	VOD 86.7	VOD 74.0	TS 3.7	TS 7.0	DT 131.1	DT 138.5	TN 45.9	TN 90.5	VOD 22.8	VOD 23.8	DT 7.3	TS 47.1	TS 37.8	VOD 86.9	FT 21.1	FT 31.5	TN 61.9	TS 33.6	VOD -19.9	VOD -45.6
3	FT 79.9	FT 66.9	DT -7.0	DT -7.6	TEF 109.3	TN 96.6	TS 17.7	TS 29.7	FT 20.0	FT 18.9	TS 4.0	DT 25.6	TN 38.4	FT 88.1	DT 31.8	DT 45.0	TEF 54.3	FT 26.7	DT -23.6	FT -48.8
4	TN 60.6	TEF 63.0	TN -9.5	VOD -14.6	TS 59.7	TS 77.4	DT 7.4	DT 5.6	TS 8.5	TEF 16.5	VOD 1.0	TN 4.8	VOD 66.6	TEF 88.2	TN 47.6	TN 124.0	FT 52.1	TEF 26.9	FT -25.4	TS -50.2
5	TS 52.9	TS 56.6	VOD -12.7	TN -15.7	TN 50.7	TEF 68.1	FT -12.3	FT -24.6	TN 6.2	TS 12.5	TN 0.3	VOD 4.4	FT 67.0	TS 97.6	TS 59.8	TS 158.2	VOD 43.6	VOD 23.7	TEF -27.4	TEF -50.5
6	TEF 46.4	TN 51.1	FT -13.0	FT -16.3	FT 49.9	FT 37.6	TEF -41.2	TEF -37.7	TEF 3.3	TN 6.5	FT -1.1	FT -5.5	DT 70.7	DT 102.5	TEF 67.5	TEF 326.1	DT 36.8	DT 13.2	TS -33.8	DT -64.1

Distance Between the TSPs' Aggregate Foreign Market Portfolios and Their Home Markets															
	Macro-Risk (High to Low)			Population (M) (High to Low)			GDP/CAP (US $) (High to Low)			Subs/100 (Low to High)			CAG (%) (High to Low)		
Rank	2002	2007	Ch.	2002	2007	Ch.	2002	2007	Ch.	2002	2007	Ch.	2002	2007	Ch.
1	DT -0.1	DT -5.6	TEF 18.7	VOD 83.2	VOD 225.7	VOD 142.5	DT 4.0	DT -4.1	TEF -2.6	TEF -60.4	VOD -31.6	VOD -14.9	TS 67.4	TN 56.0	TEF 15.3
2	VOD -5.9	VOD -17.0	TS 3.8	TEF 68.6	TN 91.9	TN 45.8	VOD -3.9	TEF -15.8	DT -8.1	TS -50.3	TN -24.5	DT -14.1	TN 44.2	TS 27.9	DT 14.2
3	FT -11.5	FT -22.8	DT -5.5	TS 52.3	TS 69.6	TS 17.3	FT -4.4	VOD -21.1	VOD -17.2	TN -44.9	TEF -22.0	FT -4.2	FT 6.1	FT 19.2	VOD 13.2
4	TN -34.9	TEF -23.4	VOD -11.1	DT 48.5	DT 55.9	DT 7.4	TEF -13.2	FT -22.7	TS -16.9	VOD -16.7	TS -16.6	TN 20.4	TEF 3.8	TEF 19.1	FT 13.1
5	TS -39.9	TS -36.1	FT -11.3	TN 46.1	TEF 23.8	FT -14.4	TS -18.6	TS -35.5	FT -18.3	DT -1.0	DT -15.1	TS 33.7	VOD 2.6	VOD 15.8	TN 11.8
6	TEF -42.1	TN -46.4	TN -11.5	FT -9.7	FT -24.1	TEF -44.8	TN -36.0	TN -74.8	TN -38.8	FT 2.4	FT -1.8	TEF 38.4	DT -11.4	DT 2.8	TS -39.5

Key: TEF = Telefonica; DT=Deutsche Telekom; FT = France Telecom; VOD = Vodafone; TN = Telenor; TS = TeliaSonera

spect to the other measures, it only scored high on the number of proportionate subscribers. Telenor ranked in the top ten on all four measures, but only scored high on the percentage of proportionate subscribers outside the domestic market. TeliaSonera did not rank in the top ten on the number of proportionate subscribers, but ranked in the top ten on the other three measures. The company's highest ranking was on the number of countries.

Curwen and Whalley did not look at how the degree of internationalization rankings changed over time, an analysis which was possible using the 2002-2007 data collected in this study and is presented in Table 5. Following are some observations: In 2007 Vodafone ranked first (along with France Telecom) in the number of foreign markets and the number of foreign proportionate subscribers, and second in the percentage of foreign proportionate subscribers. However, because it was already so highly internationalized in 2002, Vodafone did not rank high when the change in the degree of internationalization from 2002 to 2007 is considered. In fact, Vodafone is the only TSP in the study which operated in fewer foreign markets in 2007 compared to 2002, and the company ranked sixth in the percentage increase in the number of foreign proportionate subscribers and sixth in the increase in percentage of foreign proportionate subscribers to the total number of proportionate subscribers. The company that advanced the most in the rankings from 2002 to 2007 was Telefonica. In 2002 it ranked fifth in the number of foreign markets in which it operated, fourth in the number of foreign proportionate subscribers, and sixth in the percentage of foreign proportionate subscribers. Although it still ranked third in the number of foreign markets in 2007, it had added the most new foreign markets (representing the second highest percentage increase), added the second highest number of foreign proportionate subscribers (representing the largest percentage increase in this category), and ranked first in the change in percentage of foreign proportionate subscribers (representing the highest percentage increase in this category as well.) France Telecom, which ranked second to Vodafone in both 2002 and 2007 in the number of foreign markets, dropped from fourth to fifth in the percentage of foreign proportionate subscribers, although it still ranked third in the number of foreign proportionate subscribers. Deutsche Telekom, which had ranked sixth in 2002 in the number of foreign markets, second in the number of foreign proportionate subscribers, and fifth in the percentage of foreign proportionate subscribers, lost ground in 2007 dropping to fourth in the number of foreign proportionate subscribers and sixth in the percentage of foreign proportionate subscribers. However, its addition of six foreign markets represented the largest percentage increase in that category. Telenor also added six foreign markets from 2002 to 2007, ranking third in both change and percentage change with respect the number of foreign markets. In doing so, the company experienced a substantial increase in the number of foreign proportionate subscribers, ranking third among the six companies in the increase in the number of foreign subscribers over the time period (second in percentage increase). In 2007 Telenor ranked first in the percentage of foreign proportionate subscribers. Finally, like Telenor, TeliaSonera had a relatively low base of foreign proportionate subscribers in 2002. While it ranked fifth in the number of foreign markets added from 2002 to 2007 and sixth in the number of foreign subscribers added, TeliaSonera experienced the second largest change, and second largest percentage change, in the percentage of foreign proportionate subscribers.

In terms of the growth, profitability, and risk characteristics considered in this paper, an attractive market is one possessing high growth (high 5-year CAG, low number of subscribers per 100), high profitability (large population, high GDP per capita), and low risk (high country risk score) (Abell & Hammond, 1979). In Table 5 the TSPs are ranked according to their scores with respect to each characteristic in descending order of attractiveness. For example, in 2007 Vodafone's foreign market portfolio ranked higher in attractiveness with respect to the risk characteristic than Telefonica's, but lower than Deutsche Telekom's, since the aggregate country risk score of Vodafone's foreign market portfolio in 2007 is higher than Telefonica's (74.0 versus 63.0), but lower than Deutsche Telekom's (74.0 versus 84.8).

Table 5 indicates that on an aggregate basis Vodafone's foreign market portfolio in 2007 ranked second in attractiveness with respect to country risk, first with respect to population, second with respect to GDP/capita, second with respect to subscribers per 100, and fifth with respect to 5-year subscriber CAG. Except for subscribers per 100, where it jumped from fourth to second, its rankings changed very little from 2002, although the aggregate scores did - improving significantly in population, but worsening in growth and risk. Deutsche Telecom's rankings in 2007 remained the same as 2002: high in country risk and profitable markets, but low in growth. In fact, on an aggregate basis its markets became more attractive from a profitability perspective, but less attractive from a growth and risk perspective. Another company whose rankings remained relatively invariant from 2002 to 2007 is France Telecom, which retained its ranking of third in risk, sixth in population, and third in GDP/capita, but rose in its growth ranking (from fifth to second in subscribers per 100 and fourth to third in 5-year CAG). However, considering aggregate score France Telecom's markets decreased in growth and also became riskier and less profitable. Telenor also improved its growth ranking (from third in subscribers per 100 to first and from second to first in 5-year CAG) and in population from fifth to third rank. However, the company's risk ranking worsened from 2002 to 2007 (from fourth to sixth) as did its GDP per capita ranking (from fifth to sixth) ranking. Although Telenor's markets also became riskier on the basis of aggregate score, they became more profitable, especially on the basis of population. However, while Telenor's 5-year CAG actually increased, the only company of the six to experience an increase in 5-year CAG, this was offset considerably by a significant increase in market penetration. On the other hand, Telefonica's and TeliaSonera's growth rankings worsened: Telefonica fell from first in 2002 in subscribers per 100 to fourth in 2007 and from third in 5-year CAG in 2002 to fourth in 2007; and TeliaSonera dropped from second in 2002 in subscribers per 100 to fifth in 2007 and from first in 5-year CAG in 2002 to second in 2007. The decrease in attractiveness with respect to subscribers per 100 for both companies on an aggregate basis was substantial. While Telefonica's population rank also fell from 2002 to 2007 (from third to fifth), it did improve its other profitability measure, GDP per capita (from sixth to fourth), as well as its risk rank (from sixth to fourth). On the basis of aggregate score the attractiveness of Telefonica's markets with respect to risk and GDP per capita improved noticeably, but worsened significantly with respect to growth and population. Although TeliaSonera retained its ranking of fourth in population from 2002 to 2007, its ranking in GDP per capita dropped from fourth to fifth, and its risk ranking failed to improve from fifth. On the basis of aggregate score TeliaSonera's risk did not change very much from 2002 to 2007, although its profitability characteristics improved and its growth characteristics worsened.

Table 5 also ranks the six TSPs in 2002 and 2007 in terms of the difference (or distance) between the companies' foreign market portfolios and their home markets (Ghemewat, 2007). Distance is measured with respect to a given market characteristic and is calculated here by subtracting the score of the company's home market from the aggregate score of the foreign market portfolio. The distance between a company's foreign market portfolio and its home market with respect to a particular attractiveness characteristic is a dynamic quantity which depends on two variables that can change over time. As was the case with the TSPs' foreign market portfolios, which varied in their attractiveness from company to company and which changed from 2002 to 2007, the characteristics of the TSPs' home markets (Spain, Germany, France, UK, Norway, and Sweden/Finland) differed and changed over time. For example, in 2002 Germany's population was quite a bit higher than the other markets, while the populations of Sweden/ Finland and Norway were considerably lower; Spain had a lower GDP per capita than the other markets; France's number of subscribers per 100 was quite a bit lower than the other countries; and the 5-year CAG of subscribers in Spain was the highest of the home markets, closely followed by France, while the 5-year CAG of the UK, Norway, and Finland/Sweden was a good deal lower. On the other hand, in 2002 the country risk characteristics of all the home markets were very similar. From 2002 to 2007 there was a significant increase in the markets' GDP per capita and number of subscribers per 100, as well as a decrease in their subscriber 5-year CAG, although the country risk and population of the home markets remained fairly constant.

In Table 5 companies were ranked with respect to a given market characteristic according to the size of the difference between their aggregate foreign market portfolio and home market scores in order of decreasing attractiveness, that is, descending order of distance with respect to risk, population, GDP per capita, and 5-year CAG (the four market characteristics which are more attractive the higher the score) and ascending order with respect to subscribers per 100 (the only market characteristic which is more attractive the lower the score). Note that the distance between a company's foreign market portfolio and its home market is negative with respect to a particular attractiveness characteristic if the home market is more attractive than the aggregate foreign market portfolio with respect to that characteristic.

Table 5 shows that, for the most part, the distance rankings of the six TSPs were the same as the companies' attractiveness rankings. Some notable exceptions occurred in the population rankings, where Deutsche Telekom was lower ranked in distance than attractiveness, the GDP per capita rankings, where TEF was higher ranked in distance than attractiveness, and subscribers per 100, where France Telecom was ranked lower in distance than attractiveness. Telefonica's risk distance score changed the most among the six TSPs from 2002 to 2007 and did so in the positive direction—unlike all the other companies except TeliaSonera. Vodafone, France Telecom and Telenor experienced the largest decreases in risk distance. The population distance scores changed significantly from 2002 to 2007 with Vodafone experiencing an extremely large positive change. The positive change for Telenor was smaller, but still significant, as was the negative change for Telefonica. From 2002 to 2007 the changes in the GDP per capita distance scores were all negative, Telenor experiencing the largest negative change, Telefonica the least. The changes in the number of subscribers per 100 distance scores showed more variability than the other characteristic distance scores from 2002 to 2007: three companies' distance scores increasing in attractiveness and three decreasing in attractiveness. Vodafone's and Deutsche Telekom's subscriber per 100 distance scores experienced the largest increase in attractiveness; Telefonica and TeliaSonera the largest decrease. On the other hand, all the companies except TeliaSonera increased their 5-year CAG distance scores; TeliaSonera's reduction in the score being a significant one.

CONCLUSION AND DIRECTIONS FOR FUTURE RESEARCH

This paper has addressed the international expansion efforts in the wireless telecommunications industry from 2002 to 2007 of six European telecommunications service providers, Telefonica, Deutsche Telekom, France Telecom, Vodafone, Telenor and TeliaSonera, all of which operate in highly competitive, mature home markets. The six companies have pursued different internationalization strategies, a fact which is reflected in differences in the growth, profitability, and risk characteristics of their foreign market portfolios. Vodafone pursued a global strategy, with investments in Asia, Australia, Africa, the United States, and Europe. Although it has been contracting its footprint somewhat, it has increased its investments in emerging high-growth, large-sized markets like India. Overall, in 2007 it was the most internationalized of the six companies: tied for first in terms of the number of foreign markets (24); first in the number of foreign proportionate customers (242.0 million); and second in the percentage of foreign proportionate subscribers (92.9%). From 2002 to 2007 both Telefonica and France Telecom continued to pursue expansion strategies into less developed, high-growth, high-risk foreign markets with a cultural affinity to their home markets—in Telefonica's case, Latin America, in France Telecom's, French-speaking Africa—while also investing in slower growth, less risky developed markets in Europe, their home region. Telefonica, which added ten new markets and increased its percentage of foreign proportionate subscribers in 2007 to 84.4% from 43.5% in 2002, increased its extent of internationalization the most during the time period. Compared with Telefonica, France Telecom's internationalization was less dramatic, changing less in the number and percentage of foreign proportionate subscribers, although in 2007 the company was tied for first in the number of foreign markets. From 2002 to 2007 Telenor and TeliaSonera also continued to expand into less developed, high-growth, high-risk foreign markets, principally in Eastern Europe and Asia, although these markets did not have a great deal of cultural affinity to either Telenor's or TeliaSonera's home markets. The companies also expanded selectively into less risky developed markets in Western Europe. Although Telenor added fewer foreign markets to its portfolio than Telefonica from 2002 to 2007 (six compared with nine), like Telefonica it experienced a very large increase in the number of foreign subscribers, and in 2007 ranked first in the percentage of foreign proportionate subscribers (96.4%). From 2002 to 2007 TeliaSonera added fewer foreign subscribers than the other five companies, although it did rank third in 2007 in the percentage of foreign proportionate subscribers. Finally, Deutsche Telekom stayed close to home in its expansion efforts, expanding into Eastern Europe from 2002 to 2007. Although Deutsche Telekom did not rank very high in 2007 in any of the internationalization categories, it did rank first in the percentage change of foreign markets added (120.0%) from 2002 to 2007.

The paper also investigated the attractiveness of the companies' foreign market portfolios from a growth, profitability, and risk perspective. Attractive markets are high growth, very profitable, and possess little risk. Unfortunately, very few markets fit this description entirely. High growth markets are often emerging markets, which are relatively risky and poor. As markets mature, their subscriber density increases and 5-year CAG decreases, offsetting gains in country risk and GDP per capita that generally accompany maturity. In 2002 Vodafone's foreign market portfolio was positioned in relatively mature, low-risk, but low-growth markets. However, with additional investments in high-growth markets such as India during 2002-2007, at the end of 2007 Vodafone seemed to possess the most balanced foreign market portfolio with respect to growth, market profitability, and risk of the six companies in the study. With its investment in O_2, Telefonica, on the other hand, improved its rather weak 2002 profitability and risk position, but at the expense of growth. In 2002 France Telecom was in a similar position as Telefonica, although with a less risky, but low market size (population) portfolio of foreign markets. However, with additional investments in Africa among other places, by 2007 its risk and market size position had worsened, although with respect to growth its position was equivalent to Telefonica's. In 2002 Telenor and TeliaSonera scored low in risk, although not as low as Telefonica, relatively low in the profitability characteristics (population and GDP per capita), but very well in the growth characteristics (subscriber per 100 and 5-year CAG). With its continued expansion into Europe and Asia from 2002 to 2007, Telenor's risk position worsened, but its profitability position with respect to population improved and it improved its position relative to the other companies with respect to growth. On the other hand, TeliaSonera's risk position remained the same, its profitability position with respect to population and GDP per capita improved, but its growth position worsened, especially with respect to subscribers per 100 for which the company dropped from second to fifth place. Deutsche Telekom, on the other hand, was in an excellent risk and relatively strong market profitability position in 2002, but a very poor growth position. After making several investments in Eastern Europe, in 2007 the company was still in a relatively strong position with respect to market profitability, but its growth position had not improved, and its risk position had worsened, although it was still the strongest of the six companies in the study.

Finally, the paper investigated the distance in 2007 compared with 2002 between the TSPs' foreign market portfolios and their home markets with respect to the growth, profitability, and risk characteristics described above. Although the TSPs' home market characteristics generally all moved in the same direction (their populations grew slowly, their riskiness remained low, their GDP per capita increased, their number of subscribers per 100 increased significantly, and their five-year subscriber CAG decreased significantly), the size of the change in distance from 2002 to 2007 between the TSPs' foreign market portfolios and their home markets varied from company to company depending upon the specific characteristics of each company's home market and the markets in the company's foreign market portfolio. For example, from 2002 to 2007 the distance with respect to the number of subscribers per 100 increased in the direction of greater attractiveness for Vodafone and Deutsche Telekom, changed little for France Telecom, but decreased significantly in the direction of lesser attractiveness for Telefonica, TeliaSonera, and Telenor. However, there was some uniformity in the direction of distance change from 2002 to 2007 with respect to certain market characteristics: for example, the distance increased in the negative direction of attractiveness with respect to GDP per capita for all six companies; the distance increased in the positive direction of attractiveness with respect to five-year subscriber CAG for five companies; and the distance increased in the positive direction of attractiveness with respect to population for four companies.

There are several possible directions for future research, four of which will be mentioned here. One direction for future research is to collect more data on the six TSPs investigated in this paper so that they can be studied over a longer time period. A second research direction would be to enlarge the sample of TSPs, expanding it beyond Europe. Prominent wireless service providers that might be examined include MTS, America Móvil, and NTT DoCoMo. A third direction for future research is to investigate the companies' strategies and portfolios in greater detail and to analyze their effect on performance. The framework introduced by Ghemewat (2007), which was described in the introduction, can be helpful in analyzing the strategies. Among the market attractiveness and competitive strength factors that can be investigated are degree of competition, regulation, market demographics, economic factors, market share, and technological capability. Lastly, a fourth research direction is to investigate whether other weighting schemes and distance measures yield the same results as those obtained in this analysis.

REFERENCES

Abell, D. F., & Hammond, J. S. (1979). *Strategic market planning: Problems and analytical approaches*. Englewood Cliffs, NJ: Prentice-Hall.

Curwen, P., & Whalley, J. (2006). Measuring internationalization in the mobile telecommunications industry. *International Business Review*, *15*, 660–681. doi:10.1016/j.ibusrev.2006.09.002

Deutsche Telekom. (2002-2007). *Annual and 20-F reports*. Bonn, Germany: Author.

Dunning, J. H., & Lundan, S. M. (2008). Multinational enterprises and the global economy (2nd ed.). Cheltenham UK: Edward Elgar.

Euromoney International Investor. (2003-2008). *Country risk–results & methodology*. London: Author.

France Telecom Orange. (2002-2007). *Annual and 20-F reports*. Paris: Author.

Gerpott, T., & Jakopin, N. (2005). The degree of internationalization and the financial performance of European mobile network operators. *Telecommunications Policy*, *29*, 635–661. doi:10.1016/j.telpol.2005.06.003

Ghemawat, P. (2007). *Redefining global strategy*. Boston: Harvard Business School Press.

Ghoshal, S. (1987). Global strategy: An organizing framework. *Strategic Management Journal*, *8*, 425–440.

International Telecommunications Union. (2002-2008). *Indicators*. Retrieved from http://www.itu.int/ITU-D/ict/statistics

Rugman, A. M., & Verbeke, A. (2004). A perspective on regional and global strategies of multinational enterprises. *Journal of International Business Studies*, *35*, 3–18. doi:10.1057/palgrave.jibs.8400073

Sarkar, M. B., Cavusgil, S. T., & Aulakh, P. S. (1999). International expansion of telecommunications carriers: The influence of market structure, network characteristics, and entry imperfections. *Journal of International Business Studies*, *30*(2), 361–382. doi:10.1057/palgrave.jibs.8490074

Steinbock, D. (2003). Globalization of wireless value system: From geographic to strategic advantages. *Telecommunications Policy*, *27*, 207–235. doi:10.1016/S0308-5961(02)00106-4

Telefonica and Telefónica Moviles. (2002-2007). *Annual and form 20-F reports*. Madrid, Spain: Author.

Telenor. (2002-2007). *Annual and form 20-F reports, 2002-2007*. Fornebu, Norway: Author.

TeliaSonera. (2002-2007). *Annual and form 20-F reports, 2002-2007*. Stockholm, Sweden: Author.

Vodafone. (2002/2003-2007/2008). *Annual and form 20-F reports*. Newbury, UK: Author.

World Bank. (2008). *World development indicators database*. Washington, DC: Author.

This work was previously published in International Journal of Interdisciplinary Telecommunications and Networking, Volume 1, Issue 4, edited by Michael R. Bartolacci, and Steven R. Powell, pp. 16-36, copyright 2009 by IGI Publishing (an imprint of IGI Global).

Chapter 14
Revisiting the Gatekeeping Model:
Gatekeeping Factors in European Wireless Media Markets

Vassiliki Cossiavelou
University of the Aegean, SOC/CI, Greece

Philemon Bantimaroudis
University of the Aegean, SOC/CI, Greece

ABSTRACT

Mediation in news industries has received significant attention by researchers for more than half a century. Gatekeepers decide which information should be delivered to different audiences. The Shoemaker/Reese Gatekeeping Model identifies five different filters of content processing: individual influences, professional routines, the organization, extra-media influences and ideology. Journalism practices, intra-organization and extra-media-related procedures and strategic alliances, including culture and ideology, add more complexity in the contemporary globalized media landscape. Gatekeeping is being processed through out all the above mentioned pillars. ICT technologies related to the media have influenced the interactivity among the pillars and wireless technologies have influenced the digital media landscape. The European Union has experienced dramatic changes in its regulation environment and spectrum resources allocation. In this paper, the authors examine the impact of wireless technology on gatekeeping practices in the context of EU news markets.

INTRODUCTION

For more than 50 years, gatekeeping has been an influential paradigm in the field of communication theory. The term "gatekeeping" has been coined by Kurt Lewin (1947) and ever since it has been utilized by various professional fields and academic disciplines. The literature describes gatekeepers as individuals who decide what kind and how much information should be allowed to be delivered to a particular audience. A seminal gatekeeping study is attributed to David Manning

DOI: 10.4018/978-1-60960-505-6.ch014

White who applied Lewin's concept in the field of journalism. White's (1950) study, also known as the "Mr. Gates" study, examined the personal news selection habits of a newspaper editor. White observed what news items were thrown away and which were retained. The editorial criteria, that determined what constituted newsworthiness, became subject to observation. This gatekeeping process analyzed by White was limited to one individual editor and his news selection practices.

Since the 1960s, the gatekeeping process has been refined and expanded to encompass not only information but also, raw materials, goods, services, technologies, and so forth. The definition of the "gatekeeper" does not pertain only to the individual but includes institutions and organizations of different kinds—even government bodies and international authorities. In the context of media communication, gatekeepers make quantitative as well as qualitative decisions that determine the flow of information and thereby influence public knowledge, perceptions and even behaviours related to different issues. Thereby, people's understanding of world news has been dependent on global media gatekeeper decisions on what constitutes news and how much of the available news content should become public.

Gatekeeping is a complex, multidimensional process involving different players at different levels who become subject to various institutional influences. Analyzing this interplay of functions requires a systemic understanding and knowledge of the institutions. Gatekeeping is useful for understanding different communication processes at different stages, from message creation to dissemination and public consumption. Such an assessment requires a systemic conceptualization combined with available empirical evidence to enhance scholar understanding of contemporary media industries in a global digital, and even more, wireless environment.

THE GATEKEEPING LITERATURE

The "Mr. Gates" study focused on the news filtering criteria of the individual – the editor. Editors exercise a great deal of power in deciding what constitutes news. Manning described his editor's decisions as "highly subjective." One-third of the rejected news items were discarded because the editor decided they were not suited for publication, while the other two-thirds were rejected because of limited newspaper space (White, 1950, p. 386). In that case study, the editor rejected 90% of the wire news that reached him, demonstrating the degree of authority a newspaper editor exercises on what news should become public. Similar studies examining the role of individual news editors displayed similar results (Cohen, 2002; Dimitrova, Connolly-Ahern, Williams, & Paul, 2003; Hollifield, Kosicki, & Becker, 2001; Peterson, 1981; Plaisance, & Skewes, 2003; Singer, 1997; Snider, 1967; Wanta & Craft, 2004). Those initial gatekeeping studies emphasized the ability of the individual gatekeeper to shape news content. Later works, however, demonstrated that gatekeeping is a complex process involving different actors and practices. According to Dimitrova et al. (2003), "some practices that reduce uncertainty in making news decisions include: accepting the news definition of opinion leaders within a newsroom or on a particular beat; adopting of a group consensus through daily professional interaction; keying on output of a reference institution, such as the AP or The New York Times; accepting key sources' definition of news; and using attitudes and values of reference groups other than those in the newsroom" (p. 402). Researchers realized that individual gatekeeping decisions should be examined in the context of institutional as well as cultural environments. Thus, the gatekeeping research tradition encompassed research questions and hypotheses superseding the narrow cycle of the individuals.

While outlining the origins of the media sociology tradition, Reese and Ballinger (2001) argued that the European media research tradition has favoured analyzing institutions, societies and ideologies in contrast to the American tradition which traditionally focused on individuals and practices (p. 641). Thereby, they suggested that a holistic research approach to gatekeeping should examine evidence from the micro to the macro level. The first significant deviation from analyses focused on individuals is Warren Breed's (1955) work entitled "Social Control in the Newsroom." The author examined processes of socialization in a newsroom environment. Breed's work demonstrated how an organization acculturates its employees on accepted and unaccepted behaviours and practices. The socialization process is subtle, not direct, but organizations employ different means such as "blue-pencilling—teaching reporters which objectionable phrases to omit in the future, occasional reprimands, internal house organ papers, and rare explicit policy decisions" (Breed, 1955). Breed discovered that reporters were "sensing policy" and conformed to the cultural standards of the organization.

Although the initial gatekeeping research dealt with newspaper newsrooms, subsequent work examined gatekeeping processes in television (Berkowitz, 1990; Carroll, 1985) and recently the Internet (Dimitrova et al., 2003; Singer, 1998). A recent trend toward digital media analysis is expected and justified, as traditional media, print, sound and spectacle, converge in digital media platforms. Thereby, scholars reassess gatekeeping models and processes while generating evidence derived from digital media industries, and especially wireless media outlets.

A comprehensive gatekeeping model by Shoemaker and Reese (1996) has successfully mapped gatekeeping processes outlying five dimensions of content filters: individual influences, professional routines, organizational influences, extra-media influences and the ideological environment. The authors in their influential book "Mediating the Message" generated convincing evidence in regards to the five dimensions that affect the gatekeeping process.

While assessing influences on media content derived from individual factors Shoemaker and Reese (1996) argued "the communicators' characteristics (such as gender, ethnicity, and sexual orientation) and their personal backgrounds and experiences (such as religious upbringing and their parents' socioeconomic status) not only shape the communicators' personal attitudes, values and beliefs, but also direct their professional background and experiences (such as whether the communicator goes to journalism or to film school). These professional roles and ethics have a direct effect on mass media content, whereas the effect of personal attitudes, values and beliefs on mass media content is indirect, operating only to the extent that the individuals hold power within their media organizations sufficient to override professional values and/or organizational routines" (p. 65).

In the second level of influences, Shoemaker and Reese (1996) describe journalistic routines and their influence on content. For example, reporters have developed a set of news values that govern news selection and the framing of the stories. Also defensive routines such as "objectivity" allow journalists to protect themselves from accusations regarding their perceived biases. Story structures and narratives are routinized. News stories have the same structure and format. Furthermore, the pack routine—journalists covering stories together and comparing notes—as well as deadline requirements force journalists to work within certain boundaries (1996, p. 105).

As already discussed, Warren Breed's (1955) work is important for understanding organizations and their influence on media content. Organizations do not always need to define explicitly what constitutes acceptable and unacceptable behaviour. However, there are indirect ways to reward "appropriate" coverage and punish "inappropriate" journalistic behaviour. Every media organization is unique. They have distinct images. They enforce diverse policies, and they promote diverse sets of priorities.

In terms of extra-media influences, Shoemaker and Reese (1996) refer to interest groups, public relation campaigns, advertisers, government controls, the market competition, and the influence of technology. All these elements affect media organizations and their content selection priorities (1996, p. 175).

Finally, ideology—"the symbolic mechanism that serves as a cohesive and integrating force in society" (Shoemaker & Reese, 1996, p. 221)—influences media content in various ways. Among those, Hall (1982) recognizes "the media's ability to 'define situations' and label groups and individuals as 'deviant'" (p. 307). In this context, the works of Gitlin (1998, 2003, 2007) and Chomsky and Herman (2002) are geared toward ideology. They present evidence of ideological influences on media content—a bias that exists to support established institutions and their interests.

A NEED FOR MODEL UPGRADING

The design of the Shoemaker and Reese model reflects the media environment prior the age of the Internet and the other new media, even more, prior the wireless news media. Although in the 1950s, media technologies, such as electronic newsgathering tools, changed the news media industry in a profound way, the model does not identify technology as a distinct pillar. Neither this paper suggests, as is considering "the recently involved" gatekeeping factor as influential every traditional level of analysis. Arguably, the nature of analogue technologies provided a great deal of gatekeeping power at the source level rather than at the receiver's. Media sourcesindividuals and organizations exercised absolute authority over the collected news content and subsequently determined how much of news material should become public. Indeed, all information was received, filtered, processed and disseminated in the realm of established media organizations. Thereby, the Shoemaker and Reese model did not need to identify technology as a separate filtering factor. Analogue technology has been controlled at large by institutions employing carefully socialized individual gatekeepers. Only in rare cases, individuals with audiovisual equipment would record and make public information not collected by professional reporters affiliated with news organizations.

Williams and Delli Carpini (2004) argue that digital media have practically rendered the gatekeeping paradigm inapplicable since media elites have no control on the flow of information toward their publics. "The new media environment creates a multiplicity of gates through which information passes to the public both in terms of the sheer number of sources of information (i.e., Internet, cable television, radio), the speed with which information is transmitted, and the types of genres the public uses for political information (i.e., movies, music, docudramas, talk shows)" (Williams & Delli Carpini, 2004, p. 1213). While discussing the challenge that blogs present to mainstream journalism, Singer (2005) argues that since almost anyone can disseminate information on the Web, the role of professional journalists shifts from "story selection and toward news judgement" (2005, p. 179). Although there is increasingly a trend towards advocacy journalism, Singer argues that mainstream elite media retain their gatekeeper function while blogs offer additional challenges to professional journalists, most notably the values of "accountability and transparency" (2005, p. 179).

Robinson (2006) argues that blogs run by mainstream media constitute a professional response toward independent news blogs. In such on-line settings, professional journalists exercise a degree of authority in terms of news control, paradigm repair while defending the values of the journalistic profession. Nevertheless, Robinson acknowledges a continuous change of journalism as she states that "blogs are a new form of journalism destined to evolve as time and broadband capabilities advance" (2006, p. 80).

In the current project, the authors argue that a sixth factor needs to be added to the Shoemaker and Reese Gatekeeping Model. The digital world has shifted dramatically every principle of gatekeeping taught in journalism schools up to the 1990s. Nowadays, media institutions cannot exercise any control whatsoever on what type and quantity of information is collected and subsequently launch it on the Web or in wireless mobiles. Any individual, group, non-media institution or national entity, which has been traditionally out of the media agenda, can tell their story without traditional mediators intervening. There is no doubt that in the digital world, technology is one of the primary factors that empowers citizens outside the established media organizations to promote their version of reality. The wireless technology is making this potentiality to have the attributes of omnipresence, and therefore giving to the user the features of eyewitness and ear witness.

However, digital technology, and wireless technology in particular, does not represent a separate, distinct factor in the previous map of gatekeeping filters. Technology represents an additional gatekeeping factor affecting each one of the five established dimensions. Furthermore, the revised model does not describe those five filters only as media sources but as nodes of media networks becoming sources or receivers or both, and in every possible interconnection direct or indirect. It becomes apparent that a new generation of empirical evidence is needed to assess how digital and wireless platforms (maybe innovations and their implementations) interact with each filter.

At the individual level, a multitude of technologies are expected to affect individual content creation and consumption behaviours. All citizens, consumers, users of the world have editor—gatekeeper capacities as they manage Semantic Web metadata, become members of social networks like Facebook, Twitter, and so forth, or uploading files utilities, such as Flickr, YouTube, and so forth, engage in distance learning and master Web authoring techniques, they use cognitive and language/speech processing technologies, and so forth. Individuals are not bound by organizational pressures on what to select unless they are media professionals in established organizations. Even in those cases their flexibility in selecting and comparing sources is unprecedented in the history of time. Technology does not minimize individual characteristics addressed in the traditional gatekeeping model. As a matter of fact gender, cultural background, education, and religions constitute significant factors in current content selections. However, the variety of content available, the ability to verify sources of information and the manner individuals interact with the content and other individual's leads to the creation of a digital techno-culture and new generations of gatekeepers (see Figure 1).

Professional routines have also been affected by technology. Although professionals still apply established professional practices, a set of new professional demands is presented to new professionals. Familiarization with ICT technologies affects all stages of content gathering, packaging and dissemination. Furthermore, digital technologies lead to new professional practices, professional standards at every stage of news making and at codes of ethics, as well. For example, if anyone can present his/her version of reality what is the effect on the objectivity routine, one of the most traditional values in western journalism? Although organizations still socialize their personnel "teaching" them what constitutes appropriate behaviour, they realize their employees are socialized by technology, as well (see Figure 1).

Furthermore, the organization relies on technology savvy individuals to carry out its mission. The proliferation of digital media limits the organizational degree of control on the information. What an organization chooses to discard, various

Figure 1. Revisiting the shoemaker and reese media gatekeeping model

competing outlets may choose to publicize. Furthermore, organizational dependence on technology creates a constant urgency in establishing standards and values in order to maintain a strategic advantage. The ownership patterns are affected too by technology adoption—the transition to IPTV, MVB, podcasting, mobile casting and so forth. affected the subscription models and policies, the broadcasting rights management emerge as a new factor of those policies, and the global nature of joint ventures among media industries, but of adjacent industries, as well, turn the M&A schemas subject of their R&I strategies (see Figure 1).

In the fourth dimension, influences originating from extra media sources have proliferated. Not only advertisers, public relations officers, governments and lobbyists but also ICT corporations, software and hardware designers as well as Research, Development & Innovation firms have mounted unprecedented pressures to adopt new means of communications and to make choices about technologies of the future. Not only established institutions but informal networks of Internet surfers put additional pressures as their behaviours diminish the time previously allocated for mainstream news media consumption, storage and retrieval. In the context of the media industry, technology-driven national regulatory agencies or international organizations such as WTO, ITU, EU, FCC, ICANN have become dominant players as the affect media industries and their related lobbies (see Figure 1).

Finally, ideology remains a significant factor combined with the diffusion of digital media. For the first time in history, the average citizen-user has global exposure and interaction. Multiculturalism and multilingualism are both widely experienced as citizens easily interact with one another. In this global context, a significant gatekeeping factor remains the rate of new technologies adoption. ICT products need to conform to global standards supporting more than one language and training personnel in dealing with culturally diverse consumers (see Figure 1).

The current revised set of influences represents examples of digital gatekeeping processes. Empirical evidence is needed to map out accurately how those gatekeeping processes are affected by ICT technologies. Both qualitative evidence assessing different roles of individuals and the routines they employ in the context of modern newsrooms, as well as quantitative evidence derived from organizational and social environments are necessary to establish revised relationships among the five factors and digital and wireless technology as a new emerging gatekeeping factor (see Figure 2).

Figure 2. Proposal for the revised gatekeeping model

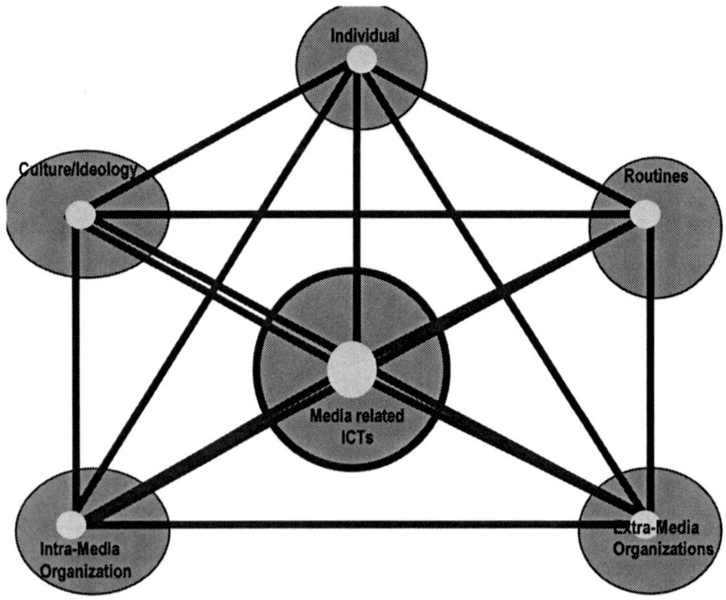

Methodology

The central research question of the current project is as follows:

RQ: *To what extent digital technology, including wireless technology, affects content selection processes as information goes through five different filters, via changing the attributes of each pillar and of its interdependence on each other?*

The authors argue that although the Shoemaker/Reese gatekeeping model was constructed during an era of technological significance—electronic news gathering and satellite feeds had a tremendous impact on the packaging and flow of information —the current digital environment needs to be accounted for as a significant factor in an upgraded gatekeeping model suited for current analyses. The authors however, do not propose a separate technological dimension added onto the previous five-dimensional model, but rather a technological factor interplaying in distinct ways with every one of the five gatekeeping dimensions described by Shoemaker and Reese.

To empirically support this proposal, the authors conducted a global on-line survey of news gatekeepers representing a wide spectrum of organizations. Those institutions were targeted because of their significant status as global gatekeepers as well as their expertise on current ICT platforms widely used for the management and dissemination of information. An online questionnaire was designed and uploaded at the following URL: http://www.cossiavelou.gr.

The questionnaire was sent all over the world to more than 5000 organizations such as International News Media, National News Agencies, Media Content Providers, Designers, Information Science & Technology corporations, Olympic Games Authorities, Broadcasting Authorities, International News Correspondents, Embassies, European Union Institutions, Wireless Service Operators, Law & Regulation Authorities and Universities. The gatekeeper concept was not treated in narrow fashion, in the context of media workers (journalists and editors only) but it encompassed different institutions, as both media and extra-media workers have gatekeeping capacities. As already established in the literature, virtually anyone can acquire gatekeeper capacities, which represents a challenge for professional gatekeepers.

The online database collected data from 230 participants working in 28 different countries, 86 of which hold a PhD, and their affiliations included all the above mentioned industries, where academia, Telecom operators and CE providers and the global media corresponded about the 60% of the replies and the rest is concerning research centres and institutions, EU organs, governmental bodies, international ICT experts' organizations and corporations, news agencies, journalistic unions and reporters, media groups and broadcasters, R/TV Channels (national or regional), print media, new media, technology clusters, law and regulation groups, Olympic organizers and broadcasters and cultural industries.

The questionnaire was distributed electronically by direct e-mail. A database kept detailed records of all returned questionnaires and was updated automatically. The database removed automatically respondents who replied from the pool of those targeted. Respondents, who indicated they did not wish to participate in the survey, were automatically removed from the list. Repeated, electronic messages were sent in a time frame of a year and a half, informing a pool of ICT, media and telecom specialists of the survey and urging for their participation. The database built to support the online questionnaire kept accurate records of the overall visitation, the organization affiliated with every visitor and the duration of each visit.

The questionnaire was designed to include six sets of questions addressing each of the five levels of analysis plus the proposed addition, while assessing how technology affects news selection functions at each level. The questionnaire incorporated the Likert scale as a standard measurement device utilized in similar descriptive survey projects. A convenience sample of 230 respondents representing an array of experts participated in the survey. Thereby, a generalization of the findings to a larger population is not intended but rather establishing new gatekeeping principles and needs, which characterize trends in the global news media industry.

ON LINE RESEARCH FINDINGS

Findings on All Media Gatekeeping Pillars

Because of the descriptive nature of the current research, frequency counts and cross-tabulations were deemed adequate measures of analysis. As already argued the intent of the current study is not generalizing results to a larger population, but rather generating initial evidence of the impact of digital technology, on modern gatekeepers. The current project in particular deals with wireless technologies.

The current project dealt with the five levels of analysis as described in the Shoemaker/Reese model in order to assess its applicability in the contemporary digital environment. As respondents were asked to display their degree of agreement with research statements pertaining to gatekeeping functions from the individual (micro level) to the ideology (macro level).

THE INDIVIDUAL LEVEL

The authors attempted to assess whether individuals employed in newsrooms need a continuous updating of their ICT capabilities. This statement assesses the need for additional training and capabilities that modern gatekeepers need to possess beyond traditional qualifications, such as writing and research skills. They asked respondents to display their degree of agreement with the following statements.

H-1: *Career evolution into the media organizations is related to the amount of continuous training of the personnel working in news-gathering and news production on new digital, especially wireless and mobile technologies.*

This is a significant statement because it defines individual roles in the context of the newsroom. Assessing whether gatekeepers should rely on continuous updating of ICT skills and knowledge defines the kinds of gatekeepers that modern newsrooms search to offer employment to. Prior to the 1990s, gatekeepers with strong writing skills and relative depth of knowledge would possess adequate credentials to work for the media.

The respondents had six choices: N/E (Non Expertise, meaning they did not feel qualified to answer), strongly disagree, disagree, neutral, agree and strongly. From those who responded, 1.6% strongly disagreed, 10.9% disagreed, 31.1% were neutral, 30.6% agreed and 9.8% strongly agreed with the statement. 40.4% of the respondents acknowledged that continuous ICT training is essential for individuals working as gatekeepers.

H-2: *The news industry professionals are increasingly dependent on digital and especially wireless technologies in terms of production and delivery.*

The question assesses the degree of dependence on people with particular technological skills or access to decision-making points in R&D/I industries in ICT and related sectors. Such dependence may be indicative of alternating power relations in the newsroom, as individuals who lack particular skills will be seeking help from trained colleagues. From the respondents, 0.5% strongly disagreed, 9.8% disagreed, 16.1% were neutral, 50.8% agreed with the statement and 15% strongly agreed with the statement. 65.8% of the respondents indicated that in certain circumstances professional gatekeepers rely on NPDT professionals for the production of news.

THE PROFESSIONAL ROUTINES LEVEL

The second level of influences in the Shoemaker and Reese model is described as the routines level, identifying professional practices, standards, and codes of ethics, which every profession subscribes to. In the context of the current project, the authors examined a set of two "routines" variables as an initial evaluation of the revised gatekeeping model, attempting to assess whether digital technologies influence a routine shift in newsrooms environments.

H-3: *The influence of News Production and Delivery Technologies (including wireless, after 1990's) on news industry ethics (claiming objectivity, giving conflicting evidence, presenting supportive facts, quoting) increases rapidly.*

Objectivity and presentation of conflicting evidence are two significant professional routines in the field of journalism. Respondents are asked whether digital technologies affect the nature of those routines. 1% of the respondents strongly disagreed with the statement, while 8.8% disagreed, 20.2% were neutral, 39.9% agreed and 22.8% strongly agreed. A total of 62.7% either agreed or strongly agreed with the statement.

H-4: *Target audiences are formulated according to their familiarity with news-receiving gadgets/ equipment (mobile phones, decoders, etc.).*

Knowing an audience and designing content appropriate for it constitutes perhaps the most significant routine. Previous gatekeeping research has demonstrated that editors are keen on selecting content appropriate for their audience, while developing an instinct about what audiences tolerate and what they do not. Familiarity with digital devices, especially the wireless, can be a significant gatekeeping factor on what messages become public and those which are discarded.

People followed similar response patterns, as there were no strong disagreements. 10.4% disagreed with the statement, and 25.4% were neutral, 40.4% agreed and 15.5% strongly agreed, the combined agreement percentage being 58.2%.

THE ORGANIZATION LEVEL

Every organization represents a unique cultural environment. Warren Breed's research has demonstrated the degree of influence the organization exercises on individual decision-making, as the corporate culture dictates what constitutes appropriate behaviour. "Appropriate" behaviour is not always dictated through explicit, formal actions against the personnel but in most cases through tacit interventions. Very often the culture of the collective dictates how the individual should behave. A set of two statements were presented two respondents in the context of organizational influences:

H-5: *Corporate policies, and ownership patterns, have changed because of telecommunications—especially the wireless—penetration into the nature of news reporting and production.*

The statement assesses a degree of change in corporate culture as a direct outcome of technological penetrations into the organization. The literature indicates that ownership patterns depend on technological innovations, implementations and integrations as managers take into account the compatibility factor. From those who responded, 2.1% strongly disagreed with the statement, 11.9% disagreed, 26.4% were neutral, while 36.3% agreed and 8.8% strongly agreed, thus the combined agreement equalled 45.1%.

H-6: *The autonomy of remote offices in the news industries has increased due to the globalized telecommunications—especially the wireless—infrastructure.*

The statement assesses the role of technology as a factor providing a degree of autonomy in remote offices of news industries. 2.6% of the respondents strongly disagreed with the statement, 11.4% disagreed, 19.7% were neutral, while 33.2% agreed and 15.5% strongly agreed, thus the combined agreement equalled 48.7%.

THE EXTRA-MEDIA LEVEL

H-7: *The news industry is increasingly dependent on R&D departments and their external technological partners for news gathering.*

The media increasingly rely on other organizations while establishing symbiotic relationships with them. Advertisers, public relations agencies and government bodies exercise a great deal of influence on media content. In a wide list of extra-media influences, there are new additions as technological organizations research, innovate, design and develop different information technologies. The question assesses the degree of influence of R&D on media content. 1.6% of the respondents strongly disagreed with the statement, 8.3% disagreed, 17.6% were neutral, while 46.6% agreed and 11.4% strongly agreed with the statement, thus the combined agreement equalled 58%.

H-8: *The degree of conformity to international standards for news documentation, compression and transmission depends on international institutions (IEEE, ITU, ICANN, etc.) mandates.*

The role of international institutions dictating technological standards constitutes an extra-media influence on media content. The statement assesses the degree of influence of those international bodies on media organizations and their content. The respondents had six choices: very low, low, neutral, high, and very high. The responses were as follows: 4.1% very low, 5.7% low, 26.9% neutral,

Figure 3. The role of digital technology in gatekeeping: Survey results

	The Role of Technology		
Level of Analysis	Low	Medium	High
Individual	11,40%	23,60%	53,10%
Routines	10,10%	22,80%	59,30%
Organization	14,00%	23,05%	46,90%
Extra-Media	09,85%	22,25%	54,15%
Culture / Ideology	09,60%	21,50%	63,70%

N= 193

Percentages based on average responses from each level of analysis

31.6% high and 18.7% very high, % were neutral, while 46.6% agreed and 11.4% strongly agreed with the statement, thus the combined agreement equalled 58%.

IDEOLOGY/CULTURE

H-9: *The degree to which new digital, most of it wireless, technologies are adopted by an audience influences the content of the media.*

The statement assesses the cultural/social factor as different societies are geared toward innovation while other cultural groups resist innovation in content and/or in medium to access the content. 1.6% strongly disagreed with the statement, 8.8% disagreed, 20.7% were neutral, while 49.2% agreed and 15.5% strongly agreed, thus the combined agreement equalled 64.7%.

H-10: *The impact of new digital technologies, especially the wireless, on media content is a crucial factor for understanding the nature of the content's effects on people and society.*

The statement assesses the impact of technology not only to media content but to societal perceptions. People responded as follows: 1% strongly disagreed, 7.8% disagreed, 22.3% were neutral, while 49.2% agreed and 13.5% strongly agreed, thus the combined agreement equalled 62.7%.

Figure 3 provides comprehensive evidence in regards to all five levels of analysis as the average percentages of every level of analysis were calculated to assess differences among the five levels.

Although there is a degree of conformity among for all five levels of gatekeeping, as respondents demonstrate that technology has altered irreversibly the nature of modern information gatekeeping and thereby rendering technology as a significant sixth filter as the current project proposes. Table 1 seems to indicate that technological influences seem to less significant in the context of the organization. Less that 50% of the respondents agreed that technology affects how information is filtered in the context of the organization. Approximately 54% indicated an average or a low impact of technology within the organization.

On the other hand, technology was presented as a major gatekeeping factor for the other four levels: individual, routines, extra-media and ideology. More than 50% of all respondents recognized different ways through which technology affects each of those filters while recognizing a macro/societal influence centred on technology (63%).

EUROPE WIRELESS MEDIA GATEKEEPING FACTORS

During the past decade, the European news markets experienced tremendous changes in the field of regulation. Twenty-three of the 27 countries,

Figure 4. Chart, Europe wireless media markets' gatekeeping factors

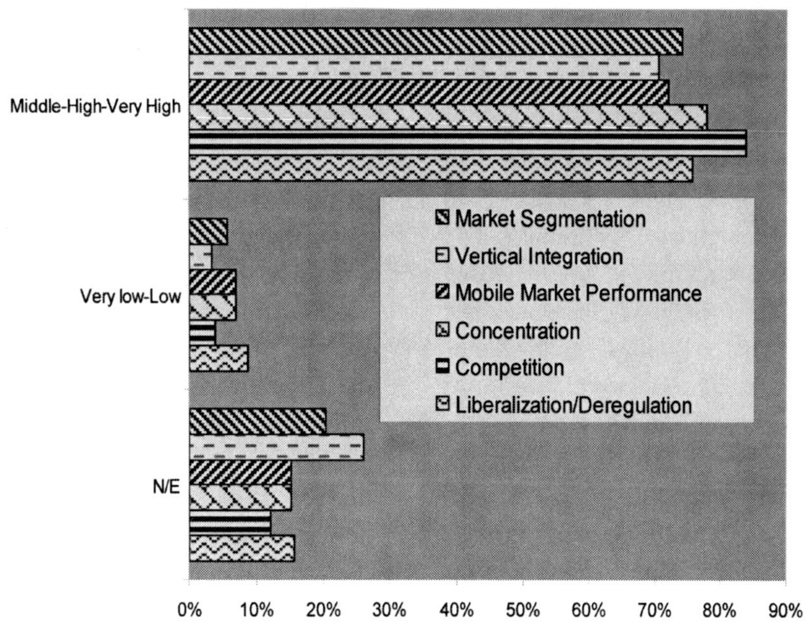

member states which have already ratified the Lisbon Treaty (2007) have revised their business development models related to new digital technologies. Convergence between e/m-communications, broadcasting and communications 2.0 not only have blurred the boundaries between network and content regulation, but also present new challenges in the spheres of deregulation and liberalization, thereby setting new rules in terms of competition, segmentation, verticalization and concentration. In this new regulatory environment, new principles will be imposed in terms of radio spectrum management and digital dividend allocation by 2012.

The current study gathered some preliminary survey data on wireless European media and their applications in the context of gatekeeping (See Figure 4). In the statement 31.a.: "The structure of the news market in terms of news products and services is influenced by Liberalization in Europe," 52.60% of the respondents agreed that the effect of deregulation of European news markets was important.

In the past few years, European markets have experienced dramatic changes that are going to affect the twelve accession countries, with different intensity and different policies of management of the radio spectrum. The "raw material' in short supply in wireless Europe is the radio spectrum, a 'key resource' for, among other everyday services involves mobile, wireless communications, TV and radio broadcasting, radio location, Global Positioning System (GPS)/Galileo, Broadband Wireless Access (BWA) technologies together with enhanced 3G mobile communications, and so forth. In the past, it was treated as a "scarce resource," used by a few public sectors only (like defence and broadcasting). The related policy now changes rapidly, following the technology innovation and investment on applications that contributed toward the "Lisbon goals." The Radio spectrum involves different pan-European networks and services that deal with the Lisbon goals which supports a single European communications market and focuses on a European "digital inclusion."

In the question 31.b.: "The structure of the news market in terms of news products and services is influenced by Competition in Europe," 69.50% of the respondents agreed that the structure of the European market has been influenced considerably by competition, while 83.80% agreed in some degree with the statement. 3.9% disagreed while a 12.2% did not answer. Preliminary data demonstrate that the technology has increased the competition in the context of European news markets. In Europe where digitalization constitutes a significant objective, digital media technologies are often synonymous with wireless technologies, because of the already existing wireless operators' networks.

In statement 31.c.: "The structure of the news market in terms of news products and services is influenced by Concentration in Europe," 50% of the respondents strongly agreed that the structure of European market has been influenced considerably by the concentration, while the 77.80% agreed with the statement. Seven percent disagreed, while 15.2% did not answer. The data indicate an ianic effect of ICT. Wireless markets show a significant degree of convergence in traditionally separated industries, as wireless operators become content providers and advertisements slots operators.

In the statement 31.d.: "The structure of the news market in terms of news products and services is influenced by mobile market performance in Europe" 52.20% of the respondents agreed that the structure of European market has been influenced considerably by the use of wireless technology, while the 71,80% agree in some degree with the statement. 7% disagreed, while a 15.2% did not answer. The influences of wireless technologies seem more influential in Europe than in different parts of the world.

In the statement 31.e.: "The structure of the news market in terms of news products and services is influenced by vertical integration in Europe," 43.40% of the respondents agreed that the structure of the European market has been influenced considerably by verticalization, while the 70.40% partially agreed with the statement. 3.4% disagreed with the statement, while a 26.1% did not answer. Verticalization is often related to concentration in news markets. In the context of the European news markets most of the respondents (77.80%) agreed that the structure of the news market is influenced by the vertical integration factor.

In the statement 31.f.: "The structure of the news market in terms of news products and services is influenced by market segmentation in Europe," 48.70% of the respondents agreed that the structure of the European market has been influenced considerably by segmentation, while 73.90% partially agreed with the statement; 5.6% disagreed, while a 20.4% did not answer. The new media (Internet and mobile phones enabled) fragment the communities, producing niche media markets, on a multi-platform approach, that is being more effective via a concentrated business model. Additionally, the national policies on the spectrum, the resource for wireless applications, could "fragment" the EU single media and telecom market, as the coordination is still in processing at a national level.

DISCUSSION

The authors recognize the usefulness of the established media gatekeeping model. However, they propose its upgrading to identify the role of different wireless media technologies in global creative industries. The revised model should provide space for continuous change in a dynamic fashion as technology changes rapidly.

Because of the nature of the data—convenience sample—and the acknowledged difficulty to instigate participation in on-line surveys of such scope, the authors are hesitant to generalize these results to a larger population. However, initial evidence from this global survey, based on a convenience sample of 230 media content and networks related gatekeepers working in several culture environments of some dozens of countries all over the world, indicates the need to account technological innovations, in the context of various creative industries. This new design should be supported by multidisciplinary studies accumulating empirical evidence at all levels to lead to a continuously updated media gatekeeping model, technology sensitive, in terms of content and networks.

The current preliminary study demonstrates the obtrusive nature of digital media in content and network selection. At the individual level, individual career evolution seems to be dependent on continuous training in the NPDTs. The data suggest that life-long training on NPDTs is of critical significance and not just a matter of option. Furthermore, individual gatekeepers seem to be dependent on other workers familiar with new technologies when it comes to utilizing them for news production. The same principle applies to gatekeepers depending on experts on news storage and information retrieval and other management.

There is also preliminary evidence that technology affects traditional news gathering and processing routines, as NPDTs have an impact on industry ethics, the notion of objectivity, presentation of facts, keeping up with deadlines, making decisions on presenting the news content, and so forth. Even deciding on target audiences is largely dependent on the audience's familiarity with ICTs such as mobile phones, palmtops, PDAs, and so forth.

Cross-tabulating those six variables with the age, gender, years of professional experience and the level of education yielded some differences demonstrating that at the individual and routines level, such characteristics should be taken into account. For example, the data indicate different gender perceptions on the need for dependence on technologically literate colleagues, and that people with extensive professional experience in the field and advanced education perceive technology as more important than their peers with lesser experience and lower educational status.

Although this initial evidence is quite convincing, additional empirical evidence, drawn from different organization/institutional environments is needed to complement the current evidence. Furthermore, qualitative evidence, utilizing in-depth interviews or testimonies can enhance the existing quantitative evidence providing an accurate picture of processes and practices.

The current study attempted to draw a global convenience sample from different media & technology related organizations and corporations, in different cultural settings to assemble preliminary evidence in regards to media gatekeeping factors in the digital and particularly in the wireless world. Thus, it answers a broad research question dealing with the need for upgrading, if not changing, a comprehensive media gatekeeping model. The EU media markets present a unique field for observation of changing media gatekeeping factors. Specifically, while examining extra-media influences, in the context of the Lisbon goals urging for regulation of the radio spectrum and a digital dividend allocation, researchers notice a changing digital terrain that triggers an array of reformations in EU media markets. These extra-media redefinitions present new challenges to scholars as gatekeeping practices are altered affecting media markets in Europe and in the world.

ACKNOWLEDGMENT

To Pennsylvania State University (PSU), IST (USA) for the trust and the opportunities that gave me with invited speeches and honoraria based on this research. Particularly, to Associate Professor of IST Michael R. Bartolacci at Penn State, USA, for his outstanding and valuable support and plenty of time he spent with me to discuss research issues and facilitating my access to the resources and digital libraries of PSU, and encouraging me to publish papers in journals and participate in international conferences. I cannot suppress the enthusiastic in-depth dialogues I had on wireless issues many times with Assistant Professor Dong Hee Shin at PSU Berks. Last but not least I would like to thank Abdullah Konak, Associate Professor of IST and Mahdi Nasereddin Assistant Professor of IST, for their friendly inclusiveness into the PSU academic community and the inspiration for continuing this research.

To George Carayannis, General Director of Research and Innovation Centre in Information, Communication and Knowledge Technologies "Athena" and Professor NTUA, Greece, with my deep respect as this research could not be conducted at the same level, and be noticed in public without the ILSP scholarship that award me and his kind personal help in some critical turning points of this academic work.

REFERENCES

Berkowitz, L. (1990). On the formation and regulation of anger and aggression. A cognitive-neoassociationistic analysis. *The American Psychologist*, *45*(4), 494–503. doi:10.1037/0003-066X.45.4.494

Breed, W. (1955). Social control in the newsroom: A functional analysis. *Social Forces*, *33*(4), 326–335. doi:10.2307/2573002

Carroll, R. L. (1985). Content values in TV news programs in small and large markets. *The Journalism Quarterly*, *62*(4), 877–938.

Chochliouros, I., & Spiliopoulou, A. (2003). Innovative horizons for Europe: The new European telecom framework for the development of modern electronic networks and services. *The Journal of the Communications Network*, *2*(4), 53–62.

Chochliouros, I., Spiliopoulou, A., Chochliouros, S., & Doukoglou, T. (2007) European challenges towards forming and promoting an innovative radio-spectrum policy in a fully converged electronic communications market. *The Journal of the Communications Network*, *6*(2), 60-67. Retrieved January 15, 2009, from http://theitp.org/UserData/root/Files/2007Vol6Prt2_08.pdf

Cohen, E. (2002). Online journalism as market-driven journalism. *Journal of Broadcasting & Electronic Media*, *46*(4), 532–548. doi:10.1207/s15506878jobem4604_3

Department of Trade and Industry (DTI). (1996). *Spectrum management: Into the 21st century* (DTI White Paper). London: Author.

Dimitrova, D., Connolly-Ahern, C., Williams, A., & Paul, K. (2003). Hyperlinking as gatekeeping: Online newspaper coverage of the execution of an American terrorist. *Journalism Studies*, *4*(3), 401–414. doi:10.1080/14616700306488

European Commission. (1998). *Green paper on radio spectrum policies, in the context of European community policies such as telecommunications, broadcasting, transport, and R&D*. (Rep. COM1998 596). Brussels, Belgium: Author.

European Commission. (1999). *Next steps in radio spectrum policy: Results of the public consultation on the green paper* (Rep. COM1999 538 final). Brussels, Belgium: Author.

European Commission. (2005). *Communication to the Spring European Council–working together for growth and jobs: A new start for the Lisbon Strategy* (Rep. COM2005 24). Brussels, Belgium: Author.

Gitlin, T. (1998). Public sphere or public spectacles. In T. Liebs & J. Curran (Eds.), *Media, ritual, and identity* (pp. 168-174). London, Routledge.

Gitlin, T. (2003). *The whole world is watching: Mass media in the making & unmaking of the new left*. Los Angeles: University of California Press.

Gitlin, T. (2007). *Media unlimited: How the torrent of images and sounds overwhelms our lives*. New York: Henry Holt.

Hall, S. (1982). Ideology. In E. Barnouw (Ed.), *The international encyclopedia of communication* (Vol. 2, pp. 307-311). New York: Oxford University Press.

Herman, E., & Chomsky, N. (2002). *Manufacturing consent: The political economy of the mass media*. New York: Pantheon Books.

Hollifield, C., Kosicki, G., & Becker, L. (2001). Organizational vs. professional culture in the newsroom: Television news directors' and newspaper editors' hiring decisions. *Journal of Broadcasting & Electronic Media*, *45*(1), 92–117. doi:10.1207/s15506878jobem4501_7

Peterson, S. (1981). International news selection by the elite press: A case study. *Public Opinion Quarterly*, *45*(2), 143–163. doi:10.1086/268647

Plaisance, P., & Skewes, E. (2003, October). Personal and professional dimensions of news work: Exploring the link between journalists' values and roles. Paper presented at the Annual Meeting of the International Communication Association, San Diego, CA.

Radiocommunications Agency (RA). (2002). *The economic impact of radio* (Reg. No. 876/2002, official journal L138, 1-8). Brussels, Belgium: Council of the European Union.

Reese, S. (2001). Understanding the global journalist: A hierarchy of influences approach. *Journalism Studies*, *2*(2), 173–187. doi:10.1080/14616700118394

Reese, S., & Ballinger, J. (2001). Roots of a sociology of news: Remembering Mr. Gates and social control in the newsroom. *Journalism & Mass Communication Quarterly*, *78*(4), 641–658.

Robinson, S. (2006). The mission of the j-blog: Recapturing journalistic authority online. *Journalism*, *7*(1), 65–83. doi:10.1177/1464884906059428

Shoemaker, P. J., & Reese, S. D. (1996). *Mediating the message: Theories of influences on mass media content*. New York: Longman.

Singer, J. (1997). Still guarding the gate? The newspaper journalist's role in an on-line world. *Convergence: The international journal of research into new media technologies*, *3*(1), 72-89.

Singer, J. (1998). Online journalists: Foundations for research into their changing roles. *Journal of Computer-Mediated Communication*, *4*(1).

Singer, J. (2005). The political j-blogger: 'Normalizing' a new media form to fit old norms and practices. *Journalism*, *6*(2), 173. doi:10.1177/1464884905051009

Snider, P. (1967). Mr. Gates revisited: A 1966 version of the 1949 case study. *The Journalism Quarterly*, *44*(3), 419–427.

Wanta, W., & Craft, S. (2004). Women in the newsroom: Influences of female editors and reporters on the news agenda. *Journalism & Mass Communication Quarterly*, *81*(1), 124–138.

White, D. M. (1950). The "Gate-Keeper": A case study in the selection of news. *The Journalism Quarterly*, *27*, 383–390.

Williams, B. A., & Delli Carpini, M. X. (2004). Monica and Bill all the time and everywhere, the collapse of gatekeeping and agenda setting in the new media environment. *The American Behavioral Scientist*, *47*(9), 1208. doi:10.1177/0002764203262344

Wireless Federation. (2007). *EU to open up mobile spectrum*. Retrieved June 25, 2009, from http://wirelessfederation.com/news/8794-eu-to-open-up-mobile-spectrum

This work was previously published in International Journal of Interdisciplinary Telecommunications and Networking, Volume 1, Issue 4, edited by Michael R. Bartolacci, and Steven R. Powell, pp. 37-53, copyright 2009 by IGI Publishing (an imprint of IGI Global).

Chapter 15
Trust-Based SAODV Protocol with Intrusion Detection, Trust Management and Incentive Cooperation in MANETs

Floriano De Rango
DEIS Department, University of Calabria, Rende (CS), Italy

ABSTRACT

Distributed wireless systems present many security issues, and offering security preserving network resources is often a challenging effort. Offering security through only cryptography is not always a suitable solution if the high dynamic context of MANET is considered and the trust mechanism that reduces the computationally intensive number of security operations becomes strategic. In this work, a well-known routing protocol, called SAODV, has been considered. To improve its performance and offer more resilience to attack from malicious nodes authenticated by the network, some additional features are added. A preventive approach based on a cryptographic mechanism and a reactive approach to detect anomalous and malicious nodes are considered. An extension of SAODV offering an Intrusion Detection Mechanism (IDM) and a Trust-based Mechanism (TBM) to promote the collaboration of the cooperating nodes and penalize the selfish nodes are proposed. Simulation results were performed under a NS-2 simulator to show the effectiveness of our proposal compared with AODV and SAODV. Simulation parameters, such as Normalized Routing Overhead, Data Packet Delivery Ratio, average number of signatures and throughput of collaborating and malicious nodes, have been considered.

INTRODUCTION

In MANET the security issue is a hot and challenging task, considering the distributed nature of the communication and the absence of any underlying fixed infrastructure (Royer & Toh, 1999). Mobile host join on the fly and create a network on their own. With the network topology changing dynamically and the lack of any centralized network management functionality, these networks tend to be vulnerable to a number of attacks. These attacks can be assessed by node external to the networks or by node already authenticated in the network (internal attacks). This means that

DOI: 10.4018/978-1-60960-505-6.ch015

for this kind of network a distributed monitoring system to verify the actions of mobile nodes and a signature mechanism to verify the identity and authenticity of nodes become necessary. However, it is important to observe that any security mechanism introduced in the node features need to be scalable and not too complex due to the limited resources of mobile nodes. In literature there are many routing protocols for MANET that offer some security features (Wu, Chen, Wu, & Cardei, 2006). However, these protocols consider just some specific aspects of security. In this case, we tried to consider more features like intrusion detection, trust establishment among nodes and incentive cooperation to reduce the selfish behavior that can reduce the network performance. In order to reduce the computational complexity of mobile nodes, only a preventive mechanism such as cryptography can be unsuitable. In this last case, Intrusion Detection Mechanism (IDM) and Trust-based Mechanisms (TBM) can become essential (Anantavalee & Wu, 2006; Pirzada, McDonald, & Datta, 2006). To combat malicious behavior coming of nodes authenticated by the network, reactive approaches based on the observation node activity can become important and appealing. Often, both approaches (IDM and TBM) can request a supplementary deployment of resources and an additional overhead not present in the traditional routing protocols (Anantavalee & Wu, 2006; Pirzada et al. 2006). Owing to this overhead, both of them are always considered separately. However, we think that through a joint and efficient use of both solutions, a greater degree of security can be offered for the right working of a routing protocol in a MANET.

Our proposal will reduce the complexity of signature mechanism the hop counting procedure from source and destination in the route discovery process. Moreover, a trust management based on trust levels will be proposed. Through this approach it will possible to associate different security actions. Additionally to this feature, also a collaboration grade index is defined in order to promote the cooperation among nodes and to individuate selfish nodes in the network. In our proposal, selfish behavior is discouraged through penalty actions.

The paper structure will be the following: some related works on security protocols, Intrusion Detection mechanisms and Trust Management over MANETs are presented in the next section; the SAODV protocol will be examined in another section (Zapata, 2006); later the additional security features introduced in SAODV and trust-level management are presented; finally, simulations results and conclusions are respectively presented.

RELATED WORK

In the literature many attacks (Wu et al., 2006) have been actuated in a MANET to produce Denial of Service (DoS). Regarding on-demand protocols, the route discovery phase can be a weak point of the protocol. For example, during the RREQ propagation a node can modify some fields of the packet in order to determine the selection of the path that goes through itself. In this section we briefly refer to some well-known secure routing protocol and then the attention will be focused on the Intrusion Detection Mechanism proposed in literature and the Trust Management policies.

Security-Based Protocols

Protocols that make use of the public key cryptography to generate signature, or protocols that use symmetric cryptography or hash functions are proposed (Perrig, Johnson, & Hu, 2004).

The first routing solutions are represented by protocols such as SAODV (Zapata, 2006), which will be discussed in the next section, SRP (Papadimitratos & Haas, 2002) that represents an extension of the protocol DSR for some security mechanism support, SEAD (Hu, Johnson, & Perrig, 2002a) and Ariadne (Hu, Johnson, & Perrig 2002b) that introduce a security mechanism in DSDV and DSR, ARAN (Kravets, Yi, & Naldurg, 2001) that proposes security schemes to apply on-demand routing protocols such as SAODV. All these protocols have been studied in the literature and their performance have been evaluated. Among them, SAODV resulted in a good compromise between reactive info exchange and security mechanisms based on an on-demand authentication mechanism and control overhead. For this reason, we focused on this protocol that will be briefly described in the next sub-section.

Intrusion Detection

An intrusion in a network system is defined as "any set of actions that attempt to compromise the integrity, confidentiality or availability of resources." In literature many preventive techniques to avoid these bad actions have been considered such as authentication and encryption. However, preventive techniques are not always sufficient especially in a distributed wireless context where every node can potentially behave maliciously. At this purpose in these last years many intrusion detection mechanisms (IDM) based on proactive or on demand monitoring have been proposed for MANETs. IDMs can be categorized into *misuse detection* and *anomaly detection*.

Examples of misuse detection based IDMs are IDIOT (Ilgun, Kemmerer, & Porras, 1995) and STAT (Kumar & Spafford, 1995) that use patterns of well-known attacks or weak spots of the system to math and identify known intrusions. The main advantages of misuse detection are the capability to accurately and efficiently detect instances of known attacks. The drawback is the lack to detect novel attacks in the networks.

On the other hand, anomaly detection systems like IDES (Lunt et al., 1992), try to monitor the system activities to detect behaviors that deviate significantly from the established normal use profiles. The main advantage of anomaly detection is that they do not need of a priori knowledge of intrusion and they can thus detect new forms of intrusions. The main drawback is the incapability to describe the type of attack and they can have high false positive rate (Mishira, Nadkarni, & Patcha, 2004; Zhang & Lee, 2000).

A distributed and cooperative architecture for better intrusion detection was proposed in (Zhang & Lee, 2000). A statistical anomaly detection approach is used and a local detection mechanism is implemented in each node.

Trust Management

Trust and reputation have been recently suggested as an effective security mechanism for open distributed wireless systems such as MANET. Considerable research has been done on modeling and management of trust and reputation applied to an Internet context. However, concerning the Mobile Ad Hoc Networks (MANET), traditional trust policies are not suitable. The bandwidth limitation of MANET and the dynamic topology make the trust and reputation management very challenging.

Ren, Li, Wan, Bao, Deng, and Kim (2004) proposed a certification-based trust establishment scheme. They addressed the trust establishment with dynamically joining and leaving nodes under the assumption of existing sparse social relationships among nodes.

Theodorakopoulos and Baras (2004) proposed a trust evaluation mechanism for MANETs without the need for a centralized infrastructure. In the authors' mechanism trust evaluation is viewed as a generalized shortest path problem on a weighted directed graph where vertices represent nodes and weighted edges correspond to the opinions that one end node has about the other end node in the edge direction.

Liu and Issarny (2004) considered another reputation mechanism for MANETs, where service reputation and recommendation reputation is differentiated. By the joint use of recommendation and reputation, this mechanism is able to distinguish between truth-telling and lying nodes and thus to be resilient against attacks of defamation and collusion.

Boukerch, Xu, and El-Khatib (2007) proposed a trust and reputation management scheme for sensor networks in order to reduce the overhead in terms of extra messages and time delay. The authors' idea is to use mobile agents in each mobile node in order to administrate the trust and reputation of its hosting node. Moreover, each node has a trust and reputation information stored in the forms of *t*-instruments and *r*-certificates.

Yan, Zhang, and Virtanen (2004) introduced in their proposal a rational security mechanism into the ad hoc networks by simulating human being's decision-making procedure. This approach permits to reach a good average security level based on accumulated knowledge and experience such as trust relationship established and modified.

Pirzada and McDonald (2004) proposed a trust management in a pure ad hoc networks (a network without any infrastructure and created on the fly). In particular, the authors, after defining a way to derive, to quantify and to compute the trust, extend the DSR protocol behavior in order to manage also security and trust. Thus, the DSR path selection is based on the trust value rather than minimum hop count.

The contributions of this paper are listed below:

1. A reduction of number of signature and a simplification in the hop counting management have been proposed with a version of SAODV called SAODV-DO.
2. A multilevel trust mechanism based on the concept of entropy has been introduced in SAODV-DO in order to reduce the signature and verification of mobile node in the considered MANET.
3. An intrusion detection mechanism and a credit based management system to promote the node cooperation and discourage the selfish behavior have been also introduced in SAODV-DO and the enhanced protocol version is called SAODV with Selfish Detection + Trust Management and Destination only reply (SAODV-SDDO).
4. Performance evaluation under malicious nodes attacks and selfish nodes behavior have been proposed in NS2 simulator.

SAODV Protocol

The SAODV protocol (Zapata, 2006), provides security mechanisms based on non-invertible hash functions and public key cryptography use applied to the on-demand routing protocol AODV. The node authenticity is guaranteed through the knowledge of the public key in each node of the network. An underlying key distribution mechanism is supposed to exist in the network.

Route Discovery Phase

In this phase a source node broadcasts a route discovery request (RREQ) toward a destination. The RREQ travels through the network until an intermediate node, which knows the path toward the destination, is met or until the destination node is reached. In SAODV, hash chains are applied for the hop count authentication so that each node, at every hop, can verify that the hop count metric was not maliciously decreased. A Max_Hop_Count is set typically to the TTL value of the message, a hash chain is calculated starting from a seed and applying itself Max_Hop_Count times; a node verifies that the hop count metric has not been maliciously changed applying (Max_Hop_Count – Hop_Count) times the hash function to the Hash field and comparing the result with the Top_Hash value.

In order to protect the immutable field of routing messages, each node generating a message includes a digital signature generated through its private key and all fields with the exception of the Hop_Count field. For details see (Potlapally, Ravi, Raghunathan, & Jha, 2006).

When a node receives a message, it verifies its authenticity, checking that the digital signature content, decrypted through the public key of the node originating the message, is equal to the fields of the received packets. In this way, if an intermediate node modifies the message content, this modification can be detected by the next node receiving the message and the packet can be discarded.

When the request reaches the destination, after verifying its authenticity, a reply with the addition of the destination signature is sent in unicast on the reverse path forwarding.

Two modalities for the working of the protocols can be observed:

- Destination only reply (DO): it does not allow intermediate nodes, which know a path toward the destination, to reply on behalf of destination. This approach increases the overhead associated with the RREQ propagation, because it denies the intermediate node a reply with a path toward destination.
- Route cache reply: it permits replying by intermediate nodes that have the path toward the destination in their cache. This behavior permits reduction of the RREQ propagation, but a double signature for each node and for each reply operation is necessary.

It is important to observe how in the first mechanism (DO) just a signature is necessary, whereas, in the second modality, each intermediate node needs to store a signature of the reply coming from the destination and the second signature of the requests coming from the source.

Route Maintenance Phase

In this phase the protocol provides a mechanism for the active link maintenance among the network nodes. In order to advertise that a specific node is no longer reachable the interested node generates an RERR message toward its neighbors.

SAODV assumes that all these messages will be verified and signed for each hop because they can be affected by variations along the path (differently from RREQ and RREP that are signed at the bottom of the communication).

Signature and Hop Count Mechanism Simplification

In SAODV signature and verifications operations are essential to mark and verify the identity and authenticity of the source and destination nodes. However, these two operations can be power and time consuming and some simplifications could be necessary is possible. Also the hop counting mechanism that is included in the hash chain can be simplified in the route selection. In this section a simplification on double signature mechanism and hop counting mechanism is proposed.

Considerations on Double Signature Mechanism in SAODV

Considering the studies performed for the evaluation of average energy consumption in the execution of signature, verification and key generation operations through different asymmetric algorithms (Potlapally et al., 2006), it is possible to note the asymmetry in the resource deployment during the verification and signature of RSA. In particular, the consumption associated with the signature operation is much greater than the verification operation. Now, if it is assumed that the key generation phase could be applied just one time, its weight in the power consumption computation can be omitted. On the other hand, the most often executed operation is the verification (especially on the RREQ broadcasted in the network during the route discovery phase) that is less incisive in terms of computation. Thus, the most heavy and expensive operation is the *signature*.

SAODV uses a *double signature* mechanism to allow an intermediate node to reply to a Route Discovery request on behalf of destination. This mechanism is necessary because an intermediate node could answer to a request (RREQ) also without knowing a path toward the destination. Moreover, source nodes, to allow an intermediate node to reply to their request, have to generate and include in the RREQ (RREQ double signature) two signatures: one for the intermediate node and the other one for the destination (the latter is to be included in the *gratuitous reply*).

Thus, in order to reduce the complexity and computational overhead of double signature, the Destination Only (DO) modality is considered and the signature phase is executed just twice (once for the source node and another time for the destination node). Instead, in the other modality (intermediate node reply) the signature is executed 4 times (twice for the source node and twice for the destination node). In this second case the resource consumption of each node for the cryptographic operations would become very expensive. This suggests that in situation where the available resources are limited, such as can be the context of wireless ad hoc networks, an intermediate node spontaneously avoids replying on behalf of the destination.

Another issue of double signature mechanism regards security. An intermediate node stores the destination node signature (in the case of RREP) or the supplementary source node signature (in the case of RREQ) in the routing table for the *reverse path* toward the node from which the message is received. Thus, a node could permanently store the signature in a way to answer to a request also in the case it did not have a path toward the destination.

Hop Counting Mechanism Simplification in the Route Reply Phase

The hop count in SAODV is applied in the RREQ broadcast phase and in the unicast RREP (Figure 1 (a)). However, in SAODV-SDDO, differently from SAODV, in order to reduce the computational complexity, the hop count is executed just in the RREQ broadcast phase (route discovery phase), whereas in RREP forwarding the hop count value is fixed to the value extracted from the request received by the destination node and it is signed with the other packet fields. In this way it is not possible at each intermediate node to increase the hop count in RREP message without producing packet discarding at the next hop. Moreover, it is erased by RREP the *Max_Hop_Count*, *Top_Hash* and *Hash*, which in the case of SHA-1 use, occupied 160 bits for each of them. Thus, a reduction of computational complexity in the transmission phase is obtained, because the intermediate nodes will not make any hash verification of hop counts, because this is signed as the other fields of the packet.

The intermediate nodes will update the distance in terms of hops to destination, subtracting the number of hops previously stored at the RREQ arrival remaining to reach the destination, such as shown in Figure 1 (b). This solution improves the forwarding phase of RREP in terms of security and efficiency. As is well known, the hop count authentication through hashing presents two problems consisting in the possibility of not increasing or decreasing the hop count; moreover, the proposed technique reduces the message size and avoids the verification and hashing operations of hop count for each node sending the RREP.

Trust Management

The number of RREQ packets is much higher than other routing messages in SAODV. This means that the number of verification operations on these messages is absolutely the highest in comparison with signature and verifications of other messages. For this reason it has been considered suitable and necessary to implement a trust based mechanism among nodes in order to reduce the number of RREQs verifications.

Each node makes use of a supplementary table that stores useful info on the neighbors met in the

Figure 1. (a) Hop counting in SAODV;(b) Hop counting in SAODV- SDDO

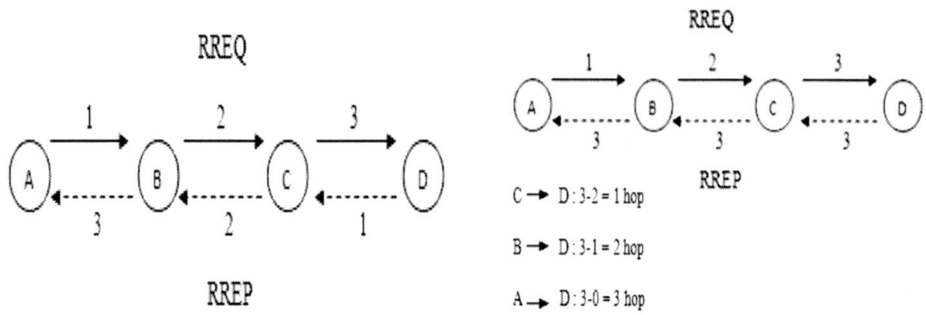

network evolution. For each new neighbor the node will initialize a table with the following entries:

- *trust_id*: address of neighbor node
- N_{RREQ}: the number of RREQ correctly received (at the beginning $N_{RREQ} = 0$)
- N_{RREP}: the number of RREP correctly received (at the beginning $N_{RREP} = 0$)
- T: trust level (at the beginning it is set to 1)
- N_f: number of consecutive messages whose verification is failed. It permits reduction of the trust level.
- th_{Nf}: maximum number of signature failures before putting the trust level to 0 and temporally exclude the node by communication.

The node maintains the counting of received and correctly verified RREQs and RREPs number for each neighbor and it regulates the trust level on the basis of this counting. The computational overhead is not so high because this mechanism is activated locally to each node.

Trust is a relationship established between two entities for a specific action. In particular, according with (Sun, Yu, Han, & Liu, 2006), one entity trusts the other entity to perform an *action*. In this work, the first entity is called the *subject*, the second entity is called the *agent*. We introduce the notation *{subject: agent; action}* to describe a trust relationship. Trust is a function of uncertainty. In particular, if the subject believes that the agent will perform the action for sure, the subject fully "trusts" the agent to perform the action and there is no uncertainty; however, if some malicious or not cooperating behavior is detected the trust value is decreased and more control is applied to the node with higher uncertainty grade. For more details about the trust definition associated to the entropy concept to see (Sun et al., 2006).

Let T{*subject: agent, action*} denote the trust value of the trust relationship {*subject: agent, action*}, and P{*subject: agent, action*} denote the probability that the agent will perform the action from the subject's point of view. The entropy-based trust value is defined as:

$$T_v\left\{subject : agent, action\right\} = \begin{cases} 3 - \mathrm{H}(\mathrm{p}) & \text{for } 0.5 \le p \le 1 \\ 1 + H(p) for\ 0 \le p < 0.5 \end{cases} \quad (1)$$

where

$$H(p) = -p\log_2(p) - (1-p)\log_2(1-p)$$

is the entropy function and

$$p = P\left\{subject : agent, action\right\}$$

This entropy based trust value is similar to that presented in (Sun et al., 2006) but it is translated to offer positive value and greater than 1. It is necessary for our purpose to associate probability *signature verification grade* to trust levels such as will be explained later. In general, trust value is positive when the agent is more likely to perform the action ($p>0.5$), and is negative when the agent is more likely not to perform the action ($p<0.5$).

It is assumed that X's behaviors in the past N trials and in the future $(N+1)^{th}$ trial are independent but governed by the same Bernoulli distribution such as in (Sun et al., 2006). Thus, if for example we consider as action the sending of RREQs from a node A to a node B, a trust value associated to the RREQ forwarding *action* can be expressed as follows:

$$P\left\{A:B,RREQ\_forwarding\right\}=\frac{k_{RREQ}+1}{N_{RREQ}+2} \quad (2)$$

For details about the math formulation of the Equation (2) above refer to (Sun et al., 2006). In general, considering a node A as subject, a node X as agent and a specific *action*, a good estimate of trust probability P can be calculated as follows. At every *HELLO* interval, node A observes the action of its neighbors. For example, if at time t_j, node A observes that node X performs the action k_j times upon the request of performing the action N_i times, the trust value associated with the specific action is expressed as below:

$$P\left\{A:X,action\right\} \quad (3)$$

where t_c represents the current time when this calculation is performed, β is a memory factor ($0 \leq \beta < 1$) such as in (Sun, 2006). Our considered action are the RREQ, RREP and DATA forwarding observation. On the basis of the value calculated in (3) it is possible to associate a trust value T_v such as expressed in (1). Then, after getting the T_v it is possible to calculate the respective discrete trust level. In particular, having *l* trust levels where t_l is the specific trust level *i* with i=1,2,..l, the association to the trust level is determined considering *l* intervals [1+ε; (3-ε)/l], [(3-ε)/l; 2*(3-ε)/l],...., [(l-1)*(3-ε)/l; 3-ε] (with $\varepsilon \to 0^+$. This value is adopted because $H(p) \to 0$ if $p \to 0^+$) and, on the basis of the current trust value, it is determined which interval the current T_v belongs to. To each interval is associated a discrete value specifying the trust level. Let tl_i indicate the trust level value associated to the *i*-th interval. If we have *l* trust levels, their values are *1,2,...,l*, where the lower trust level is 1 (the trust level 0 is assigned just in case of totally untrustworthiness about a specific node). This allows the definition of a *signature verification grade* (*sfg*) as follows:

$$sfg = 1 - \frac{tl_i}{\sum_{j=1}^{l} tl_j} \quad (4)$$

where $tl_i > 1$ and $0 < sfg \leq 1$. Then, on the basis of trust level, the node probabilistically determines a number of verifications to apply on a certain number of messages coming from each specific neighbor. At the beginning, the trust level is set to 1 and this means that a node will verify every message coming from its neighbor. For increasing grade of trust levels the verification operations are reduced.

In the level greater than first trust level, the RREQ verification frequency is determined through the *sfg* value. The mechanism consists of a random function $f_r(p)$ and a probability *p* randomly selected in the interval [0;1].

The probability of making a verification is associated with the trust levels and in particular is extracted a value from $f_r(p)$; if this value is lower than *sfg* the signature is applied otherwise not. In particular if the trust level is high, lower verification are needed and a low *sfg* value close to 0 is calculated. On the other hand, if the trust level is low (e.g., it could be due to malicious or not cooperating behavior) more verifications will be performed and a *sfg* value close to 1 is calculated. When a node detects a number of messages verification failures greater than N_f, it resets the counting of messages correctly received from that node and it reduces the trust level by 1.

Moreover, the number of received packets whose verification is failed, is stored. If this number exceeds a specific *threshold* th_{Nf}, the trust level is put to 0 and no more packets coming from that node are considered for a certain amount of time t_w.

Credit Management and Node Collaboration Grade

Different solutions for monitoring exist in the literature (Pirzada et al., 2006). Most of them perform neighbor monitoring and some penalty action in the case of protocol packet or missing data forwarding. In our case, the credit of nodes collaborating correctly are promoted and increased, and the credit of nodes requesting route discoveries is decreased.

Credits Management to Promote Cooperation

A node can decide to cooperate during routing operations to save its own network resources and continue to use network services making requests and generating traffic. If a lot of nodes assume this behavior, the network performance can sensibly degrade and the cooperating nodes could manage the traffic in an unbalanced way. In a general case, a node can avoid participating in routing operations discarding both RREQ and RREP. Inside our proposed protocol a mechanism was implemented to mitigate this selfish behavior and in particular the actions of selfish nodes are penalized and the actions of collaborating nodes are rewarded.

For these reasons not all penalty actions are applied to all the nodes whose credit is reducing but, on the basis of credit state, it is decided whether to:

- accept all RREQs coming from a neighbor
- accept or reject with a specific frequency RREQs
- reject all RREQs

The mechanism consists of monitoring neighbor node activity concerning data packet forwarding operation and routing message generation. The basic idea is the following:

1. A positive credit is assigned to each node
2. When a node makes a Route Discovery request generating an *RREQ*, its credit is decreased by an amount *X*.
3. When it forwards a request generated by another node its credit is increased by an amount *Y*.
4. Each node that generates or forwards an *RREQ* is able to receive *RREQ* that the neighbor nodes broadcast. Before discarding this previously sent packet, each node increases by an amount *Z* the credit of neighbors that correctly executed the operation
5. During the *RREP* forwarding a node, after forwarding a message, is in a *promiscuous modality* and it assures that the next node sends RREP. If this sending is not detected, the node decreases the credit of the neighbor node by an amount *Q*.

Let α_p^i and α_n^i be the credit associated to the positive operations (collaborative of node *i*) and negative operations, respectively. They can be defined as: $\alpha_p^i = Y + Z$ and $\alpha_n^i = X + Q$. It is possible to define the ratio between the number of collaborative actions associated to a specific node *i* and the number of negative actions associated to the same node. In particular:

$$\alpha(i) = \frac{\alpha_p^i - \alpha_n^i}{\alpha_n^i} \tag{5}$$

Figure 2. Credit Management Diagram associate to a node

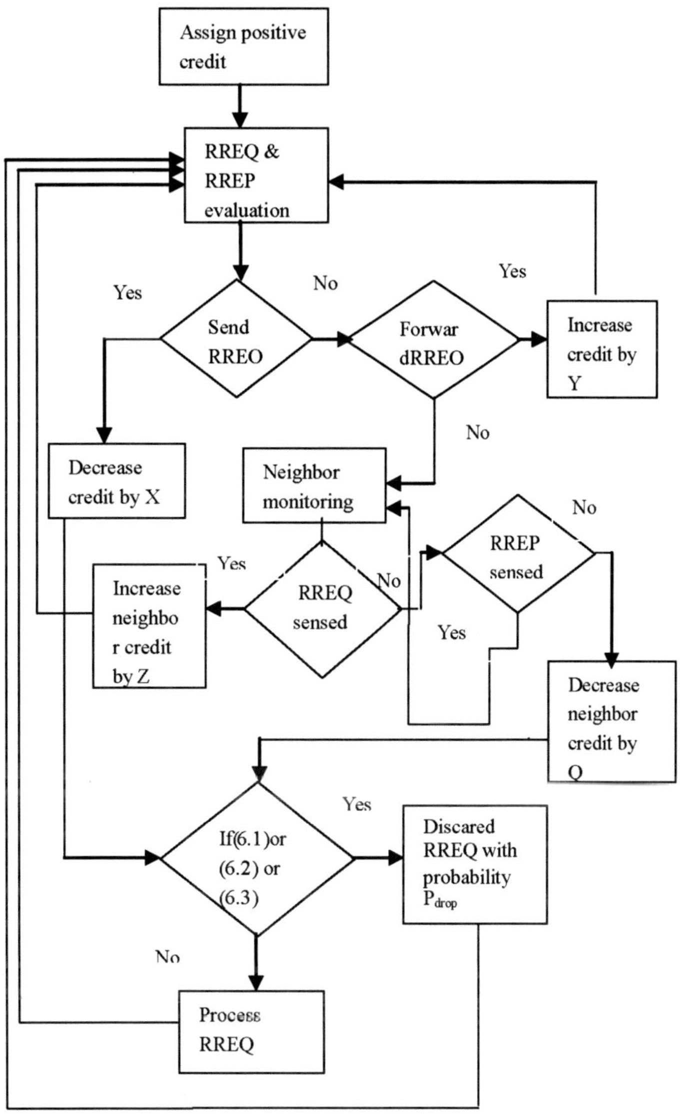

α(i) expresses the collaboration grade of each node *i*. In this way, a probability by each neighbor to forward the RREQ on behalf of each specific node can be evaluated on the basis of this index. The idea is to penalize not collaborating nodes and to exclude these nodes from the network operations for a while if they are not cooperating and they are maintaining a malicious behavior. Because α can assume different values we use the mapping between α and the probability P_{drop} to drop a packet coming from a not collaborating node. In particular:

$$\begin{cases} If\alpha < p_{th} \,\&\&P_{drop} = 1 \Rightarrow discard(RREQ)(6.1) \\ If(p_{th} < \alpha < 0 \,\&\&P_{drop} \in [p_d;1]) \Rightarrow discard(RREQ)(6.2) \\ If(0 < \alpha < 1 \,\&\&P_{drop} \in]0;p_d]) \Rightarrow dicard(RREQ)(6.3) \\ (\alpha > 1 \,\&\&P_{drop} = 0) \Rightarrow process(RREQ)(6.4) \end{cases} \quad (6)$$

where p_{th} is a threshold for negative collaborative grade. Under this threshold the node is isolated tor a time *t*. After this time the variable *X*, *Q*, *Y*, *Z* are reset to 0 and the *Credit Management* monitoring is activated again; p_d is a threshold to determine the packet dropping frequency.

A node selects a number P_{drop} through a uniform distribution in the range [0;1], if this number belongs to the range $[p_d;1]$ the packet is dropped otherwise it is considered (processed and forwarded) by neighbor nodes. The same rationale is for the other interval; the value P_{drop}=0 means that no packet is dropped by neighbors because the node is highly collaborative. All cases not expressed in (6) determines a *RREQ* processing by the node. In this way nodes that erroneously produce the reduction of their own credit toward their neighbors *will not be* excluded definitely from network operations and they will have some chance of recovering their credit.

In order to apply the previously described mechanism in the routing operations, a node should know its neighbors' state otherwise it could not be able to detect which node forwarded or otherwise a RREQ and it could not react through penalty or promotion of a neighbor. This kind of mechanism applied in SAODV would require a HELLO periodical messaging mechanism activation for neighbor updating. The applied approach avoids this issue not introducing particular computational overhead because nodes observation is local such as decisions to react to some selfish behavior.

Performance Evaluation

In this section simulation scenario where some attacks and malicious behavior is adopted by some nodes is considered. Moreover, some simulation parameters have been considered in order to compare our security features introduced in the SAODV protocol.

Simulation Parameters and Scenario

In order to validate the efficacy of the implemented mechanisms ns-2 simulator was adopted (Fall & Varadhan, 2000). SAODV and SAODV-SDDO were developed on the basis of AODV present in the simulator and with the addition of OpenSSL library for the adoption of the cryptography functions (openssl). The metrics evaluated for the simulations are:

1. *Data Packet Delivery Ratio*
2. *Normalized Control Overhead (NRO)*
3. *Throughput*
4. *Average number of signatures*

Let us define two kinds of nodes: malicious and selfish nodes. A malicious node is a node that tray to not execute right protocol operations in order to disrupt the network. A selfish node is a node that, in order to save its resources, does not want to collaborate assuming a selfish behavior.

The parameters adopted in the simulation are reported in Table 1. First of all, the effect of malicious nodes in the network were considered; then, the presence of selfish nodes was evaluated. In the different simulations, AODV and SAODV with route cache reply and without (DO) and the proposed protocol SADOV-SDDO were considered.

The following graphics show the protocol performance for increasing attacking nodes. The

Table 1. Simulation scenario

nodes	50	Grid	800mx400m
Simulation time	400	Traffic type	CBR 512 byte
Max speed	20m/s	Traffic source	15
Max pause time	2 sec.	Transmission range	200 m
N_f	5	th_{Nf}	10
Trust levels	4	p_{th}	-0.5
p_d	0.5	t_w	5 sec.
X	2	*Y*	4
Z	3	*Q*	4

percentage of malicious nodes are changed in the range [2, 10] on a total of 50 network nodes and they are configured as follows:

- A half of malicious nodes substitute the destination IP address in the RREQ packet.
- A half of malicious nodes increases the broadcast id in the RREQ packets before forwarding them.

Moreover, also selfish behavior was investigated and the selfish nodes were changed in the range [10%, 50%] of the total network nodes. This selfish behavior was modeled in the following way:

- a node with a probability 1/2 decide to discard the RREQ.
- A node with a probability of 1/2 decide to generate RREQs to consume resources.

Simulation Results

As it is possible to observe, the absence of security mechanisms reduces the performance of AODV owing to the increase of malicious nodes (Figures 3 and 4). On the other hand, SAODV and SAODV-SDDO maintain constant performance thanks to the integrity check of routing messages. However, SAODV-SDDO outperforms SAODV

Figure 3. Packet delivery ratio vs number of malicious nodes

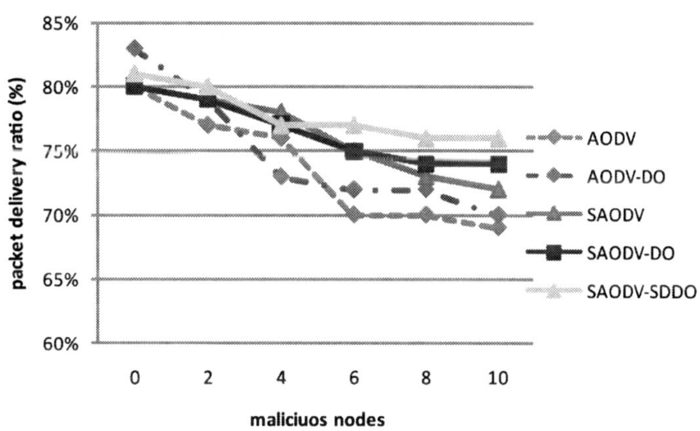

Figure 4. Normalized routing overhead vs number of malicious nodes

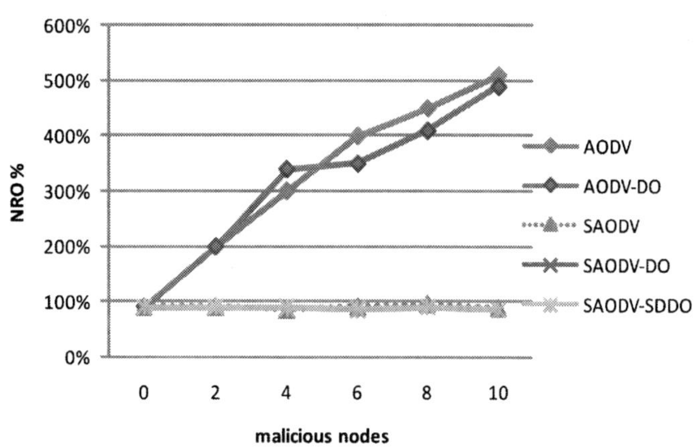

for the reduction on the computational overhead associated to the double signature mechanism and for the capacity to isolate the malicious and selfish nodes. Moreover, it is possible to note the high increase of the routing overhead of AODV for a high number of malicious nodes. In Figure 4 it is possible also to observe as SAODV and the respective enhanced versions (SAODV-DO and SAODV-SDDO) are robust under malicious nodes attacks and they maintain low the NRO.

The following results (Figures 5 and 6) show the detection mechanism for not collaborating nodes and TBM penalizes the action of selfish nodes. In particular, let us see how the selfish node throughput (it is considered as the capacity to forward data packet) is reduced after the detections of neighbor nodes.

It is possible to observe how in SAODV-SDDO selfish nodes are penalized and they are forced to generate a greater number of RREQ in order to be able to make their connection. This is true for both cases of RREQs and RREPs discarding. In Figure 7 is shown the case of the RREQs.

Through this mechanism, selfish nodes do not preserve their own resources and they are penalized inside the network. In this way an incentive to the cooperative behavior is provided. From Figure 8 it is possible to see how without malicious nodes, on the basis of trust mechanisms, SAODV-SDDO requires a lower number of routing message verifications.

In the presence of malicious nodes the verifications of SAODV-SDDO could increase because the slower increase of trust levels due to the

Figure 5. Selfish node throughput vs time

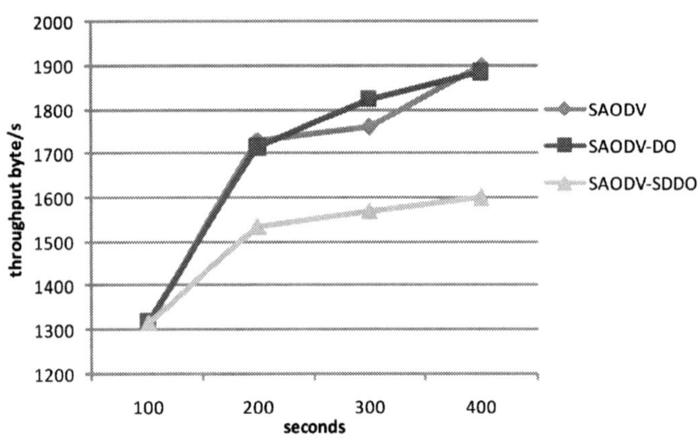

Figure 6. Average throughput vs percentage of selfish nodes in the network

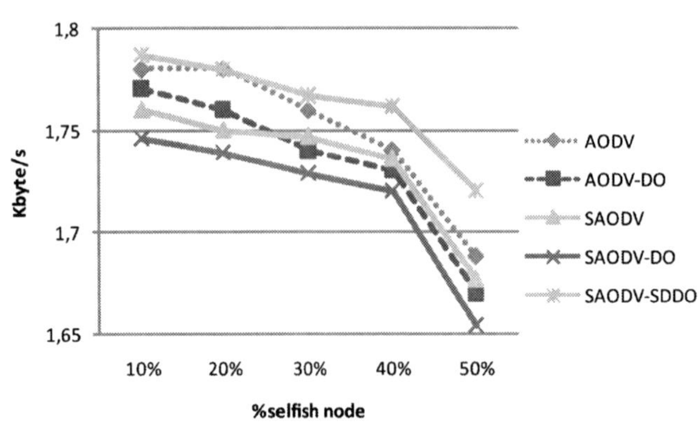

verification failure of packets coming from these nodes. Simulations under node mobility have not been shown due to space limitations. However, also under node mobility SAODV-SDDO outperformed SAODV and AODV under conditions of the malicious nodes attacks and selfish behavior. Details about this protocol performance under mobility will be offered in further contributions.

CONCLUSION

Additional security features to improve SAODV protocol have been introduced. The enhanced SAODV version is called SAODV-SDDO. Some mechanisms such as reply for intermediate nodes with double signature have been also evaluated in comparison with the single signature with reply only at the destination. Moreover, a trust mechanism to evaluate the trust of neighbor nodes through the observation of specific actions such as RREQ and RREP forwarding have been proposed and evaluated. Selfish nodes have been introduced in the MANET to see if the trust mechanism is able to reduce the trust value of malicious nodes and some penalty actions can be performed. Extensive simulation with the additional mechanism of SAODV with IDM and TBM has also been investigated. SAODV-SDDO proves to outperform SAODV when internal attacks are effectuated.

Figure 7. Number of dropped RREQs vs percentage of selfish nodes in the network

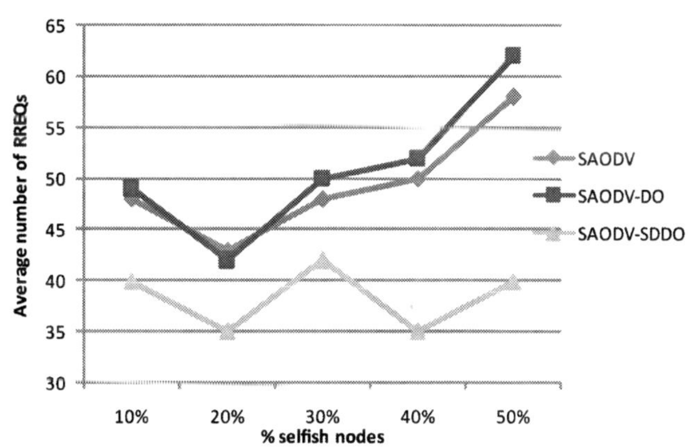

Figure 8. Average number of verifications vs time

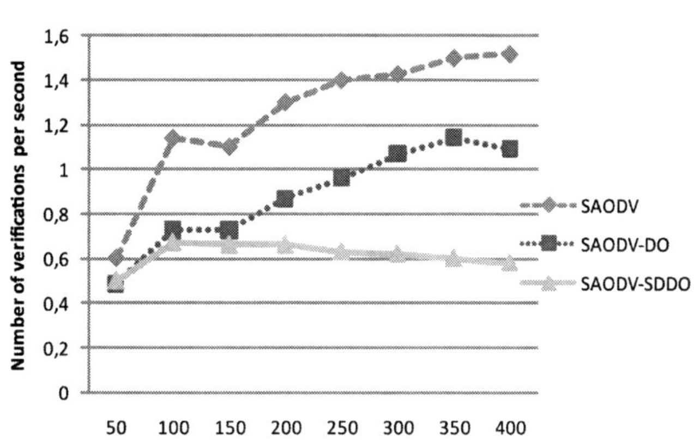

Performance evaluation has been evaluated under NS-2 in terms of data throughput, Normalized Routing Overhead, Data Packet Delivery Ratio and Average Number of Signatures.

ACKNOWLEDGMENT

We thank Eng. Giulio Pulitano for implementing the first version of SAODV in NS-2 in our Research Lab. Through his first contribution, it has been possible to extend the SAODV features and to propose an enhanced version based on the IDM, TBM and incentive cooperation.

REFERENCES

Anantvalee, T., & Wu, J. (2006). A survey on intrusion detection in mobile ad hoc networks. In Y. Xiao, X. Shen, & D.-Z. Du (Eds.), *Wireless mobile networks security* (pp. 170-196). New York: Springer.

Boukerch, A., Xu, L., & El-Khatib, K. (2007). Trust-based security for wireless ad hoc and sensor networks. *Computer Communication Journal*, *30*, 2412–2427.

De Rango, F. (2009, April 22-24). *Improving SAODV protocol with trust levels management, IDM and incentive cooperation in MANET*. Paper presented at the Wireless Telecommunication Symposium (WTS'09), Prague, Czech Republic.

Djahel, S., Naït-Abdesselam, F., Zhang, Z., & Khokhar, A. (2008). Defending against packet dropping attack in vehicular ad hoc networks. *Security and Communication Networks Journal*, *1*(3), 245–258. doi:10.1002/sec.25

Fall, K., & Varadhan, K. (Eds.). (2000). *Ucb/lbnl/vint Network Simulator–ns(version2)*. Retreived from http//www-mash.cs.berkeley.edu/ns

Feeney, L. M., & Nilsson, M. (2001, April 22-26). Investigating the energy consumption of a wireless network interface in an ad hoc networking environment. In *Proceedings of the 20th Annual Joint Conference of the IEEE Computer and Communications Societies (INFOCOM 2001)*, Anchorage, AK (Vol. 3, pp. 1548-1557). IEEE.

Hu, Y. C., Johnson, D. B., & Perrig, A. (2002a, June). SEAD: Secure efficient distance vector routing for mobile wireless ad hoc networks. In *Proceedings of the 4th IEEE Workshop on Mobile Computing Systems and Applications (WM-CSA'02),* Callicoon, NY (pp. 3-13). IEEE.

Hu, Y. C., Johnson, D. B., & Perrig, A. (2002b, September 23-26). Ariadne: A secure on-demand routing protocol for ad hoc networks. In *Proceedings of the 8th Annual International Conference on Mobile Computing and Networking (Mobicom'02)*, Atlanta, GA (pp.12-23). ACM Publishing.

Ilgun, K., Kemmerer, R. A., & Porras, P. A. (1995). State transition analysis: A rule-based intrusion detection approach. *IEEE Transactions on Software Engineering*, *21*(3), 181–199. doi:10.1109/32.372146

Kravets, R., Yi, S., & Naldurg, P. (2001, October 4-5). *A security-aware routing protocol for wireless ad hoc networks*. Paper presented at the ACM Symposium on Mobile Ad Hoc Networking and Computing (MobiHoc 2001), Long Beach, CA.

Kumar, S., & Spafford, E. H. (1995, October). A software architecture to support misuse intrusion detection. In *Proceedings of the 18th National Information Security Conference* (pp. 194-204).

Liu, J., & Issarny, V. (2004, March). Enhanced reputation mechanism for mobile ad hoc networks. In *Proceedings of the 2nd International Conference on Trust Management,* Oxford, UK (Vol. 2995, pp. 48-62). Springer-Verlag.

Lunt, T., Tamaru, A., Gilham, F., Jagannathan, R., Jalali, C., Neumann, P. G., et al. (1992). *A real-time intrusion detection expert system (IDES).* Menlo Park, CA: Computer Science Laboratory, SRI International.

Michiardi, P., & Molva, R. (2000). *Core: Collaborative reputation mechanism to enforce node cooperation in mobile ad hoc networks* (Research Rep. No. RR-02-062). Sophia-Antipolis, France: Institut Eurécom.

Mishira, A., Nadkarni, K., & Patcha, A. (2004). Intrusion detection in wireless ad hoc networks. *IEEE Wireless Communications Magazine*, *11*(1), 48–60. doi:10.1109/MWC.2004.1269717

Papadimitratos, P., & Haas, Z. J. (2002, January). *Secure routing for mobile ad hoc networks*. Paper presented at the SCS Communication Networks and Distributed Systems Modeling and Simulation Conference (CNDS 2002), San Antonio, TX.

Perrig, A., Johnson, D. B., & Hu, Y. C. (2004). A survey of secure wireless ad hoc routing. *IEEE Security & Privacy*, *2*(3), 28–39. doi:10.1109/MSP.2004.1

Pirzada, A. A., & McDonald, C. (2004, January). Establishing trust in pure ad-hoc networks. In *27th Australian Computer Science Conference*, Dunedin, New Zealand (Vol. 26, pp.47-54). Australian Computer Society.

Pirzada, A. A., McDonald, C., & Datta, A. (2006). Performance comparison of trusted-based reactive routing protocols. *IEEE Transactions on Mobile Computing*, *5*(6), 695–710. doi:10.1109/TMC.2006.83

Potlapally, N. R., Ravi, S., Raghunathan, A., & Jha, N. K. (2006). A study of the energy consumption characteristics of cryptographic algorithms and security protocols. *IEEE Transactions on Mobile Computing*, *5*(2), 128–143. doi:10.1109/TMC.2006.16

Ren, K., Li, T., Wan, Z., Bao, F., Deng, R. H., & Kim, K. (2004). Highly reliable trust establishment scheme in ad hoc networks. *Computer Networks: The International Journal of Computer and Telecommunications Networking*, *45*(6), 687–699.

Royer, E. M., & Toh, C.-K. (1999). A review of current routing protocols for ad hoc mobile wireless networks. *IEEE Personal Communications*, *6*(2), 46–54. doi:10.1109/98.760423

Sanzgiri, K., Dahill, B., Levine, B., Shields, C., & Belding-Royer, E. (2002, November 12-15). A secure routing protocol for ad hoc networks. In *Proceedings of the 10th IEEE International Conference on Network Protocols (ICNP),* Paris (pp. 78-87). IEEE.

Sun, Y. L., Yu, W., Han, Z., & Liu, K. J. R. (2006). Information theoretic framework of trust modeling and evaluation for ad hoc networks. *IEEE JSAC Special Issue on Security in Wireless Ad Hoc Networks*, *2*(24), 305–317.

Theodorakopoulous, G., & Baras, J. S. (2004, October). Trust evaluation in ad-hoc networks. In *Proceedings of the ACM Workshop on Wireless Security,* Philadelphia, (pp.1-10). ACM Publishing.

Wu, B., Chen, J., Wu, J., & Cardei, M. (2006). A survey on attacks and countermeasures in mobile ad hoc networks. In Y. Xiao, X. Shen, & D.-Z. Du (Eds.), *Wireless mobile networks security* (Ch.12). New York: Springer.

Yan, Z., Zhang, P., & Virtanen, T. (2004). *Trust evaluation based security solution in ad hoc networks* (Nokia Tech. Rep.) Retrieved from http://www.nokia.com/library/files/docs/

Zapata, M. G. (2006). Secure ad hoc on-demand distance vector (SAODV) routing. *ACM SIGMOBILE Mobile Computing and Communications Review*, *6*(3), 106–107. doi:10.1145/581291.581312

Zhang, Y., & Lee, W. (2000, August 6-12). Intrusion detection in wireless ad-hoc networks. In *Proceedings of the 6th Annual Conference on Mobile Computing and Networking (Mobicom 2000)*, Boston (pp. 275-283). ACM Publishing.

This work was previously published in International Journal of Interdisciplinary Telecommunications and Networking, Volume 1, Issue 4, edited by Michael R. Bartolacci, and Steven R. Powell, pp. 54-70, copyright 2009 by IGI Publishing (an imprint of IGI Global).

Chapter 16
Modeling and Simulation of Self-Similar Traffic in Wireless IP Networks

Dimitar Radev
University of Rousse, Bulgaria

Izabella Lokshina
SUNY Oneonta, USA

Svetla Radeva
CAE, UACEG, Bulgaria

ABSTRACT

The paper examines self-similar properties of real telecommunications network traffic data over a wide range of time scales. These self-similar properties are very different from the properties of traditional models based on Poisson and Markov-modulated Poisson processes. Simulation with stochastic and long range dependent traffic source models is performed, and the algorithms for buffer overflow simulation for finite buffer single server model under self-similar traffic load SSM/M/1/B are explained. The algorithms for modeling fixed-length sequence generators that are used to simulate self-similar behavior of wireless IP network traffic are developed and applied. Numerical examples are provided, and simulation results are analyzed.

INTRODUCTION

The growth of broadband networks and the Internet has been exponential in terms of users and user-end systems as well as in traffic in recent years. High-speed communications networks are able to support a wide range of multimedia applications, such as audio, video and computer data that differ significantly in their traffic characteristics and performance requirements (Hayes & Ganesh Babu, 2004).

One possible goal for developers is to build a unified high-speed communication network platform capable of carrying diverse traffic and supporting diverse levels of Quality of Service (QoS). Recent studies of real telecommunications

DOI: 10.4018/978-1-60960-505-6.ch016

network traffic data have shown that teletraffic exhibits self-similar (or fractal) properties over a wide range of time scales (Boxma & Cohen, 2000; Marie, Blackledge, & Bez, 2007).

The properties of self-similar telecommunications network traffic are very different from properties of traditional models based on Poisson, Markov-modulated Poisson, and related processes (Giambene, 2005). The use of traditional models in networks characterized by self-similar processes can lead to incorrect conclusions about the performance of analyzed networks (Jeong, 2002). Traditional models can lead to over-estimation of the network performance (Faraj, 2000), insufficient allocation of communication and data processing resources, and hence difficulties in ensuring the QoS. Then, full understanding is that the self-similar nature in teletrafic is an important issue.

Self-similar teletraffic is observed in LAN and WAN, where superposition of strictly independent alternating ON/OFF traffic models whose ON- or OFF-periods have heavy-tailed distributions with infinite variance can be used to model aggregate network traffic that exhibits self-similar (or long-range dependent) behavior typical for measured Ethernet LAN traffic over a wide range of time scales (Kushner, 2001).

In ATM network traffic self-similar traffic arriving at an ATM buffer results in a heavy-tailed buffer occupancy distribution, and buffer cell loss probability decreases with the buffer size, not exponentially as in traditional Markovian models, but hyperbolically.

Another implementation of traffic self-similarity is in Internet traffic, where many characteristics of WWW can be modeled using heavy-tailed distributions, including the distribution of traffic times, the distribution of user requests for documents and the distribution of WWW document sizes (Jeong, 2002).

In TCP/IP network traffic the transfer of files or messages shows that the reliable transmission and flow control mechanisms of TCP serve to mainain long range dependent structure included by heavy-tailed file size distributions (Bobbio, Horváth, Scarpa, & Telek, 2003). The relationship between self-similar traffic and network performance is defined in (Ilnickis, 2004) as captured by performance measures such as packet loss rate, retransmission rate and queueing delay, where increased self-similarity results in degradation of performance, and queueing delay exhibits a dramatic increase as self-similarity increases.

The self-similarity observed in video traffic provides possibility for developing models for Variable Bit Rate (VBR) video traffic using heavy-tailed distributions (Radev, 2005). The autocorrelation of the VBR video sequence decay hyperbolically and can be modeled using Fractional Autoregressive Integrated Moving-Average (F-ARIMA) and Fractional Gaussian Noise (FGN) self-similar processes (Radev, Rashkova, & Stankovski, 2007; Radev & Lokshina, 2009).

The impact of self-similar models on queuing performance is significant and the main trends in such findings are connected with (a) permission traffic modeling for high-speed networks, (b) efficient simulation of actual network traffic and (c) analyzing queuing models and protocols under realistic traffic scenarios (Mehdi, 2003). The traditional models of teletraffic that assume independent arrivals, based on Poisson processes, Markov-modulated Poisson processes and other related processes are not able to capture the self-similar nature of teletraffic (Hayes & Ganesh Babu, 2004).

The time series of self-similar processes exhibit burstiness over a wide range of time scales. Self-similarity can statistically describe wireless IP network traffic that is bursty on many time scales. Modeling and simulation of self-similar telecommunications network traffic can be performed with the generators of synthetic self-similar sequences (Horn, Kvalbein, Blomskøld, & Nilsen, 2007), which are divided into two practical classes: the sequential generators and the fixed-length sequence generators. The fixed-length sequence generators for simulation of self-similar wireless IP network traffic are considered in this paper.

PROPERTIES OF LONG-RANGE DEPENDENT SELF-SIMILAR PROCESSES

Self-similarity can be classified into two types: deterministic and stochastic (Stathis & Maglaris, 2000). One can distinguish two types of stochastic self-similarity. A continuous-time stochastic process Y_t is strictly self-similar with a self-similarity (or, Hurst) parameter H ($1/2<H<1$), if Y_{ct} and $c^H Y_t$ (the rescaled process with time scale ct) have identical finite-dimensional probability for any positive time stretching factor c. This definition, in a sense of probability distribution, is quite different from that of the second-order self-similar process, observed at the mean, variance and autocorrelation levels (Park & Willinger, 2000). The process X is asymptotically second-order self-similar with $0.5<H<1$, if for each k large enough $\rho_k^{(m)} \to \rho_k$, as $m\to\infty$, where $\rho_k = E[(X_i-\mu)(X_{i+k}-\mu)]/\sigma^2$. In this work exact or asymptotic self-similar processes are used in an interchangeable manner, which refers to the tail behavior of the autocorrelations (Popescu, 2001).

In modern telecommunications networks the most striking feature of some second-order self-similar processes is that the accumulative functions of the aggregated processes do not generate as the non-overlapping batch size m increasing to infinity. Such processes are known as Long-Range Dependent (LRD) processes (Tan, 2009). This is in contrast to traditional processes used in models of telecommunications network traffic, all of which include the property that the accumulative functions of their aggregated processes degenerate as the non-overlapping batch size m increasing to infinity, that is, $\rho_k^{(m)}\to 0$ or $\rho_k^{(m)}=0$ ($|k|>0$), for $m>1$. The equivalent definition of long-range dependence is given as:

$$\sum_{k=-\infty}^{\infty} \rho_k = \infty \qquad (1)$$

Another definition of LRD is presented as:

$$\rho_k \sim L(t)k^{-(2-2H)}, \quad \text{as} \quad k \to \infty\,, \qquad (2)$$

where $½<H<1$ and $L(\cdot)$ slowly varies at infinity, that is, for all $x>0$ it could be determined as:

$$\lim_{t\to\infty} \frac{L(xt)}{L(t)} = 1\,. \qquad (3)$$

The Hurst parameter H characterizes the relation in (2), which specifies the form of the tail of the accumulative function. One can show that is true for $1/2<H<1$, is given:

$$\rho_k = \frac{1}{2}\left[(k+1)^{2H} - 2k^{2H} + (k-1)^{2H}\right]. \qquad (4)$$

For $0<H<1/2$ the process is Short Range Dependent (SRD) and could be presented as:

$$\sum_{k=-\infty}^{\infty} \rho_k = 0\,. \qquad (5)$$

For $H=1$ all autocorrelation coefficients are equal to one, no matter how far apart in time the sequences are. This case has no practical importance in real telecommunications network traffic modeling. If $H>1$, then is true:

$$\rho_k = \begin{cases} 1 & \text{for} \quad k=0 \\ \frac{1}{2}k^{2H}g(k^{-1}) & \text{for} \quad k>0 \end{cases} \qquad (6)$$

where:

$$g(x) = (1+x)^{2H} - 2 + (1-x)^{2H}\,. \qquad (7)$$

One can see that $g(x)\to\infty$ as $H>1$. If $0<H<1$ and $H\neq 1/2$, then the first non-zero term in the Taylor expansion of g(x) is equal to $2H(2H-1)x^2$ and is true:

$$\rho_k / (H(2H-1)k^{2H-2}) \to 1, \quad \text{as} \quad k \to \infty . \tag{8}$$

In the frequency domain, an essentially equivalent definition of LRD for a process X with given spectral density:

$$f(\lambda) = \frac{\sigma^2}{2\pi} \sum_{k=-\infty}^{\infty} \rho_k e^{ik\lambda} , \tag{9}$$

is that in the case of LRD processes, this function is required to satisfy the following property:

$$f(\lambda) \sim c_{f_1} \lambda^{-\gamma}, \quad \text{as} \quad \lambda \to 0 , \tag{10}$$

where c_{f_1} is a positive constant and $0<\gamma<1$, $\gamma=2H-1<1$.

As a result, LRD manifests itself in the spectral density that obeys a power-law in the vicinity of the origin. This implies that $f(0)=\Sigma_k \rho_k=\infty$. Consequently, it requires a spectral density, which tends to $+\infty$ as the frequency λ approaches 0. For a Fractional Gaussian Noise (FGN) process, the spectral density $f(\lambda,H)$ is given by:

$$f(\lambda, H) = 2c_f (1 - \cos(\lambda)) B(\lambda, H) , \tag{11}$$

with $0<H<1$ and $-\pi\leq\lambda\leq\pi$, where (12) is true:

$$\begin{aligned} c_f &= \sigma^2 (2\pi)^{-1} \sin(\pi H) \Gamma(2H+1) \\ B(\lambda, H) &= \sum_{k=-\infty}^{\infty} | 2\pi k + \lambda |^{-2H-1}, \end{aligned} \tag{12}$$

and $\sigma^2=\mathrm{Var}[X_k]$ and $\Gamma(\cdot)$ is the gamma function.

The spectral density $f(\lambda,H)$ in (11) obeys a power-law at the origin, as shown in:

$$f(\lambda, H) \to c_f \lambda^{1-2H}, \quad \text{as} \quad \lambda \to 0 \tag{13}$$

where $1/2<H<1$.

STEADY – STATE SIMULATION OF SELF – SIMILAR QUEUING SYSTEMS

There is a significant difference in the queueing performance between traditional models of wireless IP network traffic such as Poisson processes and Markovian processes, and those exhibiting self-similar behavior. More specifically, while tails of the queue length distributions in traditional models of wireless IP network traffic decrease exponentially, those of self-similar traffic models decrease much slower. Let us consider the potential impacts of traffic characteristics, including the effects of self-similar behavior on queuing and network performance, protocol analysis, and network congestion controls.

The steady-state simulation of self-similar queueing systems include generation of self-similar traffic, simulation of queing process and simulation of overflow probability. This can be demonstrated with buffer overflow simulation for SSM/M/1/B queueing systems ($B<\infty$, i.e., queuing systems with the finite buffer capacity) at self-similar queuing processes. In this case, the difference from M/M/1/B queuing system is that the arrival rate λ_j into SSM/M/1/B queuing system is not a constant value. It depends on the sequential number of time-series i, the total number of observation n and the Hurst parameter H, which determine the rate of self-similarity. The analyzed SSM/M/1/B queuing system has exponential service times with constant rates $1/\mu$ as is shown on Figure 1.

Figure 1. State transition diagram for a SSM/M/1/B queuing system

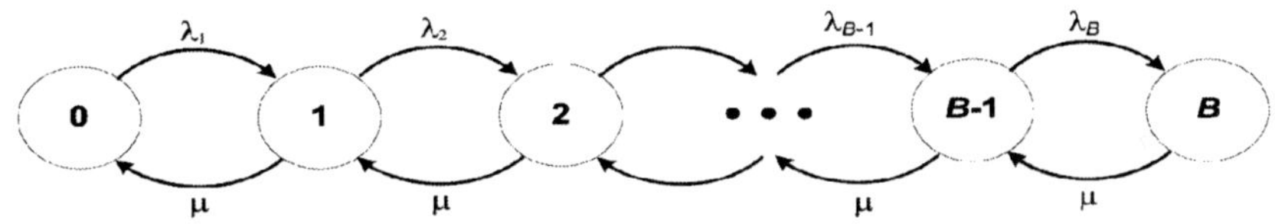

The flow balance equations are (Trivedi, 2001):

$$\begin{aligned} \lambda_j &= \lambda(i,n,H); \qquad j = 1,2,\ldots,B \\ \lambda_j &= 0; \qquad j \geq B+1 \\ \mu_j &= \mu; \qquad j = 1,2,\ldots,B+1 \end{aligned} \tag{14}$$

This system is stable whether the throughput $\rho = \frac{\lambda(i,n,H)}{\mu} < 1$. Let us consider two separated cases: ρ=1, and $\rho \neq 1$. For $j = 0,1,2,\ldots,B$ the distribution of number of customers in system is $P_j = \rho^j P_0$, which is determined according to

$$\begin{aligned} P_j &= \frac{\rho^j(1-\rho)}{1-\rho^{B+1}}; \quad \rho \neq 1 \\ P_j &= \frac{1}{B+1}; \quad \rho = 1 \end{aligned} \tag{15}$$

Hence, the rate at which the customers are lost (blocked) is λP_B. The queuing process is described with the steady-state simulation scheme, presented on Figure 2. The self-similar traffic was generated and the sequence of arrivals was received. The fixed length of self-similar traffic is extracted by fixing the number of observations. As the service process is Markovian, the sequence of departures has exponential distribution, generated with inverse transform generator. The next step is the calculation of buffer size. If the service size is greater than the size of arrivals, than the buffer size B=0 (e.g., it is impossible to have negative buffer size). In cases when the buffer is overflow L, for example, $B>L$ the traffic is lost, that's why is making an assumption that $B=L$.

The simulation is performed with the RESTART method (Georg & Schreiber, 1996), which is a variant of splitting where any chain is split by a fixed factor when it hits a level upward, and one of the copies is tagged as the original for that simulation level (Georg & Fuss, 1999). When any of those copies hits that same level downward, if the copy is the original it just continues its path, otherwise it is killed immediately (L'Ecuyer, Demers, & Tuffin, 2006). This rule applies recursively, and the method is implemented in a depth-first fashion, as follows: whenever there is a split, all the non-original copies are simulated completely, one after the other; then the simulation continues for the original chain (Villen-Altamirano & Villen-Altamirano, 2006). The reason for eliminating most of the paths that go downward is to reduce the work. The calculation of the buffer size for all sequences gives possibilities for determining the overflow probability using the RESTART with the Limited Relative Error (LRE) simulation.

LIMITED RELATIVE ERROR SIMULATION OF BUFFER OVERFLOW

The limited relative error (Bucklew, 2004) provides possibility to determine the complementary cummulative function of arrivals at single server buffer queues with Markov processes. In order to describ the LRE principles for steady-state simulation in disctere-time Markov chains, let us consider homogenious two-node Markov chain, which is extended to regular disctere-time Markov chain,

Figure 2. Steady-state simulation of queuing process

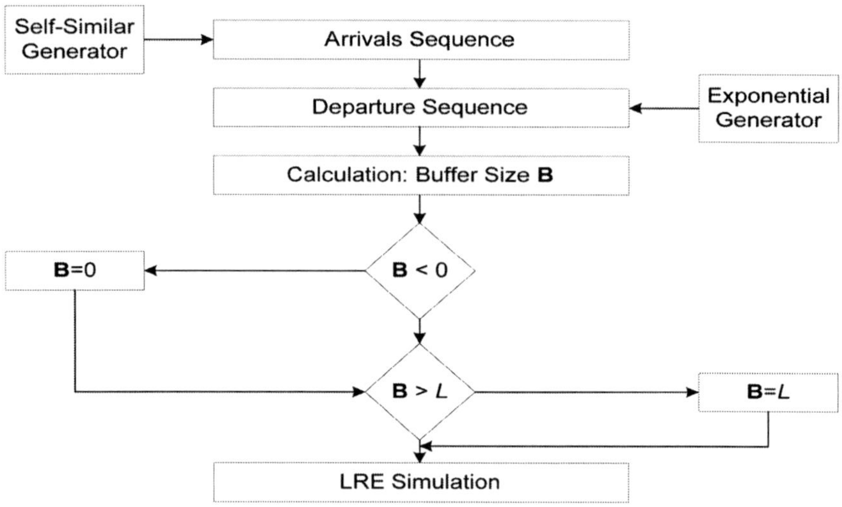

consisting of $(k+1)$ nodes with states, respectively $S_0, S_1, \ldots, S_k$, as shown in Figure 3.

We obtain the random generated sequence $x_1, x_2, \ldots, x_t, x_{t+1}, \ldots$ for $x=0,1,\ldots, k$, for which a transition for state S_j at the time t exists, for example, $x_t=j$ and there are no constraints to the parameters of the transition probabilities:

$$p_{ij} = \mathrm{P}\left(j \mid i\right); \quad \left(i, j\right) = 0, 1, \ldots, k; \quad \sum_{j=1}^{k} p_{ij} = 1 \tag{16}$$

There are no absorbing states S_i at $p_{ii}=1$ for all stationary probabilities P_j $j=0,1,\ldots,k$, which satisfy the constraint condition:

$$0 \le \mathrm{P}_j < 1; \quad \sum_{j=0}^{k} \mathrm{P}_j = 1. \tag{17}$$

The cummulative distribution $F(x)$ can be presented as:

$$\left.\begin{aligned} &F\left(x\right) = F_i; \quad \left(i-1\right) \le x < i; \quad i = 1, 2, \ldots, k+1; \\ &F_i = \sum_{j=0}^{i-1} \mathrm{P}_j; \quad F_0 = 0; \quad F_{k+1} = 1; \end{aligned}\right\} \tag{18}$$

To simulate the $(k+1)$ nodes of Markov chain, the complementary cummulative distribution $G(x)=1-F(x)$ that is more significant, can be determined along with the local correlation coefficient $\rho(x)$ through the limited relative error approach.

After having the two-node Markov chain defined as shown in Figure 1, with changing the states n times, an estimation of the local correlation coefficient $\hat{\rho}(x)$ can be received, which connects the number of transitions through a dividing line $a_i \approx c_i$, with the total number of observed events $l_i=n-d_i\left(\beta = 0, 1, \ldots, i-1\right)$ at left side, and d_i at right side $\left(\beta = i, i+2, \ldots, k\right)$.

The value of simulated complementary cumulative distribution $\hat{G}_i$ can be defined directly by using relative frequency d_i / n, if there is enough number of samples:

$$n \ge 10^3; \quad \left(l_i, d_i \ge 10^2\right); \left(a_i, c_i, l_i - a_i, d_i - c_i\right) \ge 10 \tag{19}$$

The posterior equations can be used for the complementary function $\hat{G}\left(x\right)$, the average number of generated values of $\hat{\beta}$, the local cor-

Figure 3. Cummulative function F(x) for (k+1)- node Markov chain

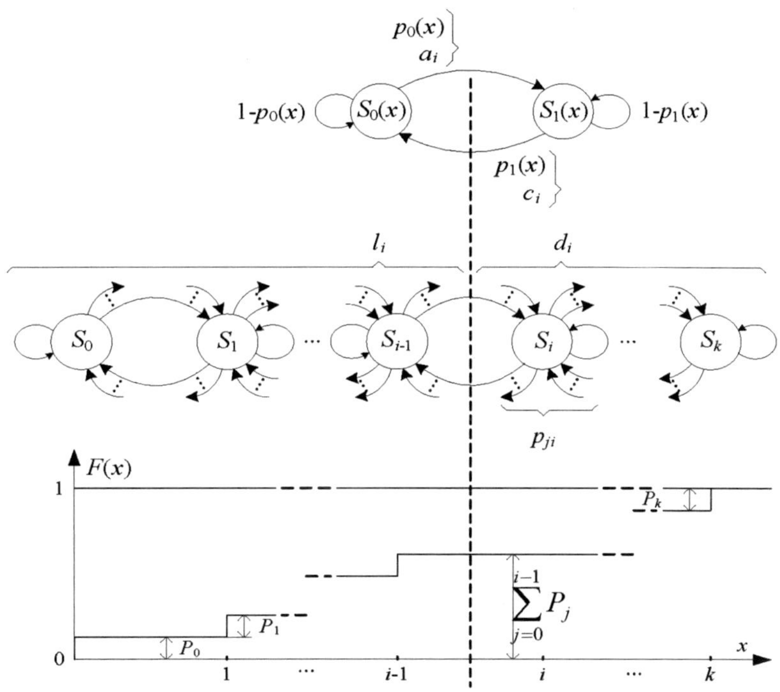

relation coefficient $\hat{\rho}(x)$, the correlation coefficient Cor[x] and the relative error RE[x]:

$$\hat{G}(x) = \hat{G}_i = d_i/n \qquad \hat{\beta} = \frac{1}{n}\sum_{i=1}^{k} d_i$$
$$\hat{\rho}(x) = \hat{\rho}_i = 1 - \frac{c_i / d_i}{1 - d_i / n} \qquad \begin{matrix} i-1 \le x < i \\ i = 1,\dots,k \end{matrix}$$
$$\mathrm{Cor}[x] = \mathrm{Cor}_i = (1+\hat{\rho}_i)(1-\hat{\rho}_i)$$
$$\mathrm{RE}[x]^2 = \mathrm{RE}_i = \frac{1 - d_i/n}{d_i} \cdot \mathrm{Cor}_i \tag{20}$$

The main advantage of this approach is that the relationships between transitions c_i are reached with routine statistical calculations. The necessary total number of simulation trails n is determined with the maximal relative error $\mathrm{RE}_{\max}\left[x\right]^2$ and with the less value of the function $G(x)$, presented as $G_{\min} = \hat{G}_k$ in approximation equation:

$$n = \frac{(1 - G_{\min})}{G_{\min}.\mathrm{RE}_{\max}\left[x\right]^2} \approx \frac{\mathrm{Cor}_k}{\hat{G}_k.\mathrm{RE}_{\max}\left[x\right]^2};$$
$$\mathrm{Cor}_k = \frac{1+\hat{\rho}_k}{1-\hat{\rho}_k} \tag{21}$$

This procedure can be described with a standard version of limited relarive error algorithm of random discrete sequences for buffer arrivals.

Algorithm 1 for limited relarive error**:**

Initialization of minimal and maximal values of the simulation parameter.
Estimation and management of the simulation time.

Cycle L_1:

Determine the current variable for calculating the Markov chain, for example, $\omega := \beta$;

Figure 4. Buffer overflow probabilities (L=41) in SSM/M/1/40 queuing system

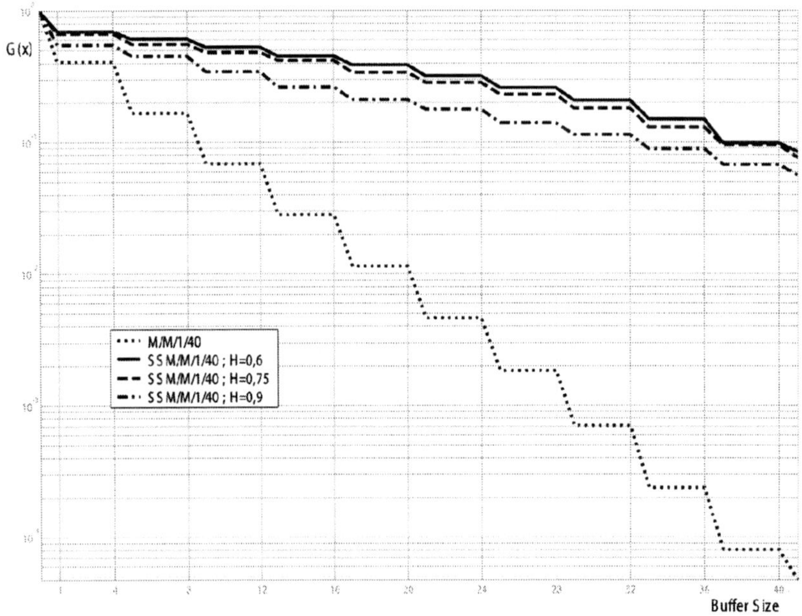

Generate a new value for β with given distribution.
Increase the number of state h_β;
If the condition $\beta < \omega$ is true, then increase the number of transitions $c_{\beta+1}$ while it reaches the value of c_ω.

Cycle L_2:

Determine the total number of events at the left part l_s and at the rigth part d_s of the Markov chain and number of transitions $a_s := c_s$;
Check on the constraint condition (18) for the index i=s. If the constraint condition (19) is true, then calculate the posterior values of the local correlation coefficient $\hat{\rho}_s$ and relative error $\mathrm{RE}\left[x\right]$ according to (20);
Calculate whether the relative error
$\mathrm{RE}\left[x\right] \leq \mathrm{RE}_{\max}\left[x\right]$;
If $s<k$, than leave the cycle L_2.
If the index s=k is reached, than leave the cycle L_1 and increase the index of the simulation time s:=s+1;

Printing out the experimental results for i=1,2,…,k. The results for the total frequency d_i are determined according to:

$$d_i = \sum_{j=1}^{k} h_j \quad \text{for } i = 0,1,\ldots,k \quad \text{where} \quad d_0 = n \tag{22}$$

The values of the complementary function $\hat{G}_i$, the local correlation coefficient $\hat{\rho}_s$ and the relative error $\mathrm{RE}\left[x\right]$ are calculated as given in (20).

LRE ALGORITHM PERFORMANCE AND SIMULATION RESULTS

As an example, the overflow probability of SSM/M/1/B queuing system was simulated with different characteristics of self-similar arrival processes. In order to demonstrate the effects of self-similarity on the buffer overflow probability, the achieved experimental results were compared with the complementary cumulative distribution

Figure 5. Buffer overflow probabilities in SSM/M/1/B with different buffer sizes

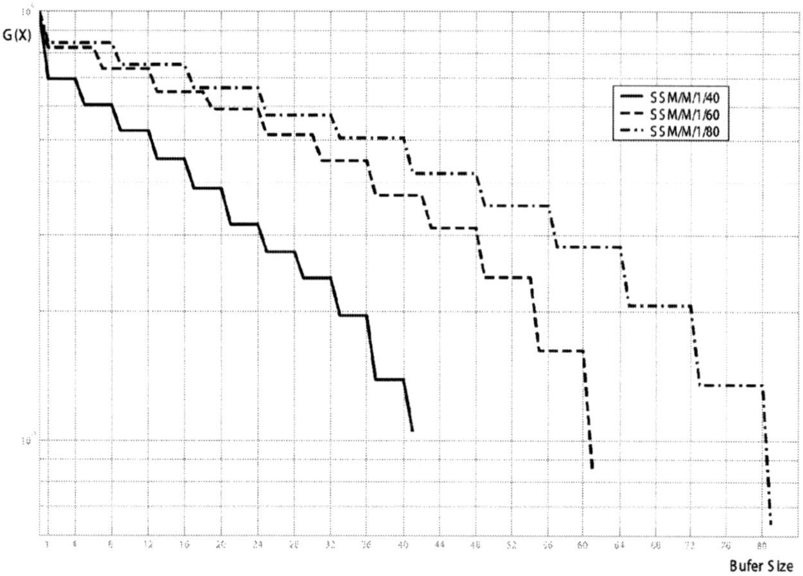

in traditional single server finite buffer queue M/M/1/B. The obtained results in logarithmic scale are shown in Figure 4.

To obtain representative and steady results the sequences of 10,000 observations were used. With the suggested LRE algorithm the values of complementary cumulative function $G(x)$ for different buffer sizes were calculated. The calculations were provided with the step m=4. As is seen in Figure 4, the increasing Hurst parameter leads to insignificant decrease in overflow probability. For example, for the value of Hurst parameter H=0.6 the overflow probability is $G(L)=10.45*10^{-2}$, and for H=0.9 it is $G(L)=5.6*10^{-2}$. On the other side the overflow probability of self-similar queuing system increases significantly in contrast to the theoretical M/M/1/B queuing system, for which $G(L)=4.79*10^{-5}$. Then, the simulation was repeated for SSM/M/1/B queuing system by using self-similar arrival process with H=0.6 and different buffer sizes. The obtained results for buffer size B=40, B=60 and B=80 are demonstrated in Figure 5.

As is seen in Figure 5, when the buffer size was increased twice, the overflow probability was changed simply in one order—from $10.45*10^{-2}$ to $6.4*10^{-3}$. It confirmed that to design the single server finite buffer model with self-similar arrival processes, the buffer size has to be increased many times in order to decrease the overflow probability.

Sequential Generators of Self-Similar Traffic in Wireless IP Networks

As confirmed above, we are able to construct a sequential Markovian model that imitates a self-similar sequence. A disadvantage of this approach is in a connection between the model parameters and its self-similar properties that is difficult to understand. Markovian models for self-similar wireless IP network traffic require including several control parameters with a wide range of input values.

As a result, controlling these values in sequential generators is much more complex than in generators of fixed-length sequences of self-similar processes with a given Hurst parameter (Law & Kelton, 2006). The following sequential generators are developed by Radev and Lokshina (2007), based on:

- Fractal-Binomial-Noise-Drivent Poisson Processes (FBNDP);
- Superposition of Fractal Renewal Processes (SFRP);
- $M/G/\infty$ Processes (MGIP);
- Pareto-Modulated Poisson Processes (PMPP);
- Spatial Renewal and FGN Processes;
- Superposition of Autoregressive Processes (SAP).

For the standard Fractal Renewal Process (FRP), inter-event times are independent random variables (Faraj, 2000). The marginal Probability Density Function (PDF) of such a FRP can be defined as:

$$f(t) = \begin{cases} 0, & t \leq A, \\ \delta A^{\delta} t^{-(\delta+1)}, & t > A \end{cases} \tag{23}$$

where $0<\delta<2$. Selecting δ in this interval proves far superior to $0<\delta<1$ for the same required values that the inter-event time PDF can be further improved (Borst & Mitra, 1998).

The improved PDF of the FRP decays as a power law, as shown in:

$$f(t) = \begin{cases} \delta A^{-1} e^{-\delta t/A}, & 0 < t \leq A, \\ \delta e^{-\delta} A t^{-(\delta+1)}, & t > A, \end{cases} \tag{24}$$

which is continuous for all t, producing smoother spectral density function.

A method, based on the FBNDP processes adds M independent and identically distributed (iid) alternating FRP to generate a Fractal Binomial Noise (FBN) process that serves as the rate function for a Poisson process.

The FBNDP requires five input parameters to generate self-similar sequences: A, δ, R, Δt, and M. The resulting Hurst parameter H assumes the value $(\alpha+1)/2$. If S is a simulation clock, which advances in time, and $S^{(j)}$ is the elapsed time of the j-th FRP sequence, then $S^{(j)}=\tau_0^{(j)}+\tau_1^{(j)}+\ldots+\tau_k^{(j)}$ for some k and $j=1,2,\ldots,M$, where $\tau_k^{(j)}$ is the inter-arrival time. The sequence of self-similar pseudo-random numbers $X_0, X_1, \ldots$ is generated through the algorithms developed by Radev and Lokshina (2007).

Fixed-Length Sequence Generators of Self-Similar Traffic in wireless IP Networks

The most frequently studied discrete-time models of self-similar traffic belong to Fractional Autoregressive Integrated Moving-Average (F-ARIMA) and FGN processes. Radev and Lokshina (2007) considered the following possible fixed-length sequence generators based on:

- F-ARIMA processes;
- Fast Fourier Transform (FFT);
- FGN and Daubechies Wavelets (DW) (Daubechies, 1992);
- Random Midpoint Displacement (RMD);
- Successive random additions.

The use of self-similar time series is very important, because the classical approach for describing the processes in telecommunications systems via generating the calls with standard statistical distributions is not valid for current implementation similar to wireless IP networks, where the LRD traffic is observed.

The shortcoming of the normal self-similar time series is related to the fact that it cannot describe at the same time different telecommunications applications. Therefore, a suggestion is made to multiply the values of self-similar time series with those standard statistical distributions that are known to be used to adequately describe different applications in communication systems. However, such approach does not work well regarding the time parameter. Various telecommunications network traffic flows are simulated with different statistical distributions that could indirectly describe the time parameter (in frequency domain). The obtained values show self-similar and LRD characteristics in wide scales and are appropriate for cluster analysis in frequency domain.

Let us consider F-ARIMA(0,d,0) method for generating self-similar sequences, where d is the fractional differencing parameter, $0<d<1/2$, and let us generate the process $X=\{X_i: i=0,1,2,\ldots,n\}$ with a normal marginal distribution, the mean of zero and the variance σ_0^2, and the Autocorrelation Function (ACF), $\{\rho_k\}$ ($k=0,\pm1,\ldots$) defined as:

$$\rho_k = \gamma_k / \gamma_0 = \frac{\Gamma(1-d)\Gamma(k+d)}{\Gamma(d)\Gamma(k+1-d)}, \qquad (25)$$

where:

$$\gamma_k = \sigma_0^2 \frac{(-1)^k \Gamma(1-2d)}{\Gamma(k-d+1)\Gamma(1-k-d)}. \qquad (26)$$

The fixed-length sequence of self-similar pseudo-random numbers could be generated through the following steps of Algorithm 2.

Algorithm 2:

Set $N_0=0$ *and* $D_0=1$. X_0, *the first pseudo random element in the output self-similar sequence, is generated from the normal distribution* $N(0,\sigma_0^2)$, where σ_0^2 is the required *variance of the* X_i.
($i=1,\ldots,n-1$). *Compute* $mean_i$ *and* var_i *of* X_i *recursively, using the following equations:*

$$N_i = \rho_i - \sum_{j=1}^{i-1} \varphi_{i-1,j}\rho_{i-j},$$

$$D_i = D_{i-1} - N_{i-1}^2 / D_{i-1}, \ \varphi_{ii} = N_i / D_i,$$

$$\varphi_{ij} = \varphi_{i-1,j} - \varphi_{ii}\varphi_{i-1,i-j}, \qquad j = 1,\ldots,i-1,$$

where φ_{ij}, $i=0$, $j=0,\ldots,n-1$, is given by:

$$\varphi_{ij} = -\binom{i}{j}\frac{(j-d-1)!(i-d-j)!}{(-d-1)!(i-d)!},$$

$$mean_i = \sum_{j=1}^{i} \varphi_{ij} X_{i-j}, \ \mathrm{var}_i = (1-\varphi_{ii}^2)\,\mathrm{var}_{i-1},$$

Generate X_i *from* $N(mean_i, var_i)$. *Increase i by* 1. *If* $i=n$, *then stop.*

It took 4 hours, 1 minute, 24 seconds on the Centrino-based HP PC (1.76 GHz) to generate the F-ARIMA traffic sample sequence with 1,048,576 numbers. Therefore, the algorithm, based on the F-ARIMA method, is too rigorous to be used to generate long sample sequences.

A new generator of pseudo-random self-similar sequences based on FGN and DW, called FGN-DW approach, is developed by Radev and Lokshina (2007). The DW are used because the generators based on DW produce more accurate self-similar sequences than those based on Haar Wavelets (HW) (Abry, Flandrin, Taqqu, & Veitch, 2000). In that case not only estimates of H obtained from the DW are closer to the true values than those from the HW, but also variances obtained from the DW are much lower. The reason behind is that the DW produce smoother wavelet coefficients that are used in the Discrete Wavelet Transform (DWT), than the HW. The HW are discontinuous, and they do not have good time-frequency localisation properties, since their Fourier transforms decay as $|\lambda|^{-1}$, $\lambda\to\infty$, meaning that resulting decomposition has a poor scale. Therefore, the DW produce more accurate coefficients than the HW.

The suggested approach for generating synthetic self-similar FGN sequences in a time domain is based on a DWT. Wavelets could provide compact representations for a class of FGN processes, because the structure of wavelets naturally matches the self-similar structure of the LRD processes.

This paper shows that the FGN-DW method is sufficiently fast being used for practical generation of synthetic self-similar sequences as simulation input data. The wavelet analysis transforms a sequence onto a time-scale grid, where the term scale is used instead of frequency, because the mapping are not directly related to frequency as in the Fourier transform. The wavelet transform generates the wavelet coefficients $d_x(i,j)$ from a sequence of given numbers. For a LRD process, the variance of the wavelet coefficients at each level i is defined by:

$$E[d_x(j,\cdot)^2 = CP_f 2^{i(2H-1)}, \quad (27)$$

where $C>0$ and is true:

$$P_f = 2(2\pi)^{1-2H} c_\gamma E(2H-1)\sin((1-H)\pi), \quad (28)$$

where $E(\cdot)$ is the Euler function and c_γ is a positive constant. The power parameter P_f that plays a major role in fixing the absolute size of the LRD process and generates effects in the applications is an independent quantitative parameter with the dimensions of variance.

A spectral estimator can be obtained though calculating time average Q_i of the $d_x(i,\cdot)$ at a given scale that could be described as:

$$Q_i = \frac{1}{n_i}\sum_{j=1}^{n_i} d_x^2(i,j), \quad (29)$$

where n_i is the number of wavelet coefficients at scale i, i.g. $n_i=2^{-i}n$, where n is the number of data points. The estimator uses a weighted linear regression as the variance of $\log_2(Q_i)$ vary with $\log_2(2^i)$. An estimated Hurst parameter $\hat{H}$ could be obtained with the linear regression:

$$y_i = \log_2(Q_i) = (2\hat{H}-1)i + c + 1/(n_i \ln 2) \quad (30)$$

where c is a constant.

The wavelet transform delivers good resolution in both time and scale, as compared to the Fourier transform, which provides only good frequency resolution. The fixed-length sequence of self-similar pseudo-random numbers could be generated with the recently developed Algorithm 3 that consists of the following steps provided below.

Algorithm 3:

Given: H, Start for i=1 and continue until i=n. Calculate a sequence of values $\{f_1,f_2,\ldots,f_n\}$ *using*

$$f(\lambda,H) = c_f \mid\lambda\mid^{1-2H} + O(\mid\lambda\mid^{\min(3-2H,2)}), \quad (31)$$

where $c_f = \sigma^2 (2\pi)^{-1}\sin(\pi H)\Gamma(2H+1)$, $O(\cdot)$ *represents the residual error and* $f_i = \hat{f}(\pi i / n; H)$,the value of the frequencies f_i corresponds to the spectral density of an FGN process for f_i ranging between $\pi/n \div \pi$.

- *Multiply* $\{f_i\}$ *by realizations of the independent exponential random variable with the mean of one to obtain* $\{\hat{f}_i\}$, *because the spectral density estimated for a given frequency is distributed asymptotically as the independent exponential random variable with the mean f(λ,H).*
- *Generate a sequence* $\{Y_1,Y_2,\ldots,Y_n\}$ *of complex numbers such that* $\mid Y_i\mid = \sqrt{\hat{f}_i}$ and the phase of Y_i is uniformly distributed between 0 and 2π. This random phase technique preserves the spectral density corresponding to $\{\hat{f}_i\}$. *It also makes the marginal distribution of the final sequence normal and produces the requirements for FGN.*
- *Calculate two synthetic coefficients of orthonormal Daubechies wavelets. The*

Figure 6. Self-similar time series with 1,000,000 values

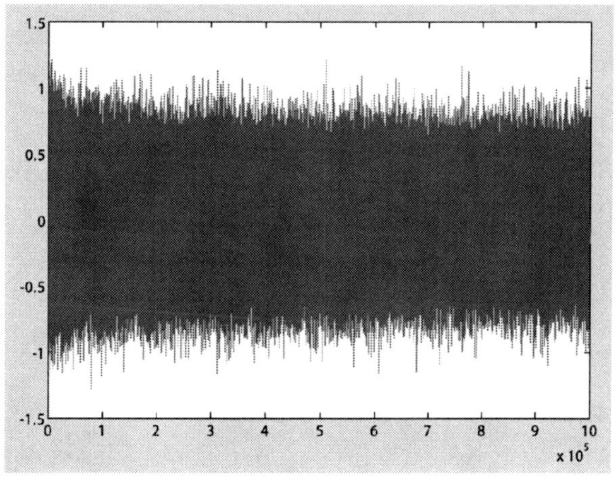

output sequence $\{X_1, X_2, ..., X_n\}$ representing approximately self-similar FGN process is obtained by applying the inverse Daubechies wavelets transformation of the sequence $\{Y_1, Y_2, ..., Y_n\}$.

It took only 2 seconds on Centrino-based HP PC (1.76 GHz) to generate a sequence of 1,048,576 numbers using this algorithm. Analysis of the mean times required to generate sequence of a given length demonstrates that the sequential generators are more attractive for practical simulation studies of the computer networks than the F-ARIMA-based generators, since they are much faster. However, these generators require more input parameters, and selecting appropriate values is a problem that remains.

FGN – DW ALGORITHM PERFORMANCE AND SIMULATION RESULTS

Let us consider generation of self-similar time series with the help of Algorithm 3 with one million values Hurst parameter H=0.5; time scaling factor equal to 10 and number of wavelet coefficients equal to 10. The obtained results are shown in Figure 6.

For the generated time series the values $Q(j_1)$ were calculated as function of j_1, as is shown on Figure 7. When linear regression (20) was implemented, better representation was obtained if to compare to the analytical values, as shown in Figure 8.

The obtained results were estimated based on correspondence criteria with parameter Q=0.16155, and time series with equivalent length of 1,000,000 values were generated with Pareto distribution and α=1.8. Then, they were multiplied with self-similar time series, generated with Algorithm 3. The implementation of wavelet criteria provided good results for $Q(j_1)$ as function of j_1, which were very similar to the results presented in Figure 7.

Considering the whole time interval (j_1, j_2)=[1,16], the optimal values were registered at (j_1, j_2)=[2,13], in addition to the fact that the correspondence criteria parameter Q increased from 0.16155 to 0.36899. The linear regression provided an opportunity to define the Hurst parameter with better precision (0.502 versus 0.504, for the exact value of 0.5).

The logarithmic diagram obtained with weighted linear regression is shown in Figure 9.

Figure 7. Comparison of self-similar time series with different scale parameters

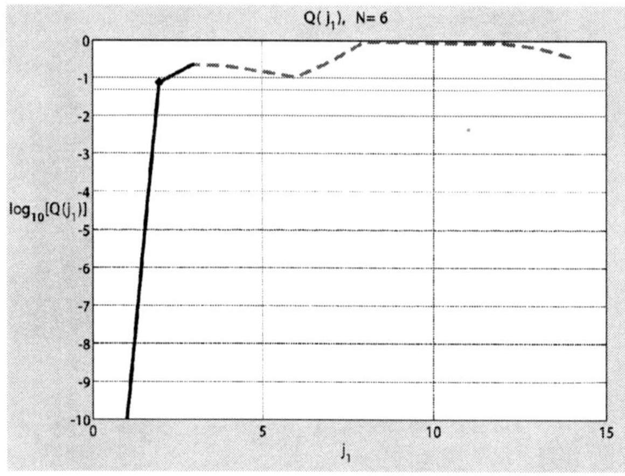

Figure 8. Comparison between generated time series and analytical values

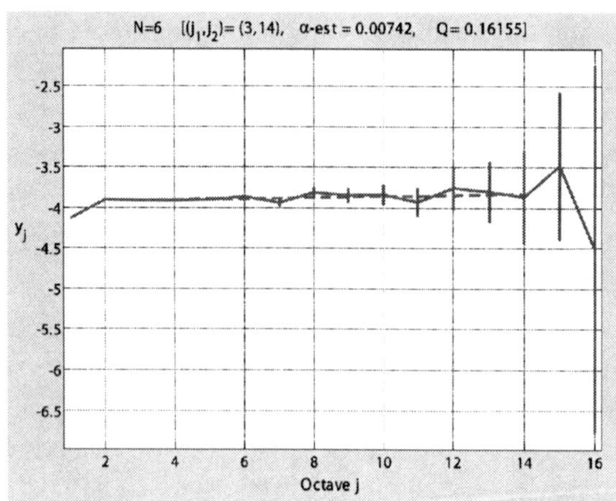

Figure 9. Comparison between generated time series and analytical values obtained with weighted linear regression

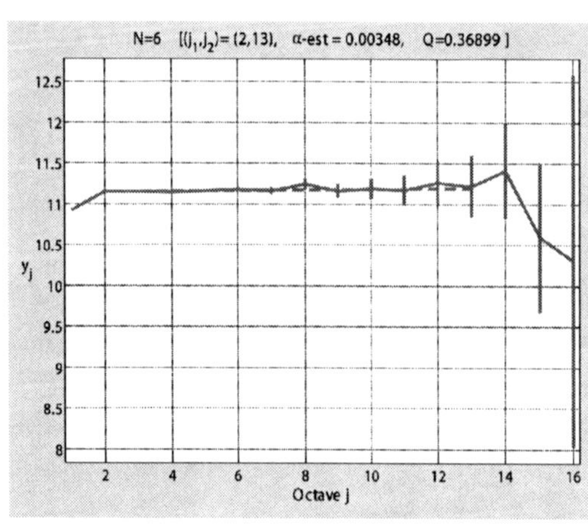

The confidential intervals are shown with use of vertical lines. They are small for j_1<13 and obtained with confidential probability of 0.95 after generation of 2^{10} (1,024) samples.

CONCLUSION

Wireless IP network traffic exhibits self-similar properties over a wide range of time scales that are very different from the properties of traditional models. Self-similar models are appropriate for telecommunications network traffic as they provide capacity to estimate the network performance/quality, allocate the resources, and ensure the QoS.

One of the problems that computer network researchers face during simulation is how to generate long synthetic sequential self-similar sequences. To solve the problem three aspects must be considered: (a) how accurately self-similar process can be generated; (b) how quickly the methods generate long self-similar sequences, and (c) how appropriately self-similar processes can be used in sequential simulation.

Steady-state simulation of self-similar queuing processes was performed using the RESTART method. The new algorithm to simulate the buffer overflow probability was developed to generate self-similar traffic with limited relative error. The buffer overflow simulation for finite buffer single server model under self-similar traffic load was described.

The implementation of the suggested simulation approach with stochastic and long range dependent wireless IP network source models was demonstrated. The models of SSM/M/1/40 queuing system were considered with different parameters of self-similar arrival processes and different buffer sizes.

The algorithms developed on the base of different methods for fixed-length sequence generators were improved and used for modeling and simulation of self-similar wireless IP network traffic. The wavelet criteria to estimate the precision of self-similar time series generation was applied. The DWT provided better results for the long range dependent self-similar processes when implementing the scaling regression. Numerical examples and simulation results were provided.

REFERENCES

Abry, P., Flandrin, P., Taqqu, M. S., & Veitch, D. (2000). Wavelets for the analysis, estimation and synthesis of scaling data. In K. Park and W. Willinger (Eds.), *Self-similar network traffic analysis and performance evaluation* (pp. 39-88). New York: John Wiley & Sons.

Bobbio, A., Horváth, A., Scarpa, M., & Telek, M. (2003). Acyclic discrete phase type distributions: Properties and a parameter estimation algorithm. *Performance Evaluation*, *54*(1), 1–32. doi:10.1016/S0166-5316(03)00044-0

Boxma, O., & Cohen, J. (2000). *Self-similar network traffic and performance evaluation.* New York: John Wiley & Sons.

Bucklew, J. A. (2004). *Introduction to rare event simulation.* New York: Springer-Verlag.

Daubechies, I. (1992). *Ten lectures on wavelets* (CBMS-NFS regional conference series in applied mathematics, vol. 61). Philadelphia: SIAM Press.

Faraj, R. (2000). *Modeling and analysis of self-similar traffic in ATM networks.* Unpublished PhD thesis, Concordia University.

Georg, C., & Fuss, O. (1999, June). Simulating rare event details of ATM delay time distributions with RESTART/LRE. In *Proceedings of the IEEE International Teletrafic Congress (ITC16),* Edinburgh, Scotland (pp. 777-786). Elsevier.

Georg, C., & Schreiber, F. (1996, December 8-11). The RESTART/LRE method for rare event simulation. In *Proceedings of the Winter Simulation Conference*, Coronado, CA (pp. 390-397). IEEE.

Giambene, G. (2005). *Queueing theory and telecommunications: Networks and applications*. New York: Springer.

Hayes, J., & Ganesh Babu, T. (2004). *Modeling and analysis of telecommunications networks*. New York: John Wiley & Sons.

Heidelberger, P. (1995). Fast simulation of rare events in queueing and reliability models. *ACM Transactions on Modeling and Computer Simulation*, *5*, 43–85. doi:10.1145/203091.203094

Horn, G., Kvalbein, A., Blomskøld, J., & Nilsen, E. (2007). An empirical comparison of generators for self similar simulated traffic. *Performance Evaluation*, *64*(2), 162–190. doi:10.1016/j.peva.2006.06.005

Ilnickis, S. (2004). Modelling telecommunication systems in case of self-similar input flow. *Scientific Proceedings of Riga Technical University*, *17*, 137–142.

Jeong, H.-D.-J. (2002). *Modelling of self-similar teletraffic for simulation*. Unpublished PhD thesis, University of Centerbury, UK.

Kushner, H. (2001). *Heavy traffic analysis of controlled queueing and communications networks*. New York: Springer.

L'Ecuyer, P. (2001). Random number generators. In S. I. Gass & C. M. Harris (Eds.), *Encyclopedia of operations research and management science* (pp. 695-702). Norwell, MA: Kluwer Academic Publishers.

L'Ecuyer, P., Demers, V., & Tuffin, B. (2006, December 3-6). Splitting for rare-event simulation. In *Proceedings of the 2006 Winter Simulation Conference,* Monterey, CA (pp. 137-148).

Law, A. M., & Kelton, W. D. (2006). *Simulation modeling and analysis* (4th ed.). New York: McGraw Hill.

Marie, R. R., Blackledge, J. M., & Bez, H. E. (2007, February). Characterization of internet traffic using a fractal model. In *Proceedings of the Fourth IASTED International Conference on Signal Processing, Pattern Recognition, and Applications* (*SPPRA 2007*), Innsbruck, Austria (pp. 156-167). ACTA Press.

Mehdi, J. (2003). *Stochastic models in queueing theory.* New York: Academic Press.

Park, K., & Willinger, W. (Eds.). (2000). *Self-similar network traffic and performance evaluation*. New York: John Wiley & Sons.

Popescu, A. (2001, June). Traffic self-similarity. In *Proceedings of the IEEE International Conference on Telecommunications (ICT 2001),* Bucharest, Romania (pp. 20-24). IEEE.

Radev, D. (2005, September 28-30). Stochastic process generation for traffic flow simulation. In *Proceedings of the IEEE Conference on Telecommunications in Modern Satellite, Cable and Broadcasting Services (TELSIKS 2005)*, Nis, Serbia (Vol. 2, pp. 559-562). IEEE.

Radev, D., & Lokshina, I. (2007, June 11-13). Modeling and simulation of self-similar teletraffic. In *Proceedings of the Industrial Simulation Conference 2007*, Delft, The Netherlands (pp. 355-359).

Radev, D., & Lokshina, I. (2009, April 22-24). Modeling and simulation of self-similar wireless IP network traffic. *Proceedings of the IEEE Wireless Telecommunications Symposium (WTS) 2009,* Prague, Czech Republic (pp. 1-6). IEEE.

Radev, D., Rashkova, E., & Stankovski, D. (2007, June). Simulation modelling of overflow probabilities in tandem queuing system. In *Proceedings of XLII International Science Conference on Information, Communication and Energy Systems and Technologies (ICEST 2007),* Ohrid, Macedonia (pp. 123-127).

Stathis, C., & Maglaris, B. (2000). Modeling the self-similar behaviour of network traffic. *Journal of Computer Networks*, *34*(1), 37–47. doi:10.1016/S1389-1286(00)00095-5

Tan, P.-H. (1999). *Statistical analysis of long-range dependent processes via a stochastic intensity approach, with applications in networking*. Unpublished PhD thesises, North-Carolina State University.

Trivedi, K. (2001). *Probability and statistics with reliability, queueing, and computer science applications*. New York: John Willey & Sons.

Villen-Altamirano, M., & Villen-Altamirano, J. (2006). On the efficiency of RESTART for multidimensional systems. *ACM Transactions on Modeling and Computer Simulation*, *16*(3), 251–279. doi:10.1145/1147224.1147227

This work was previously published in International Journal of Interdisciplinary Telecommunications and Networking, Volume 1, Issue 4, edited by Michael R. Bartolacci, and Steven R. Powell, pp. 71-87, copyright 2009 by IGI Publishing (an imprint of IGI Global).

Compilation of References

Abell, D. F., & Hammond, J. S. (1979). *Strategic market planning: Problems and analytical approaches*. Englewood Cliffs, NJ: Prentice-Hall.

Abilene Christian University. (2008a). *Mobile learning and the connected campus*. Retrieved from http://www.acu.edu/technology/mobilelearning/index.html

Abilene Christian University. (2008b). *ACU first university in nation to provide iPhone or iPod touch to all incoming freshmen*. Retrieved from http://www.acu.edu/news/2008/080225_iphone.html

Abilene Christian University. (2009). *Solutions for the 21st century university*. Retrieved from http://www.acu.edu/technology/mobilelearning/vision/solutions/index.html

Abry, P., Flandrin, P., Taqqu, M. S., & Veitch, D. (2000). Wavelets for the analysis, estimation and synthesis of scaling data. In K. Park and W. Willinger (Eds.), *Self-similar network traffic analysis and performance evaluation* (pp. 39-88). New York: John Wiley & Sons.

Accenture. (2006). *Handset industry evolution: navigating a path to high performance*. Accenture's reports on Supply Chain Management.

Aedudodla, S. R., Vijayakumaran, S., & Wong, T. F. (2005). Timing Acquisition in Ultra-wideband Communication Systems. *IEEE Transactions on Vehicular Technology*, (54): 1570–1583. doi:10.1109/TVT.2005.855495

Aedudodla, S. R., Vijayakumaran, S., & Wong, T. F. (2006, April). Acquisition of Direct Sequence Transmitted Reference Ultra-Wideband Signals. *IEEE Journal on Selected Areas in Communications*, *l*(24), 759–765. doi:10.1109/JSAC.2005.863825

Afuah, A., & Tucci, C. (2001). *Internet Business Models and Strategies* (International Editions). New York: McGraw-Hill.

Agrawal, G., Oikonomou, K. N., & Sinha, R. K. (2007). Network Performability Evaluation for Different Routing Schemes. *Proceedings of the OFC,* Anaheim, CA.

Ahonen, T. T., & Moore, A. (2005, March). *Communities Dominate Brands*. Business and marketing challenges for the 21st century. Futuretext Ltd.

Ahuja, R., Magnanti, T., & Orlin, J. (1998). *Network Flows*. Prentice-Hall.

Akhtar, N., Callejo, M. A., & Colas, J. A. (2005). Making migration easy: a key requirement for systems beyond 3G. Vehicular Technology Conference, 5, 2974-2978.

Allee, V. (2003). *The Future of Knowledge: Increasing Prosperity through Value Networks*. Butterworth-Heinemann.

Alleman, J., & Noam, E. (1999). *Real Options: The New Investment Theory and its Implications for Telecommunications Economics*. Kluwer Academic Publishers.

Alleman, J., & Rapport, P. (2006). Optimal Pricing with Sunk Cost and Uncertainty. In Cooper, R., Lloyd, A., Madden, G., & Schipp, M. (Eds.), *The Economics of Online Markets and Information Communications and Technology Networks: Contribution to Economics Series*. Heidelberg: Physica-Verlag.

Altera Inc. (2007). Data Books, http://www.altera.com.

ALU. (2008). *Creating more time for patient care with dynamic communication*. Retrieved from http://enterprise.alcatel-lucent.com/private/active_docs/EPG0550090601_Healthcare_EN_StraWhitePaper.pdf

ALU. (2009). *Healthcare solutions*. Retrieved from http://enterprise.alcatel-lucent.com/?solution=Healthcare&page=Homepage

Alvarez, G., Uysal, M., & Merchant, A. (2001). Efficient verification of performability guarantees. *PMCCS-5: The Fifth International Workshop on Performability Modelling of Computer and Communication Systems*.

Amram, M., & Kulatilaka, N. (1998). *Real Options: Managing Strategic Investment in an Uncertain World*. Harvard Business School Press.

An Analytical Cost Model-National Core Network Version 2.0. Consultative Document 2.0. (2000). Retrieved January 20, 2007, from, http://www.regtp.de/en/reg_tele/start/in_05-07-00-00-00_m/fs.html

Anantvalee, T., & Wu, J. (2006). A survey on intrusion detection in mobile ad hoc networks. In Y. Xiao, X. Shen, & D.-Z. Du (Eds.), *Wireless mobile networks security* (pp. 170-196). New York: Springer.

Anderson, C. (2006). *The Long Tail*. London: Random House.

Anderson, S. C. R. (2003). *Design of a transmitter for Ultra Wideband Radi*. Master dissertation, University of Linköpings, Linkoping.

Antonio, O. R. L. N., & Ricardo, R. (2005). Mexico's international telecommunications policy: Origins, the WTO dispute, and future challenges. *Telecommunications Policy*, *29*(5-6), 429–448. doi:10.1016/j.telpol.2005.03.005

Aven, T., & Jensen, U. (1999). *Stochastic Models in Reliability*. Springer-Verlag, 1999.

Balachandra, R., Goldschmitt, M., & Friar, J. H. (2004). The evolution of technology generations and associated markets: a double helix model. *IEEE Transactions on Engineering Management*, *51*(1), 3–12. doi:10.1109/TEM.2003.822463

Balfanz, D., Smetters, D. K., Stewart, P., & Wong, H. C. (2002). *Talking to strangers: Authentication in ad-hoc wireless networks*. In NDSS.

Banerjee, A., Faloutsos, M., & Bhuyan, L. (2008). The P2P War: Someone is Monitoring your Activities. *Computer Networks*, *52*(6), 1272–1280. doi:10.1016/j.comnet.2008.01.011

Basagni, S., Chalamtac, I., Syrotiuk, V. R., & Woodward, B. A. (1998, October). A Distance Routing Effect Algorithm for Mobility. In ACM MOBICOM'98. DREAM.

Basili, M., & Fontini, F. (2003). The Option Value of the UK 3G Telecom licenses. The journal of Policy. *Regulation and Strategy for Telecommunications*, *5*(3), 48–52. doi:10.1108/14636690310487246

Benaroch, M. (2002). Managing Investments in Information Technology Based on Real Options Theory. *Journal of MIS*, *19*(2), 43–84.

Benaroch, M., & Kauffman, R. J. (1998). A Case for using Real Options Pricing Analysis to Evaluate Information Technology Project Investments. *Information Systems Research*, *10*(1), 70–86. doi:10.1287/isre.10.1.70

Benaroch, M., & Kauffman, R. J. (2000). Justifying Electronic Banking Network Expansion Using Real Options Analysis. *Management Information Systems Quarterly*, *24*(2), 197–225. doi:10.2307/3250936

Berkowitz, L. (1990). On the formation and regulation of anger and aggression. A cognitive-neoassociationistic analysis. *The American Psychologist*, *45*(4), 494–503. doi:10.1037/0003-066X.45.4.494

Berners-Lee, T. (1989). *Tim Berners-Lee*. Retrieved from http://www.w3.org/People/Berners-Lee

Bertsekas, D. P. (1998). *Network Optimization: Continuous and Discrete Models*. Athena Scientific.

Black, F., & Sholes, M. (1973). The Pricing of Options and Corporate Liabilities. *The Journal of Political Economy*, *81*, 637–654. doi:10.1086/260062

Blazquez, R., Newaskar, P., & Chandrakasan, A. (2003, April). Coarse acquisition for ultrawideband digital receivers. In *Acoustics, Speech and Signal Processing*, 137–140, IEEE.

Bley, A., Koch, T., & Wessäly, R. (2004). Large-scale Hierarchical Networks: How to Compute an Optimal Architecture? *11th International Telecommunications Network Strategy and Planning Symposium. Networks*, *2004*, 429–434.

Bobbio, A., Horváth, A., Scarpa, M., & Telek, M. (2003). Acyclic discrete phase type distributions: Properties and a parameter estimation algorithm. *Performance Evaluation, 54*(1), 1–32. doi:10.1016/S0166-5316(03)00044-0

Bolch, G., Greiner, S., de Meer, H., & Trivedi, K. S. (2006). *Queueing Networks and Markov Chains*. Wiley. doi:10.1002/0471791571

Book, S. A., & Young, P. H. (1997). General-Error Regression for Deriving Cost-Estimating Relationships. *The Journal of Cost Analysis* (pp. 1-28).

Boukerch, A., Xu, L., & El-Khatib, K. (2007). Trust-based security for wireless ad hoc and sensor networks. *Computer Communication Journal, 30*, 2412–2427.

Boxma, O., & Cohen, J. (2000). *Self-similar network traffic and performance evaluation.* New York: John Wiley & Sons.

Bradu, D., & Mundlak, Y. (1970, March). Estimating in Lognormal Linear Models. *Journal of the American Statistical Association, 65*, 198–211. doi:10.2307/2283587

Braga, C.-A. P., Forestier, E., & Stern, P. A. (1999). Developing Countries and Accounting Rates Reform--A Technological and Regulatory El Niño? *Public Policy Journal*, (173). Retrieved May 14, 2008, from http://rru.worldbank.org/PublicPolicyJournal/Summary.aspx?id=173

Breed, W. (1955). Social control in the newsroom: A functional analysis. *Social Forces, 33*(4), 326–335. doi:10.2307/2573002

Brennan, M. J., & Schwartz, E. S. (1985). Evaluating Natural Resources Investments. *The Journal of Business, 58*, 135–157. doi:10.1086/296288

Brundick, B. (1995). *Parametric Cost Estimating Handbook.* Retrieved April 20, 2008, from http://cost.jsc.nasa.gov/PCEHHTML/pceh.htm

BT's 21CN Technology -21CN domain descriptions. (2008). Retrieved June 16, 2008, from http://www.btplc.com/21CN/Thetechnologyofthenetwork/21CNdomaindescriptions/21CNdomaindescriptions.htm

Bucklew, J. A. (2004). *Introduction to rare event simulation.* New York: Springer-Verlag.

Budde (2007). U.S.A-Mobile Market–Analysis, Statistics and Forecasts. Reports. (http://www.budde.com).

Butler, P., Hall, T. W., Hanna, A. M., Mendonca, L., Auguste, B., Manyika, J., & Sahay, A. (1997). A revolution in interaction. *The McKinsey Quarterly, 1/1997.*

Cabral, E. A., Erkut, E., & Laporte, G. (2007). The network design problem with relays. *European Journal of Operational Research, 180*(2), 834–844. doi:10.1016/j.ejor.2006.04.030

Callaway, E., Gorday, P., Hester, L., Gutierrez, J. A., Naeve, M., Heile, B., & Bah, V. (2002, August). Home networking with IEEE 802.15.4: A developing standard for low-rate wireless personal area network. *IEEE Communications Magazine*, (31): 70–77. doi:10.1109/MCOM.2002.1024418

Camp, T., Boleng, J., & Wilcox, L. (2002). *Location information services in mobile ad hoc networks* (pp. 3318–3324).

Camponovo, G., & Pigneur, Y. (2003). Business Model Analysis Applied to Mobile Business. *Presented at 5th International Conference on Enterprise Information Systems* (ICEIS 2003, Angers, 23-26 April).

Capkun, S., Hubaux, J.-P., & Buttyan, L. (2006). Mobility Helps Peer-to-Peer Security. *IEEE Transactions on Mobile Computing, 5*, 43–51. doi:10.1109/TMC.2006.12

Capkun, S., Hubaux, J.-P., & Buttyan, L. (2003). Mobility helps security in ad hoc networks. In *MobiHoc '03: Proceedings of the 4th ACM international symposium on Mobile ad hoc networking & computing*, 46–56. New York, NY: ACM Press.

Carlier, J., Li, Y., & Lutton, J. (1997). Reliability evaluation of large telecommunication networks. *Discrete Applied Mathematics*, 76.

Caronni, G. 2000. Walking the Web of Trust. *Proceedings of IEEE 9th International Workshops on Enabling Technologies: Infrastructure for Collaborative Enterprises* (pp. 153-158).

Carr, P. (1988). The Valuation of sequential exchange opportunities. *The Journal of Finance, 23*(5), 1235–1256. doi:10.2307/2328217

Carroll, R. L. (1985). Content values in TV news programs in small and large markets. *The Journalism Quarterly*, *62*(4), 877–938.

Carsello, R. D., Meidan, R., Allpress, S., O'Brien, F., Tarallo, J. A., & Ziesse, N. (1997). IMT-2000 standards: radio aspects. *IEEE Personal Communications*, *4*(4), 30–40. doi:10.1109/98.612275

CDG. (2007). Quick Market Statistics. CDMA Development Group. (http://www.cdg.org).

Chan, H., Perrig, A., & Song, D. X. (2003). Random key predistribution schemes for sensor networks. In *IEEE Symposium on Security and Privacy*, p. 197.

Chang-Joo, K. (1998, May). Adaptive hybrid acquisition of PN sequencesfor DS/SS communications. *Electronics Letters*, (34): 939–940.

Chen, J., Reilly, R. R., & Lynn, G. S. (2005). The impacts of speed-to-market on new product success: the moderating effects of uncertainty. *IEEE Transactions on Engineering Management*, *52*(2), 199–212. doi:10.1109/TEM.2005.844926

Chen, C., Lemberg, H., Philips, S., den Berg, E. V., & Zhang, T. (2002, March). *Slalom: A scalable location management scheme for large mobile ad-hoc networks.*

Chen, J., & Zhou, Z. (2005, March). Overview of Synchronization in DS-UWB. In *ISCIT 2005*, 951-954. IEEE.

Cherry, S. (2004, July). Edholm's law of bandwidth. *IEEE Spectrum*,58–60. doi:10.1109/MSPEC.2004.1309810

Chochliouros, I., & Spiliopoulou, A. (2003). Innovative horizons for Europe: The new European telecom framework for the development of modern electronic networks and services. *The Journal of the Communications Network*, *2*(4), 53–62.

Chochliouros, I., Spiliopoulou, A., Chochliouros, S., & Doukoglou, T. (2007) European challenges towards forming and promoting an innovative radio-spectrum policy in a fully converged electronic communications market. *The Journal of the Communications Network, 6*(2), 60-67. Retrieved January 15, 2009, from http://theitp.org/UserData/root/Files/2007Vol6Prt2_08.pdf

Choong, M. K., Logeswaran, R., & Bister, M. (2007). Cost-effective Handling of Digital Medical Images in the Telemedicine Environment. *International Journal of Medical Informatics*, *76*(9), 646–654. doi:10.1016/j.ijmedinf.2006.05.007

Christensen, C., & Rosenbloom, R., S. (1995). Explaining the attacker's advantage: Technological paradigms, organizational dynamics, and the value network. *Research Policy*, *24*, 233–257. doi:10.1016/0048-7333(93)00764-K

Christensen, C. M. (1997). *The Innovator's Dilemma*. Harvard Business School Press.

Chugg, K. M., & Zhu, M. (2005, May). A New Approach to Rapid PN Code Acquisition Using Iterative Message Passing Techniques. *IEEE Journal on Selected Areas in Communications*, *l*(23), 884–897. doi:10.1109/JSAC.2005.845424

Clemons, E. K. (1991). Evaluating Strategic Investments in Information Systems. *Communications of the ACM*, *34*(1), 22–36. doi:10.1145/99977.99985

Clemons, E. K., and GU, B. (2003). Justifying Contingent IT Investments: Balancing the Need for Speed of Action with Certainty before Action. *Journal of Management Information Systems*, *20*(2), 36–47.

Climaco, J. N., & Martins, E. Q. (1982). A Bicriterion Shortest Path Algorithm. *European Journal of Operational Research,* 11.

Cohen, E. (2002). Online journalism as market-driven journalism. *Journal of Broadcasting & Electronic Media*, *46*(4), 532–548. doi:10.1207/s15506878jobem4604_3

Colbourn, C. J. (1999). Reliability Issues in Telecommunications Network Planning. In B. Sansó, (Ed.), *Telecommunications Network Planning*. Kluwer Academic Press. Computational Infrastructure for Operations Research. http://www.coin-or.org/.

Committee on Facilitating Interdisciplinary Research, National Academy of Sciences, National Academy of Engineering, and Institute of Medicine. (2005). *Facilitating Interdisciplinary Research*. Washington, DC: The National Academies Press.

Corne, D., Dorigo, M., & Glover, F. (1999). *New Ideas in Optimisation*. Berkshire, England: McGraw-Hill Publishing Company.

Cost Estimating Handbook, N. A. S. A. 2008. (2008). Retrieved July 1, 2008, from http://www.nasa.gov/offices/pae/organization/cost_analysis_division.html

Cost-efficient network architecture for addressing low-income segments. (2007). Retrieved April 20, 2008, from http://www.ericsson.com/technology/whitepapers/wp_nw_arch_%20for_low_arpu_rev_a.pdf

CTIA. (2006). CTIA's Semi-Annual Wireless Industry Survey.

Curwen, P., & Whalley, J. (2006). Measuring internationalization in the mobile telecommunications industry. *International Business Review*, *15*, 660–681. doi:10.1016/j.ibusrev.2006.09.002

d'Halluin, Y., Forsyth, A.P. & Vetzal, R.K. (2002). Managing Capacity for Telecommunications Network Under Uncertainty. IEEE/ACM Transactions onNetworking, 10(4), 579-588.

Dahlman, E., Gudmundson, B., Nilsson, M., & Skold, J. (1998). UMTS/IMT-2000 Based on Wideband CDMA. IEEE Communications Magazine, 70-80. Deng, S., Johnson, B. & Sogomonian, A. Exotic Electricity Options and the Valuation of Electricity Generation and Transmission Assets. *Decision Support Systems*, *3*(30), 383–392.

Das, S. M., Pucha, H., & Hu, Y. C. (2005, March). *Performance comparison of scalable location services for geographic ad hoc routing.*

Daubechies, I. (1992). *Ten lectures on wavelets* (CBMS-NFS regional conference series in applied mathematics, vol. 61). Philadelphia: SIAM Press.

de Bijl, P. W. J., & Peitz, M. (2006). *Access Regulation and the Adoption of VoIP*. International University in Germany; Working Paper No. 43/2006.

De Rango, F. (2009, April 22-24). *Improving SAODV protocol with trust levels management, IDM and incentive cooperation in MANET*. Paper presented at the Wireless Telecommunication Symposium (WTS'09), Prague, Czech Republic.

Demetrescu, C., & Italiano, G. (2001). Fully Dynamic All Pairs Shortest Paths with Real Edge Weights. *IEEE Symposium on Foundations of Computer Science.*

Department of Trade and Industry (DTI). (1996). *Spectrum management: Into the 21st century* (DTI White Paper). London: Author.

Deutsche Telekom. (2002-2007). *Annual and 20-F reports*. Bonn, Germany: Author.

Dias, M. A. G., & Rocha, K. M. C. (1999). Petroleum Concessions with Extendible Options using Mean Reversion with Jumps to Model oil Prices. The 3rd International Conference on Real Options, the Netherlands.

Dikaiakos, M., Florides, A., Nadeem, T., & Iftode, L. (2007, October). Location-aware Services over Vehicular Ad-Hoc Networks using Car-to-Car Communication. *IEEE Journal on Selected Areas in Communications, 15*(8), October 2007.

Dimitrova, D., Connolly-Ahern, C., Williams, A., & Paul, K. (2003). Hyperlinking as gatekeeping: Online newspaper coverage of the execution of an American terrorist. *Journalism Studies*, *4*(3), 401–414. doi:10.1080/14616700306488

Dixit, A. K., & Pindyck, R. S. (1994). *Investment under Uncertainty*. Princeton University Press.

Dixit, A.K. & Pindyck, R.S. (1995). The Options Approach to Capital Investment. Harvard Business Review.

Djahel, S., Naït-Abdesselam, F., Zhang, Z., & Khokhar, A. (2008). Defending against packet dropping attack in vehicular ad hoc networks. *Security and Communication Networks Journal*, *1*(3), 245–258. doi:10.1002/sec.25

Dos Santos, B. (1991). Justifying Investment in New Information Technologies. *Journal of Management Information Systems*, *7*(4), 71–89.

Doverspike, R. D., Li, G., Oikonomou, K. N., Ramakrishnan, K. K., & Wang, D. (2007). IP Backbone Design for Multimedia Distribution: Architecture and Performance. *Proceedings of the IEEE INFOCOM*

Dunning, J. H., & Lundan, S. M. (2008). Multinational enterprises and the global economy (2nd ed.). Cheltenham UK: Edward Elgar.

Economides, N. (1999). Real Options and the Cost of the Local Telecommunications Network. In Alleman, J., & Noam, E. (Eds.), *The New Investment Theory of real options and its Implications for the Cost Models in Telecommunications*. Kluwer Academic Publishers.

Economist Intelligence Unit. (2007). *Business in motion: Managing the mobile workforce*. Retrieved from http://enterprise.alcatel-lucent.com/private/active_docs/WP-Business_in_motion-Managing_MobileWorkforce.pdf

Eisenmann, T., Parker, G., & Van Alstyne, M. (2006, August). Strategies for two-sided markets. *Harvard Business Review*.

Engadget. (2007). Nokia Music Store launches without Warner Music due to file sharing fears. http://www.engadget.com/2007/11/02/nokia-music-store-launches-without-warner-music-due-to-file-shar/. Referred 20.2.2008.

Eschenauer, L., & Gligor, V. D. (2002). A key-management scheme for distributed sensor networks. In *ACM Conference on Computer and Communications Security*, 41–47. Hubaux, J.-P., Buttyan, L., & Capkun, S. The quest for security in mobile ad hoc networks. In *MobiHoc '01: Proceedings of the 2nd ACM international symposium on Mobile ad hoc networking & computing*, 146–155. New York, NY: ACM Press.

ETH Zurich Vehicular Network Traces. (n.d.). http://lst.inf.ethz.ch/ad-hoc/car-traces/.

Euromoney International Investor. (2003-2008). *Country risk–results & methodology*. London: Author.

European Commission. (1998). *Green paper on radio spectrum policies, in the context of European community policies such as telecommunications, broadcasting, transport, and R&D*. (Rep. COM1998 596). Brussels, Belgium: Author.

European Commission. (1999). *Next steps in radio spectrum policy: Results of the public consultation on the green paper* (Rep. COM1999 538 final). Brussels, Belgium: Author.

European Commission. (2005). *Communication to the Spring European Council–working together for growth and jobs: A new start for the Lisbon Strategy* (Rep. COM2005 24). Brussels, Belgium: Author.

Fall, K., & Varadhan, K. (Eds.). (2000). *Ucb/lbnl/vint Network Simulator–ns(version2)*. Retreived from http//www-mash.cs.berkeley.edu/ns

Fan, D., Zhang, R., & Jin, L. (2007). *2007*. Extended Kalman Filter Location and Tracking Method Using Single Base Station in Cellular Network. In Wireless Communications, Networking and Mobile Computing.

Faraj, R. (2000). *Modeling and analysis of self-similar traffic in ATM networks*. Unpublished PhD thesis, Concordia University.

Faratin, P., Clark, D., Gilmore, P., Bauer, S., Berger, A., & Lehr, W. (2007). Complexity of Internet Interconnections: Technology, Incentives and Implications for Policy. *TPRC 2007: The 35th Annual Research Conference on Communication, Information and Internet Policy*. Retrieved April 20, 2008 from, http://people.csail.mit.edu/wlehr/Lehr-Papers_files/Clark%20Lehr%20Faratin%20Complexity%20Interconnection%20TPRC%202007.pdf

FCC. (2002). New public safety applications and broadband internet access among uses envisioned by fcc authorization of ultra-wideband technology. *Federal Communications Commission,* Retrieved, 2002 from http://www.fcc.gov/Bureaus/Engineering Technology/News Releases/2002/nret0203.html

FCC. (2006). FCC Annual Report on State of Competition in the Wireless Industry.

Federal Communications Commission. (2007). *Statistical Reports and FCC State-Link*. Retrieved May 1, 2007, from http://www.fcc.gov/wcb/iatd/stats.html

Feeney, L. M., & Nilsson, M. (2001, April 22-26). Investigating the energy consumption of a wireless network interface in an ad hoc networking environment. In *Proceedings of the 20th Annual Joint Conference of the IEEE Computer and Communications Societies (INFOCOM 2001)*, Anchorage, AK (Vol. 3, pp. 1548-1557). IEEE.

Feng, L., & Whalley, J. (2002). Deconstruction of the telecommunications industry: from value chains to value networks. *Telecommunications Policy*, *26*, 451–472. doi:10.1016/S0308-5961(02)00056-3

Festag, A., Fler, H., Hartenstein, H., Sarma, A., & Schmitz, R. (2004). FleetNet: Bringing Car-to-Car Communication into the Real World. *In 11th World Congress on ITS, Nagoya, Japan, 2004*.

Fette, B. (2006). *Cognitive radio technology (communications engineering)*. Newnes.

Fichman, R. G. (2004). Real Options and IT Platform Adoption: Implications for Theory and Practice. *Information Systems Research, 15*(2), 132–154. doi:10.1287/isre.1040.0021

Flatto, J. (1996) Using Real Options in Project Evaluation. working paper, Life Office Management Association (LOMA).

Flury, R., & Wattenhofer, R. (2006). *Mls: An efficient location service for mobile ad hoc networks.*

Formal Methods for Performance Evaluation of LNCS 4486. (2007). Springer.

France Telecom Orange. (2002-2007). *Annual and 20-F reports*. Paris: Author.

Galbi, A. G. (1998). Distinctive arrangements for international interconnection. *Telecommunications Policy, 22*(11), 945–951. doi:10.1016/S0308-5961(98)00058-5

Garey, M., & Johnson, D. (1978). *Computers and Intractability: A Guide to the Theory of NP-completeness.* W.H. Freeman.

Garg, V. (2001). *Wireless Network Evolution: 2G to 3G.* Prentice Hall Communications Engineering and Emerging Technology Series.

Geist, A., Beguelin, A., & Dongarra, J. (1994). *PVM: Parallel Virtual Machine: A Users' Guide and Tutorial for Network Parallel Computing.* MIT Press. GOBLIN graph library. http://goblin2.sourceforge.net/.

Georg, C., & Fuss, O. (1999, June). Simulating rare event details of ATM delay time distributions with RESTART/LRE. In *Proceedings of the IEEE International Teletrafic Congress (ITC16),* Edinburgh, Scotland (pp. 777-786). Elsevier.

Georg, C., & Schreiber, F. (1996, December 8-11). The RESTART/LRE method for rare event simulation. In *Proceedings of the Winter Simulation Conference,* Coronado, CA (pp. 390-397). IEEE.

Gerpott, T., & Jakopin, N. (2005). The degree of internationalization and the financial performance of European mobile network operators. *Telecommunications Policy, 29*, 635–661. doi:10.1016/j.telpol.2005.06.003

Gharaee, H., & Nabavi, A. (2008, April). A New TR-UWB Transceiver with Orthogonal Pulses. In *ICEE (16th Iranian Conference on Electrical Engineering)*, 167–171, Tehran, Iran.

Gharaee, H., Nabavi, A., & Etminan, J. (2008, May). OTR-UWB System. In *WTS2008*, 325-329, California, USA, IEEE.

Ghemawat, P. (2007). *Redefining global strategy.* Boston: Harvard Business School Press.

Ghoshal, S. (1987). Global strategy: An organizing framework. *Strategic Management Journal, 8*, 425–440.

Giambene, G. (2005). *Queueing theory and telecommunications: Networks and applications.* New York: Springer.

Gitlin, T. (1998). Public sphere or public spectacles. In T. Liebs & J. Curran (Eds.), *Media, ritual, and identity* (pp. 168-174). London, Routledge.

Gitlin, T. (2003). *The whole world is watching: Mass media in the making & unmaking of the new left.* Los Angeles: University of California Press.

Gitlin, T. (2007). *Media unlimited: How the torrent of images and sounds overwhelms our lives.* New York: Henry Holt.

Gomes, T., Craveirinha, J., & Martins, L. (2002). An efficient algorithm for sequential generation of failures in a network with multi-mode components. *Reliability Engineering & System Safety, 77*, 111–119. doi:10.1016/S0951-8320(02)00006-6

Gomes, T. M. S., & Craveirinha, J. M. F. (1998). Algorithm for sequential generation of states in failure-prone communication network. *IEE Proceedings. Communications, 145*(2). doi:10.1049/ip-com:19981838

Gomes, T. M. S., & Craveirinha, J. M. F. (1997). A Case Study of Reliability Analysis of a Multiexchange Telecommunication Network. In Soares, C. G. (Ed.), *Advances in Safety and Reliability.* Elsevier Science. doi:10.1016/B978-008042835-2/50269-1

Gou, L., Kim, J.-H., Chen, H.-H., Collins, J., Goodman, M., Zhang, X., et al. (2009). *MobiSNA: A mobile video social network application.* Retrieved from http://mobisna.ist.psu.edu/pods23-gou.pdf

Grant, R. (1991). The resource-based theory of competitive advantage: implications for strategy formulation. *California Management Review*, (Spring): 119–135.

Gross, F. (2005). *Smart antennas for wireless communications*. McGraw Hill.

GSMA. (2007). GSM World: Subscriber Statistics. GSM Association. Jeffery M. & Leliveld, I. Best Practices in IT Portfolio Management. *MIT Sloan Management Review*, *45*(3), 41–49.

GSMA. (2009). *Market data summary*. Retrieved from http://www.gsmworld.com/newsroom/market-data/market_data_summary.htm

Gulati, R., Nohria, N., & Zaheer, A. (2000). Strategic Networks. *Strategic Management Journal*, *21*, 203–215. doi:10.1002/(SICI)1097-0266(200003)21:3<203::AID-SMJ102>3.0.CO;2-K

Haas, Z. J., & Liang, B. (1999). Ad hoc mobility management with uniform quorum systems. *IEEE/ACM Transactions on Networking*, *7*(2), 228–240. doi:10.1109/90.769770

Hall, S. (1982). Ideology. In E. Barnouw (Ed.), *The international encyclopedia of communication* (Vol. 2, pp. 307-311). New York: Oxford University Press.

Han, T. (2005). Exploring Price and Product Information Search in e-Market. In *Proceedings of the International Conference on Information Technology* (pp. 397-402). Las Vegas, NV, U.S.A.: IEEE Computer Society Press.

Hanaro Telecom. (2007). *Multi-bell and Various Value Added Services*. Retrieved July 7, 2007, from http://www.hanaro.com/eng/iproduct/home.asp

Hanbo, Z. (2008). *Emerging Business Models of the Mobile Market*. Master's thesis. Department of Communications and Networking, Helsinki University of Technology, Finland.

Hanna, M., & Ramarao, B. (2006). *Cost optimization for transmission and backhaul technologies*. Retrieved April 18, 2008, from, http://www.accenture.com/Global/Services/By_Industry/Communications/R_and_I/acs_cost_optimization.htm

Hargadon, A. (2003). *How Breakthroughs Happen - The Surprising Truth About How Companies Innovate*. Boston, MA: Harvard Business School Press.

Harmatos, J., Juttner, A., & Szentesi, A. (1999, September). Cost-based UMTS Transport Network Topology Optimisation. *International Conference on Computer Communication*, Tokyo, Japan. Retrieved April 18, 2008, from, http://www.cs.elte.hu/~alpar/publications/ proc/ICCC99.pdf

Harms, D. D., Kraetzl, M., Colbourn, C. C., & Devitt, J. S. (1995). *Network Reliability: Experiments with a Symbolic Algebra Environment*. CRC Press.

Hayes, J., & Ganesh Babu, T. (2004). *Modeling and analysis of telecommunications networks*. New York: John Wiley & Sons.

He, N., & Tepedelenlioglu, C. (2007, March). Fast and Low Complexity Frame-Level Synchronization for Transmitted Reference Receivers. *IEEE Transactions on Wireless Communications*, *l*(54), 1014–1023. doi:10.1109/TWC.2007.05266

Heidelberger, P. (1995). Fast simulation of rare events in queueing and reliability models. *ACM Transactions on Modeling and Computer Simulation*, *5*, 43–85. doi:10.1145/203091.203094

Held, G. (2005). *Wireless mesh networking*. Taylor & Francis. doi:10.1201/9781420031263

Herman, E., & Chomsky, N. (2002). *Manufacturing consent: The political economy of the mass media*. New York: Pantheon Books.

Hoctor, R. T., & Tomlinson, H. W. (2002, January). An overview of delay-hopped, transmitted-reference RF communications. In Technical Information Series (pp. 1–29). GE Research and Development Center.

Hollifield, C., Kosicki, G., & Becker, L. (2001). Organizational vs. professional culture in the newsroom: Television news directors' and newspaper editors' hiring decisions. *Journal of Broadcasting & Electronic Media*, *45*(1), 92–117. doi:10.1207/s15506878jobem4501_7

Homier, E. A., &Scholtz, (2002). Rapid acquisition of ultra-wideband signals in the dense multipath channel. In *Ultra-Wideband Systems Technology*, 105–109, IEEE.

Horn, G., Kvalbein, A., Blomskøld, J., & Nilsen, E. (2007). An empirical comparison of generators for self similar simulated traffic. *Performance Evaluation*, *64*(2), 162–190. doi:10.1016/j.peva.2006.06.005

Horneffer, M. (1998). Methods for performance-analysis of internet access points. *Computer Networks and ISDN Systems*, *30*(16), 1607–1615. doi:10.1016/S0169-7552(98)00169-X

http://www.arraycomm.com/serve.php?page=Cooper

http://www.intel.com/technology/mooreslaw/index.htm

Hu, Y. C., Johnson, D. B., & Perrig, A. (2002a, June). SEAD: Secure efficient distance vector routing for mobile wireless ad hoc networks. In *Proceedings of the 4th IEEE Workshop on Mobile Computing Systems and Applications (WM-CSA'02),* Callicoon, NY (pp. 3-13). IEEE.

Hu, Y. C., Johnson, D. B., & Perrig, A. (2002b, September 23-26). Ariadne: A secure on-demand routing protocol for ad hoc networks. In *Proceedings of the 8th Annual International Conference on Mobile Computing and Networking (Mobicom'02)*, Atlanta, GA (pp.12-23). ACM Publishing.

Hughes, B. D. (1995). *Random Walks and Random Environments, Volume 1: Random Walks*. Oxford Science Publications.

Huynh, T., Jennings, N., & Shadbolt, N. 2006. Certified Reputation: How an Agent Can Trust a Stranger. *Proceedings of the Fifth International Joint Conference on Autonomous Agents and Multiagent Systems.*

Ilgun, K., Kemmerer, R. A., & Porras, P. A. (1995). State transition analysis: A rule-based intrusion detection approach. *IEEE Transactions on Software Engineering*, *21*(3), 181–199. doi:10.1109/32.372146

Ilnickis, S. (2004). Modelling telecommunication systems in case of self-similar input flow. *Scientific Proceedings of Riga Technical University*, *17*, 137–142.

Interdisciplinary Telecommunications Program. (2008). ITP Fast facts. University of Colorado at Boulder. Retrieved July 11, 2008, from http://telecom.colorado.edu/

International Telecommunications Union. (2002-2008). *Indicators*. Retrieved from http://www.itu.int/ITU-D/ict/statistics

Interpolation (n.d.). http://www.mathworks.com/moler/interp.pdf.

Intven, H., Oliver, J., & Sepulveda, E. (2000, November). *Telecommunications Regulation Handbook.* InfoDev program, World Bank. iSuppli. (2005). Mobile communications market will approach $200 Billion in 2009. *Market Tracker. Wireless Systems*, *Q3*, 2005.

ITU Regional Tariff Group on Asia and Oceania (TAS). TAS Cost Model. (2000). Retrieved April 16, 2008, from, http://www.itu.int/ITU-T/othergroups/tas/index.asp

ITU. (2006). *Key Global Indicators for the World of Telecommunication Service Sector.* http://www.itu.int/ITU-D/ict/statistics/at_glance/KeyTelecom99.html. Referred 20.2.2007.

ITU-Recommendation D.140; Annex A: Guidelines for the cost elements to be taken into account when determining accounting rates and accounting rate shares for the international telephone service. (1998). Retrieved April 16, 2008, from, http://www.itu.int/osg/csd/intset/itu-t/d140/d140_e_rev.html

ITU-T Study Group 3; Special Projects and Issues: Accounting Rate Reform undertaken by ITU-T Study Group 3 (2000). Retrieved April 16, 2008, from, http://www.itu.int/ITU-T/studygroups/com03/accounting-rate/index.html

Jain, R. (2006). Interconnection regulation in India: Lessons for developing countries. *Telecommunications Policy*, *30*(3-4), 183–200. doi:10.1016/j.telpol.2005.11.005

Jain, S., Fall, K., & Patra, R. (2004). Routing in a delay tolerant network. In *ACM Sigcomm*.

Jarvis, J. P., & Shier, D. R. (1996). An Improved Algorithm for Approximating the Performance of Stochastic Flow Networks. *INFORMS Journal on Computing*, *8*(4). doi:10.1287/ijoc.8.4.355

Jeffery, C., Mohamed, S., Parlett, R., & Pereda, R. (2008). *Programming with Unicon*. Unicon Project, http://unicon.org.

Jeong, H.-D.-J. (2002). *Modelling of self-similar teletraffic for simulation.* Unpublished PhD thesis, University of Centerbury, UK.

Jhunjhunwala, A., Ramamurthi, B., & Gonsalves, T. A. (1998). The role of technology in telecom expansion in India. *IEEE Communications Magazine*, *36*(11), 88–94. doi:10.1109/35.733480

Jiang, T., Guo, S., Wang, S., & Wei, R. (2007, September 21-25). Universal Study on Parametric Models in Equipment System Cost Estimation. W*iCom 2007: International Conference on Wireless Communications, Networking and Mobile Computing*, 2007 (pp. 5228–5231).

Johnson, D. B., & Maltz, D. A. (1996). Dynamic source routing in ad hoc wireless networks. In Imielinski and Korth, (Eds.), *Mobile Computing*. Kluwer Academic Publishers.

Josang, A., Hayward, R., & Pope, S. 2006. Trust Network Analysis with Subjective Logic. *Proceedings of the Twenty-Ninth Australian Computer Science Conference* (pp. 85-94).

Kalman and extended Kalman filters: Concept, derivation and properties. http://users.isr.ist.utl.pt/ mir/pub/kalman.pdf.

Karaoguz, J. (2001). High-rate wireless personal area networks. *IEEE Communications Magazine*, (39): 96–102. doi:10.1109/35.968818

Karger, D. (1995). A Randomized Fully Polynomial Time Approximation Scheme for the All-terminal Network Reliability Problem. *Proceedings of the 27th ACM STOC*.

Karp, B., & Kung, H. T. (2000, August). Greedy perimeter stateless routing for wireless networks. In ACM MOBICOM'00. GPSR.

Keilson, J. (1979). *Markov Chain Models-Rarity and Exponentiality*. Springer-Verlag.

Kemma, A. G. Z. (1993). Case Studies on Real Options. *Financial Management*, *19*(3), 259–270.

Kennet, D. M., & Sharkey, W. W. (2001). Calculating the Long Run Incremental Cost of Interconnection using a Network Cost Simulation Model. *Journal of Computing in Economics and Finance, 9*.

Kiess, W., Fler, H., Widmer, J., & Mauve, M. (2004). Hierarchical Location Service for Mobile Ad-Hoc Networks. In *ACM SIGMOBILE Mobile Computing and Communications Review, 2004*.

Kiiski, A. (2007). *Impacts of Virtual Operators on Mobile Market*. Licentiate thesis, Helsinki University of Technology, 2007.

Kilkki, K. (2007, May). A practical model for analyzing long tails. *First Monday, 12(5)*. http://firstmonday.org/issues/issue12_5/kilkki/index.html. Referred 2.7.2007.

Kim, H. J. (2006). The Economic Value of Modularity in the IPQoS Network: the Real Options Approach (ROA). [IJBDCN]. *International Journal of Business Data Communications and Networking*, *2*(1), 15–30. doi:10.4018/jbdcn.2006010102

Kim, Y. J., & Sanders, L. G. (2002). Strategic Actions in Information Technology Investment Based on Real Option Theory. *Decision Support Systems*, *33*, 1–11.

Kim, H. J. (2004). The Real Options to Technology Management: Strategic Options for 3G Wireless Network Architecture and Technologies. Ph.D. Thesis, University of Pittsburgh.

Kivi, A. (2006). *Mobile Internet Usage Measurements: Case Finland. Master's Thesis Series*. Networking Laboratory. Department of Electrical and Telecommunications Engineering. Helsinki University of Technology, Espoo, Finland, May 2006.

Kivi, A. (2007). Measuring mobile user behavior and service usage. *Presented at 6th Global Mobility Roundtable*, 1-2 June 2007, Los Angeles, USA.

Klein, J. T. (1990). *Interdisciplinarity: History, Theory, and Practice*. Detroit, MI: Wayne State University Press.

Klein, J. T. (2006). Resources for Interdisciplinary Studies. *Change*, (March/April): 52–58.

Knuth, D. (1969). The Art of Computer Programming: *Vol. 2. Seminumerical Algorithms*. Reading, MA: Addison-Wesley.

Ko, Y.-B., & Vaidya, N. H. (2000). Location-aided routing (lar) in mobile ad hoc networks. *Wireless Networks*, *6*(4), 307–321. doi:10.1023/A:1019106118419

Korea Telecomm. (2006). *Consumer Demographics White Pages*. Retrieved July 20, 2006, from http://english.ktf.com/eng/

Kowalski, J. P., & Warfield, B. (1995, December). Modelling traffic demand between nodes in a telecommunications network. *Australian Telecommunications and Networks Conference*. Retrieved March 12, 2008, from, http://citeseerx.ist.psu.edu/viewdoc/summary?doi=10.1.1.56.9702

Kravets, R., Yi, S., & Naldurg, P. (2001, October 4-5). *A security-aware routing protocol for wireless ad hoc networks*. Paper presented at the ACM Symposium on Mobile Ad Hoc Networking and Computing (MobiHoc 2001), Long Beach, CA.

Krawczyk, H. 1996. SKEME: A Versatile Secure Key Exchange Mechanism for Internet. *Proceedings of the 1996 Internet Society Symposium on Network and Distributed Systems Security* (pp. 114-127).

Kumar, R., & Hämmäinen, H. (2007, October). Competition in International Mobile Roaming: Alternative Clusters. *Communications and Strategies*, *3*(67), 151–172.

Kumar, R. L. (2002). Managing risks in IT projects: an options perspective. *Information & Management*, *40*, 63–74. doi:10.1016/S0378-7206(01)00133-1

Kumar, S., & Spafford, E. H. (1995, October). A software architecture to support misuse intrusion detection. In *Proceedings of the 18th National Information Security Conference* (pp. 194-204).

Kushner, H. (2001). *Heavy traffic analysis of controlled queueing and communications networks*. New York: Springer.

L'Ecuyer, P. (2001). Random number generators. In S. I. Gass & C. M. Harris (Eds.), *Encyclopedia of operations research and management science* (pp. 695-702). Norwell, MA: Kluwer Academic Publishers.

L'Ecuyer, P., Demers, V., & Tuffin, B. (2006, December 3-6). Splitting for rare-event simulation. In *Proceedings of the 2006 Winter Simulation Conference,* Monterey, CA (pp. 137-148).

Langlois, R. N., & Robertson, P. L. (1992). Networks and innovation in a modular system: Lessons from the microcomputer and stereo component industries. *Research Policy*, *21*(4), 297–313. doi:10.1016/0048-7333(92)90030-8

Law, A. (2007). *Simulation Modeling and Analysis* (4 ed.). McGraw-Hill Publishing.

Law, A. M., & Kelton, W. D. (2006). *Simulation modeling and analysis* (4th ed.). New York: McGraw Hill.

Lawless, M. W., & Anderson, P. C. (1996). Generational technological change: Effects of innovation and local rivalry on performance. *Academy of Management Journal*, *39*(5), 1185–1217. doi:10.2307/256996

Lee, M. G. S-J. (2001). Split multipath routing with maximally disjoint paths in ad hoc networks. In.ICC'01.

Levendovszky, J., Jereb, L., & Elek, Zs., & Vesztergombi, Gy. (2002). Adaptive Statistical Algorithms in Network Reliability Analysis. *Performance Evaluation*, 48.

Levy, Y., & Wirth, P. E. (1989). *A Unifying Approach to Performance and Reliability Objectives. Teletraffic Science for New Cost-Effective Systems, Networks and Services, ITC-12*. Elsevier Science.

Levy, P. (1994). *Collective intelligence*. New York: Basic Books.

Lewin, L. (Ed.). (1984a). *Telecommunications: An Interdisciplinary Survey. Dedham, MA: Artech House Publishers*. Dedham, MA: Artech House Publishers.

Li, J., Li, R., & Kato, J. (2008). Future Trust Management Framework for Mobile Ad Hoc Networks. *IEEE Communications Magazine*, *46*, 108–114. doi:10.1109/MCOM.2008.4481349

Li, V. K., & Silvester, J. A. (1984). Performance Analysis of Networks with Unreliable Components. *IEEE Transactions on Communications*, 32.

Li, J., Jannotti, J., Decouto, D., Karger, D., & Morris, R. (2000). A Scalable Location Service for Geographic Ad-hoc Routing. In *ACM MobiCom'00*.

Lin, X., Lu, R., Zhang, C., Zhu, H., Ho, P., & Shen, X. (2008). Security in Vehicular Ad Hoc Networks. *IEEE Communications Magazine*, *46*, 88–95. doi:10.1109/MCOM.2008.4481346

Linder, J. C., & Cantrell, S. (2000). *Changing business models: surveying the Landscape.* Working Paper, Accenture Institute for Strategic Change.

Lindqvist, K. (2007). *Scenario Analysis of Mobile Voice Market*. Master's Thesis Series. Networking Laboratory. Department of Electrical and Telecommunications Engineering. Helsinki University of Technology.

Lisnianski, A., & Levitin, G. (2003). *Multi-State System Reliability: Assessment, Optimization and Applications*. World Scientific.

Liu, J., & Issarny, V. (2004, March). Enhanced reputation mechanism for mobile ad hoc networks. In *Proceedings of the 2nd International Conference on Trust Management*, Oxford, UK (Vol. 2995, pp. 48-62). Springer-Verlag.

Lochert, C., Hartenstein, H., Tian, J., Fler, H., Herrmann, D., & Mauve, M. (2003). A Routing Strategy for Vehicular Ad Hoc Networks in City Environments. In *IEEE Intelligent Vehicles Symposium, June 2003*.

Lou, W., Liu, W., & Fang, Y. (2004, March). Spread: enhancing data confidentiality in mobile ad hoc networks. In *INFOCOM 2004, 2404–2413*. Hong Kong, China: IEEE.

Lovallo, D., & Mendonca, L. (2007). Strategy's Strategist: Interview with Richard Rumelt. *The McKinsey Quarterly*, (August): 2007.

Lucky, R. W., & Eisenberg, J. (Eds.). (2006). *Renewing U. S. Telecommunications Research*. Washington, DC: The National Academies Press, Committee on Telecommunications Research and Development, Computer Science and Telecommunications Board, Division on Engineering and Physical Sciences, National Research Council of the National Academies.

Lunt, T., Tamaru, A., Gilham, F., Jagannathan, R., Jalali, C., Neumann, P. G., et al. (1992). *A real-time intrusion detection expert system (IDES)*. Menlo Park, CA: Computer Science Laboratory, SRI International.

Luo, H., Kong, J., Zerfos, P., Lu, S., & Zhang, L. (2004). Ursa: ubiquitous and robust access control for mobile ad hoc networks. *IEEE/ACM Transactions on Networking*, *12*(6), 1049–1063. doi:10.1109/TNET.2004.838598

Luo, S., Kezunovic, M., & Sevick, D. R. (2004). Locating Faults in the Transmission Network using Sparse Field Measurements, Simulation Data, and Genetic Algorithm. *Electric Power Systems Research*, *71*(2), 169–177. doi:10.1016/j.epsr.2004.01.011

Ma, J., & Orgun, M. (2006). Trust Management and Trust Theory Revision. *IEEE Transactions on Systems, Man, and Cybernetics. Part A, Systems and Humans*, *36*, 451–460. doi:10.1109/TSMCA.2006.871628

Ma, Y., Chin, F., Kannan, B., & Pasupathy, S. (2002). Acquisition performance of an UWB communications wideband system over a multiple access fading channel. In *Ultra-Wideband Systems Technology*, 99–104, IEEE.

Magretta, J. (2002, May). Why business models matter? *Harvard Business Review*, 86–92.

Mahr, E., & Richardson, G. (2003). Development of the Small Satellite Cost Model (SSCM). *IEEE Aerospace Conference Proceedings 2003, 8*, 3831-3841.

Maitland, C. F., Bauer, J. M. B., & Westerveld, R. (2002). The European market for mobile data:evolving value chains and industry structures. *Telecommunications Policy*, *26*, 485–504. doi:10.1016/S0308-5961(02)00028-9

MapMechanics. (2005, May). *GB Traffic Volumes.*www.mapmechanics.com.

Margrabe, W. (1978). The Value of an Option to Exchange one Asset for Another. *The Journal of Finance*, *33*, 177–186. doi:10.2307/2326358

Marie, R. R., Blackledge, J. M., & Bez, H. E. (2007, February). Characterization of internet traffic using a fractal model. In *Proceedings of the Fourth IASTED International Conference on Signal Processing, Pattern Recognition, and Applications* (*SPPRA 2007*), Innsbruck, Austria (pp. 156-167). ACTA Press.

Maslow, A. H. (1943). A Theory of Human Motivation. *Psychological Review*, *50*, 370–396. doi:10.1037/h0054346

McDonald, R., & Siegel, D. (1986). The value of waiting to invest. *The Quarterly Journal of Economics*, *101*, 707–727. doi:10.2307/1884175

McDysan, D., Lundberg, T., Bjorkman, N., & Latour-Henner, A. (2000). Multiservice Networking using a Component-based Switch & Router Architecture. Proceedings of the IEEE Conference on ATM 2000.

McGahan, A. (2004). *How Industries Evolve - Principles for Achieving and Sustaining Superior Performance*. Boston: Harvard Business School Press.

Mehdi, J. (2003). *Stochastic models in queueing theory*. New York: Academic Press.

Meisl, C. J. (1993, December). Parametric Cost Model for solar Space Power and DIPS Systems. *IEEE AES Systems Magazine* (pp. 24-29).

Melody, W. H. (2000). Shaping liberalized telecom markets. *Telecommunications Policy*, *24*(10–11), 803–806. doi:10.1016/S0308-5961(00)00073-2

Melody, W. H. (2001). Assessing highly imperfect mobile markets. *Telecommunications Policy*, *25*(1–2), 1–3. doi:10.1016/S0308-5961(00)00083-5

Mendiratta, V. B. (2001). A Hierarchical Modelling Approach for Analyzing the Performability of a Telecommunications System. *PMCCS-5: The Fifth Int'l Workshop on Performability Modelling of Computer and Communication Systems*.

Messac, A., Ismail-Yahaya, A., & Mattson, C. (2003). The Normalized Normal Constraint Method for Generating the Pareto Frontier. *Structural and Multidisciplinary Optimization*, 25.

Meyer, J. F. (1995). Performability evaluation: where it is and what lies ahead. *1st IEEE Computer Performance and Dependability Symposium (IPDS)* (pp. 334-343).

Micalizzi, A. (1999). Plough Case: Valuing Pharmaceutical Development and Expansion Options. The 3rd Annual Real Options Conference, Netherlands.

Michiardi, P., & Molva, R. (2000). *Core: Collaborative reputation mechanism to enforce node cooperation in mobile ad hoc networks* (Research Rep. No. RR-02-062). Sophia-Antipolis, France: Institut Eurécom.

Mishira, A., Nadkarni, K., & Patcha, A. (2004). Intrusion detection in wireless ad hoc networks. *IEEE Wireless Communications Magazine*, *11*(1), 48–60. doi:10.1109/MWC.2004.1269717

Mo, Z., Zhu, H., Makki, K., & Pissinou, N. (2006). MURU: A Multi-Hop Routing protocol for Urban Vehicular Ad Hoc Networks. In *Int'l Conf. on Mobile and Ubiquitous Systems: Networking and Services (MobiQuitous 2006)*.

Morello, D. (2005). The IT Professional Outlook: Where Will We Go From Here? ID Number: G00130462. September 14, 2005: Gartner Inc.

Morris, R., Jannotti, J., Kaashoek, F., Li, J., & Decouto, D. (2000). CarNet: A Scalable Ad Hoc Wireless Network System. In *9th ACM SIGOPS European Workshop, 2000*.

Mudchanatongsuk, S., Ordonez, F., & Liu, J. (2008). Robust solution for network design under transportation cost and demand uncertainty. *The Journal of the Operational Research Society*, *59*(5), 652–662. doi:10.1057/palgrave.jors.2602362

Müller-Veerse, F., Kohlenbach, B., Bout, D., Singh, S., Golub, G., & Häyrynen, J. (2001). *UMTS Report - An Investment Perspective*. Durlacher Research Ltd, Eqvitec Partners Oy.

Naumov, V., Baumann, R., & Gross, T. (2006, May). An Evaluation of Inter-Vehicle Ad Hoc Networks Based on Realistic Vehicular Traces. In *ACM MOBIHOC*.

Netsize. (2008). *The Netsize Guide 2008. Mobile 2.0, you are in control*. Mobile Industry Report. http://www.netsize.com/downloads/intro.aspx. Referred 20.2.2008.

Newell, W. H. (2001). A Theory of Interdisciplinary Studies. *Issues in Integrative Studies*, *19*, 1–25.

Nokia. (2008). *Nokia launches premier mobile advertising network*. Nokia press release. February 11, 2008. http://www.nokia.com/A41360012newsid=1190110. Referred 20.2.2008.

O'Reilly, T. (2008). *Google, WalMart, and MyBarackObama.com: The power of the real time enterprise*. Retrieved from http://blogs.oreilly.com/cgi-bin/mt/mt-search.cgi?blog_id=57&tag=walmart&limit=20&IncludeBlogs=57

Odlyzko, A. (2004). The Evolution of Price Discrimination in the Transportation and its Implications for the Internet. *Review of Network Economics*, *3*(3), 323–346. doi:10.2202/1446-9022.1055

Ohayon-Dekel, A. (2009). Connected knowledge–connecting the brains of the enterprise in real time. *Enriching Communications*, *3*(1). Retrieved from http://www.alcatel-lucent.com/enrich/v3i12009/article_c1a5.html?l=en

Oikonomou, K. N., Ramakrishnan, K. K., Doverspike, R. D., Chiu, A., Martinez-Heath, M., & Sinha, R. K. (2007). Performability Analysis of Multi-Layer Restoration in a Satellite Network. In Mason, L., Drwiega, T., & Yan, J. (Eds.), *Managing Traffic Performance in Converged Networks, ITC 20 (LNCS 4516)*. Springer-Verlag. doi:10.1007/978-3-540-72990-7_11

Oikonomou, K. N. (2006). Unreliable Components with Fast Repair and Dynamic Network Restoration. In S. Raghavan & G. Anandalingam (Eds.), *Telecommunications Planning in Proceedings of the 8th INFORMS Conference on Telecommunications*. Kluwer Academic.

Oikonomou, K. N., & Sinha, R. K. (2006). A New State Generation Algorithm for Evaluating the Performability of Networks with Multimode Components. *8th INFORMS Telecommunications Conference*.

Oikonomou, K. N., & Sinha, R. K. (2008). Techniques for Probabilistic Multi-layer Network Analysis. *Proceedings of the IEEE Globecomm*.

Oikonomou, K.N., & Sinha, R.K. (2009). Improved Bounds for Performability Evaluation Algorithms using State Generation. *Performance Evaluation*.

Orndorff, A. M. (2004). *Transceiver Design for Ultra-Wideband Communication*. Master dissertation, University of Polytechnic Institute and State, Virginia.

Osterwalder, A., & Pigneur, Y. (2002). An ebusiness model ontology for modeling eBusiness. In *Proceedings of 15th Bled Electronic Commerce Conference – eReality: Constructing the eEconomy*, Bled, Slovenia, June 17–19, 2002.

Paddock, J. L., Siegel, D. R., & Smith, J. L. (1988). Option Valuation of Claims on Real Assets: The Case of Offshore Petroleum Leases. *The Quarterly Journal of Economics*, 479–508. doi:10.2307/1885541

Padhye, J., Firoiu, V., Towsley, D., & Kurose, J. (2000). Modeling TCP Reno Performance: A Simple Model and Its Empirical Validation. *IEEE/ACM Transactions on Networking*, 8.

Panayi, S., & Trigeorgis, L. (1998). Multi-stage real options: The cases of information technology infrastructure and international bank expansion. *The Quarterly Review of Economics and Finance*, *38*(3), 675–692. doi:10.1016/S1062-9769(99)80096-9

Papadimitratos, P., & Hass, Z. (2006). Secure Data Communication in Mobile Ad Hoc Networks. *IEEE Journal on Selected Areas in Communications*, *24*, 343–356. doi:10.1109/JSAC.2005.861392

Papadimitratos, P., & Haas, Z. J. (2003). Secure data transmission in mobile ad hoc networks. In *WiSe '03: Proceedings of the 2003 ACM workshop on Wireless security*, 41–50. New York, NY: ACM Press.

Papadimitratos, P., & Haas, Z. J. (2002, January). *Secure routing for mobile ad hoc networks*. Paper presented at the SCS Communication Networks and Distributed Systems Modeling and Simulation Conference (CNDS 2002), San Antonio, TX.

Papakiriakopoulos, D., Poulymenakou, A., & Doukidis, G. (2001). Building E-Business models: An analytical framework and development guidelines. In *Proceedings of 14th Bled Electronic Commerce Conference*, Bled, June 25-26, 2001.

Parametric Cost Estimating Model for Buildings. (2008). Retrieved April 16, 2008, from http://www.aacei.org/technical/BuildingModel.shtml

Park, K. A. (2001). Development of ICT Indicators of Korea. *2001 Annual Government Statistics Report*, Tokyo, Japan. Retrieved June 1, 2007, from http://www.stat.go.jp/English/info/meetings/iaos/pdf/park.pdf.

Park, K., & Willinger, W. (Eds.). (2000). *Self-similar network traffic and performance evaluation*. New York: John Wiley & Sons.

Pateli, G. A., & Giagli, M. G. (2004). A research framework for analysing eBusiness models. *European Journal of Information Systems*, *13*, 302–314. doi:10.1057/palgrave.ejis.3000513

Pateli, A. G., & Giaglis, G. M. (2007). The Impact of Value on Governance Decisions for IT-Based Alliances: Evidence from a Joint Venture in the Wireless Networks Industry. Proceedings on the 40th Annual Hawaii International Conference on System Sciences, HICSS 2007.

Pathirana, P., Savkin, A., & Jha, S. (2004). Robust extended Kalman filter based technique for location management in PCS networks. *Computer Communications*.

Perkins, C. E., & Royer, E. M. (1999, February). Ad-hoc on-demand distance vector routing. In *Proceedings of Second IEEE Workshop on Mobile Computing Systems and Applications (WMCSA '99)*, 90–100. IEEE. Qualnet. http://www.scalable-networks.com/.

Perrig, A., Johnson, D. B., & Hu, Y. C. (2004). A survey of secure wireless ad hoc routing. *IEEE Security & Privacy*, *2*(3), 28–39. doi:10.1109/MSP.2004.1

Peterson, S. (1981). International news selection by the elite press: A case study. *Public Opinion Quarterly*, *45*(2), 143–163. doi:10.1086/268647

Petrovic, O., Kittl, C., & Teksten, R. D. (2001). Developing business models for eBusiness. In *Proceedings of International Conference on Electronic Commerce 2001,* Vienna: October 31 – November 4, 2001.

Philip, S. J., & Qiao, C. (2003). *Hierarchical grid location management for large wireless ad hoc networks*. ACM SIGMOBILE Mobile Computing and Communications Review.

Phillips, S., & Westbrook, J. (1999). Approximation Algorithms for Restoration Capacity Planning. *Proceedings of Algorithms-ESA'99: 7tt Annual European Symposium,* Prague.

Phillips, S., Reingold, N., & Doverspike, R. (2002). Network Studies in IP/Optical Layer Restoration. *Optical Fiber Communications Conference*.

Pickles, E., & Smith, J. (1993). Petroleum Property Valuation: A Binomial Lattice Implementation of Option Pricing Theory. *The Energy Journal (Cambridge, Mass.)*, *14*(2), 1–26.

Pindyck, R. S. (1989). Irreversible Investment, Capacity Choice, and the Value of the Firm. *The American Economic Review*, *2*, 969–985.

Pioro, M., Juttner, A., Harmatos, J., Szentesi, A., Gajowniczek, P., & Myslek, A. (2001, September). Topological Design of Telecommunication Networks- Nodes and Links Localisation under Demand Constraints. *17th International Tele-traffic Congress*, Salvador da Balia, Brazil. Retrieved March 12, 2008 from http://www.cs.elte.hu/~alpar/publications/proc/ITC-TopDesTelNel.pdf

Pirzada, A. A., McDonald, C., & Datta, A. (2006). Performance comparison of trusted-based reactive routing protocols. *IEEE Transactions on Mobile Computing*, *5*(6), 695–710. doi:10.1109/TMC.2006.83

Pirzada, A. A., & McDonald, C. (2004, January). Establishing trust in pure ad-hoc networks. In *27th Australian Computer Science Conference*, Dunedin, New Zealand (Vol. 26, pp.47-54). Australian Computer Society.

Plaisance, P., & Skewes, E. (2003, October). Personal and professional dimensions of news work: Exploring the link between journalists' values and roles. Paper presented at the Annual Meeting of the International Communication Association, San Diego, CA.

Podhradsky, P. (2004). Migration scenarios and convergence processes towards NGN (present state and future trends): Electronics in Marine. Proceedings Elmar2004, 46th International Symposium, 39–46.

Polivka, J. (2007). Sprint Nextel Cites WiMAX Network Progress For 2007. News Releases on Sprint Nextel, January 8.

Popescu, A. (2001, June). Traffic self-similarity. In *Proceedings of the IEEE International Conference on Telecommunications (ICT 2001),* Bucharest, Romania (pp. 20-24). IEEE.

Popper, K. R. (1963). *Conjectures and Refutations: The Growth of Scientific Knowledge*. New York, NY: Routledge and Kegan Paul.

Porter, M. E. (1979). How competitive forces shape strategy. *Harvard Business Review*, (March/April): 1979.

Porter, M. E. (1980). *Competitive Strategy*. New York: The Free Press.

Porter, M. E. (1985). *Competitive Advantage*. New York: Free Press.

Porter, C. E., & Donthu, N. (2006). Using the Technology Acceptance Model to explain how Attitudes determine Internet Usage: The role of Perceived Access Barriers and Demographics. *Journal of Business Research*, *59*(9), 999–1007. doi:10.1016/j.jbusres.2006.06.003

Potlapally, N. R., Ravi, S., Raghunathan, A., & Jha, N. K. (2006). A study of the energy consumption characteristics of cryptographic algorithms and security protocols. *IEEE Transactions on Mobile Computing*, *5*(2), 128–143. doi:10.1109/TMC.2006.16

Prasad, R. (1998). An Overview of CDMA Evolution toward Wideband CDMA. *IEEE Communication Surveys*, *1*(1), 1–29.

Probability, C. (2000). Kluwer Academic.

Rabin, M. O. (1989). Efficient dispersal of information for security, load balancing, and fault tolerance. *Journal of the ACM*, *36*(2), 335–348. doi:10.1145/62044.62050

Radev, D. (2005, September 28-30). Stochastic process generation for traffic flow simulation. In *Proceedings of the IEEE Conference on Telecommunications in Modern Satellite, Cable and Broadcasting Services (TELSIKS 2005)*, Nis, Serbia (Vol. 2, pp. 559-562). IEEE.

Radev, D., & Lokshina, I. (2007, June 11-13). Modeling and simulation of self-similar teletraffic. In *Proceedings of the Industrial Simulation Conference 2007*, Delft, The Netherlands (pp. 355-359).

Radev, D., & Lokshina, I. (2009, April 22-24). Modeling and simulation of self-similar wireless IP network traffic. *Proceedings of the IEEE Wireless Telecommunications Symposium (WTS) 2009,* Prague, Czech Republic (pp. 1-6). IEEE.

Radev, D., Rashkova, E., & Stankovski, D. (2007, June). Simulation modelling of overflow probabilities in tandem queuing system. In *Proceedings of XLII International Science Conference on Information, Communication and Energy Systems and Technologies (ICEST 2007),* Ohrid, Macedonia (pp. 123-127).

Radiocommunications Agency (RA). (2002). *The economic impact of radio* (Reg. No. 876/2002, official journal L138, 1-8). Brussels, Belgium: Council of the European Union.

Radunovic, B., Gkantsidis, C., Key, P., & Rodriguez, P. (2008, April). An optimization framework for opportunistic multipath routing in wireless mesh networks. In *27th IEEE International Conference on Computer Communications (INFOCOM 2008)*, AZ, USA.

Ramirez-Marquez, J. E., & Gebre, B. A. (2007). A Classification Tree Based Approach for the Development of Minimal Cut and Path Vectors of a Capacitated Network. *IEEE Transactions on Reliability*, *56*, 2007. doi:10.1109/TR.2007.902788

Rappa, M. (2003). *Managing the digital enterprise – Business models on the Web.* http://ecommerce|.ncsu.edu/business_models.html. Referred 1.11.2007.

Rappa, M. (2008). Business Models on the Web. *Managing the Digital Enterprise, Lesson 5.* Retrieved January 9, 2008, from http://digitalenterprise.org/models/models.html

Rapport, T. (2002). Wireless Communications Principles and Practice, Prentice-Hall, Inc., 2002.

Rates for International Call to India from UK. (2008). Retrieved April 18, 2008, from http://www.lycatalk.com/res_rates.htm

Rauzy, A. (2005). An *m*log*m* Algorithm to Compute the Most Probable Configurations of a System with Multi-Mode Independent Components. *IEEE Transactions on Reliability*, *54*(1), 156–158. doi:10.1109/TR.2004.842084

Reese, S. (2001). Understanding the global journalist: A hierarchy of influences approach. *Journalism Studies*, *2*(2), 173–187. doi:10.1080/14616700118394

Reese, S., & Ballinger, J. (2001). Roots of a sociology of news: Remembering Mr. Gates and social control in the newsroom. *Journalism & Mass Communication Quarterly*, *78*(4), 641–658.

Ren, K., Li, T., Wan, Z., Bao, F., Deng, R. H., & Kim, K. (2004). Highly reliable trust establishment scheme in ad hoc networks. *Computer Networks: The International Journal of Computer and Telecommunications Networking*, *45*(6), 687–699.

Rezende, J., Ziviani, A., Fdida, S., & Duarte, O. (n.d.). *Enhancing Location Management in Mobile Ad Hoc Networks.*

Rittenhouse, G. (2008). Private communication to the author. June 24, 2008.

Riva, O., Nadeem, T., Borcea, C., & Iftode, L. (2007). Context-aware Migratory Services in Ad Hoc Networks. *IEEE Transactions on Mobile Computing*, 2007.

Robinson, S. (2006). The mission of the j-blog: Recapturing journalistic authority online. *Journalism*, *7*(1), 65–83. doi:10.1177/1464884906059428

Rocco, C., & Muselli, M. (2005). Approximate Multi-state Reliability Expressions Using a New Machine Learning Technique. *Reliability Engineering & System Safety*, 89.

Rochet, J. C., & Tirole, J. (2003). Platform Competition in Two-Sided Markets. *Journal of the European Economic Association*, *1*(4), 990–1029. doi:10.1162/154247603322493212

Rogers, E. (1962). *Diffusion of innovations*. New York: The Free Press.

Roycroft, T. R., & Anantho, S. (2003). Internet subscription in Africa: Policy for a Dual Digital Divide. *Telecommunications Policy*, *27*(1), 61–74. doi:10.1016/S0308-5961(02)00091-5

Royer, E. M., & Toh, C.-K. (1999). A review of current routing protocols for ad hoc mobile wireless networks. *IEEE Personal Communications*, *6*(2), 46–54. doi:10.1109/98.760423

Rugman, A. M., & Verbeke, A. (2004). A perspective on regional and global strategies of multinational enterprises. *Journal of International Business Studies*, *35*, 3–18. doi:10.1057/palgrave.jibs.8400073

Rumelt, R. (1991). How Much Does Industry Matter? *Strategic Management Journal*, *12*, 167–185. doi:10.1002/smj.4250120302

Rushforth, C. K. (1964, January). Transmitted-reference techniques for random or unknown channels. *IEEE Transactions on Information Theory*, (40): 39–42. doi:10.1109/TIT.1964.1053641

Saarikoski, V. (2006). *The Odyssey of the Mobile Inter*net. Doctoral Dissertation. University of Oulu, Finland.

Sahner, R., Trivedi, K., & Puliafito, A. (1996). *Performance and Reliability Analysis of Computer Systems*. Kluwer Academic.

Saleh, A., & Valenzuela, R. (1987, February). A statistical model for indoor multipath propagation. *IEEE Journal on Selected Areas in Communications*, *l*(5), 128–137. doi:10.1109/JSAC.1987.1146527

Sanchez, R. (1999). Modular Architectures in the Marketing Process. *Journal of Marketing*, 92–111. doi:10.2307/1252104

Sanchez, R. (1995). Strategic Flexibility in Product Competition. Strategic Management Journal, 16 (summer special edition), 135-159.

Sanzgiri, K., Dahill, B., Levine, B., Shields, C., & Belding-Royer, E. (2002, November 12-15). A secure routing protocol for ad hoc networks. In *Proceedings of the 10th IEEE International Conference on Network Protocols (ICNP),* Paris (pp. 78-87). IEEE.

Sanzgirl, K., LaFlamme, D., Dahill, B., Levine, B., Shields, C., & Belding-Royer, E. (2005). Authenticated Routing for Ad Hoc Networks. *IEEE Journal on Selected Areas in Communications*, *23*, 598–610. doi:10.1109/JSAC.2004.842547

Sarkar, M. B., Cavusgil, S. T., & Aulakh, P. S. (1999). International expansion of telecommunications carriers: The influence of market structure, network characteristics, and entry imperfections. *Journal of International Business Studies*, *30*(2), 361–382. doi:10.1057/palgrave.jibs.8490074

Sasson, Y., Cavin, D., & Schiper, A. (2005). A Location Service Mechanism for Position-Based Multicasting in Wireless Mobile Ad hoc Networks. In *Hawaii International Conference on System Sciences (HICSS'05)*.

Schilling, M. (2000). Toward a general modular systems theory and its application. *Academy of Management Review*, *25*(2), 312–334. doi:10.2307/259016

Schwartz, E. S., & Zozaya-Gorostiza, C. (2003). Investment Under Uncertainty in Information Technology: Acquisition and Development Projects. *Management Science*, *49*(1), 57–70. doi:10.1287/mnsc.49.1.57.12753

Seaberg, J. G., Hawn, J., Dincerler, G. E., Eugster, C. C., & Rao, N. (1997). Attackers versus incumbents: The battle for value in an IP-networked world. *The McKinsey Quarterly*, *4*, 138–153.

Seppänen, M., & Mäkinen, S. (2005). Business model concepts: A review with case illustration. *Proceedings of the Engineering Management Conference*, *2005*, 376–380.

Shamir, A. (1979). How to share a secret. *Communications of the ACM*, *22*(11), 612–613. doi:10.1145/359168.359176

Shannon, C. E. (1948, October). A mathematic theory of communications. *Bell Systems Technical Journal, 27*, 379-423 ; *27*, 623 – 656.

Shannon, C. E. (1948, January). Communications in the presence of noise. *Proc IRE, 37*, 10-21.

Shapiro, C., & Varian, H. R. (1998). *Information Rules: A Strategic Guide to the Network Economy*. Boston, MA: Harvard Business School Press.

Shier, D. R. (1991). *Network Reliability and Algebraic Structures*. Oxford: Clarendon Press.

Shier, D.R., Bibelnieks, E., Jarvis, J.P., & Lakin, R.J. (1990). Algorithms for Approximating the Performance of Multimode Systems. *Proceedings IEEE Infocom*.

Shoemaker, P. J., & Reese, S. D. (1996). *Mediating the message: Theories of influences on mass media content.* New York: Longman.

Shuster, M. S. (1998). Diffusion of Network Innovation: Implications for Adoption of Internet Services, MIT Internet Telephony Consortium Semiannual Meeting, Helsinki, Finland, 20-22.

Sigurdsson, H. M., Thorsteinsson, S. E., & Stidsen, T. K. (2004). Cost optimization methods in the design of next generation networks. *IEEE Communications Magazine, 42*(9), 118–122. doi:10.1109/MCOM.2004.1336727

Simon, D. (2001). *Kalman Filtering*. Embedded System Programming.

Singer, J. (1998). Online journalists: Foundations for research into their changing roles. *Journal of Computer-Mediated Communication*, *4*(1).

Singer, J. (2005). The political j-blogger: 'Normalizing' a new media form to fit old norms and practices. *Journalism*, *6*(2), 173. doi:10.1177/1464884905051009

Singer, J. (1997). Still guarding the gate? The newspaper journalist's role in an on-line world. *Convergence: The international journal of research into new media technologies, 3*(1), 72-89.

Smith, K. W., & Triantis, A. J. (1995). The Value of Options in Strategic Acquisitions (book chapter). Real Options in Capital Investment, Trigeorgis (ed.), Greenwood Publishing Group, 135-149.

Smura, T. (2006). Competition between Emerging Wireless Network Technologies: Case HSPA vs. WiMAX in Europe. In *17th European Regional ITS Conference*, August 22-24, 2006, Amsterdam, The Netherlands.

Snider, P. (1967). Mr. Gates revisited: A 1966 version of the 1949 case study. *The Journalism Quarterly*, *44*(3), 419–427.

Soderi, S., Iinatti, J., & Hamalainen, M. (2003, June). CLPDI algorithm in UWB synchronization. In *Workshop on UWB Systems*, 759–763, IEEE.

Somani, A. K., Mani, M., & Li, L. (2004). On Trading Wavelengths With Fibers: A Cost-Performance Based Study. *IEEE/ACM Transactions on Networking*, *12*(5), 944–951. doi:10.1109/TNET.2004.836130

Sorte, D. D., Femminella, M., & Reali, G. (2008). QoS-enabled Multicast for delivering Live Events in a Digital Cinema scenario. *Journal of Network and Computer Applications*.

Sousa, P., & Freitas, V. (1998). A Framework for the Development of Tolerant Real-Time Applications. *Computer Networks and ISDN Systems*, *30*(16), 1531–1541. doi:10.1016/S0169-7552(98)00188-3

Stabell, C. B., & Øystein, D. F. (1998). Configuring value for competitive advantage: On chains, shops, and networks. *Strategic Management Journal*, 19.

Staderini, E. M. (2002). UWB radars in medicine. *IEEE Aerospace and Electronic Systems Magazine International*, *1*(17), 13–18. doi:10.1109/62.978359

Stajano, F., & Anderson, R. J. (1999). The resurrecting duckling: Security issues for ad-hoc wireless networks. In *Security Protocols Workshop*, 172–194.

Stathis, C., & Maglaris, B. (2000). Modeling the self-similar behaviour of network traffic. *Journal of Computer Networks*, *34*(1), 37–47. doi:10.1016/S1389-1286(00)00095-5

Steinbock, D. (2003). Globalization of wireless value system: From geographic to strategic advantages. *Telecommunications Policy*, *27*, 207–235. doi:10.1016/S0308-5961(02)00106-4

Stonier, J. (1999). Airbus: Valuing Options in the Airline Industry. The 3rd Annual International Conference on Real Options.

Suarez, F. F. (2004). Battles for technological dominance: an integrative framework. *Research Policy*, *33*, 271–286. doi:10.1016/j.respol.2003.07.001

Sullivan, K., Chalasani, P., Jha, S., & Sazawal, V. (1999). Software design as an investment activity: A real options perspective. In Trigeorgis, L. (Ed.), *Real options and business strategy: Applications to decision making*. London, UK: Risk Books.

Sun, B., Osborne, L., Xiao, Y., & Guizani, S. (2007). Intrusion Detection Techniques in Mobile Ad Hoc and Wireless Sensor Networks. *IEEE Wireless Communications*, *14*, 56–63. doi:10.1109/MWC.2007.4396943

Sun, Y., Yu, W., Han, Z., & Liu, K. J. (2006). Information Theoretic Framework of Trust Modeling and Evaluation for Ad Hoc Networks. *IEEE Journal on Selected Areas in Communications*, *24*, 305–317. doi:10.1109/JSAC.2005.861389

Sun, W., Yamaguchi, H., Yukimasa, K., & Kusumoto, S. (2000). GVGrid: A QoS Routing Protocol for Vehicular Ad Hoc Networks. In *Fourteenth IEEE International Workshop on Quality of Service (IWQoS 2006)*.

Surowiecki, J. (2004). *The wisdom of the crowds: Why the many are smarter than the few and how collective wisdom shapes business, economies, societies and nations*. New York: Doubleday.

Suryanarayana, G., Erenkrantz, J., & Taylor, R. (2005). An Architectural Approach for Decentralized Trust Management. *IEEE Internet Computing*, *9*, 16–23. doi:10.1109/MIC.2005.119

Symposium, W. T. (2008). Retrieved July 11, 2008, from http://www.csupomona.edu/wtsi

Tabassi, A. (2008). Presentation at Wireless Telecommunications Symposium 2007. April 27, 2007, Pomona, CA.

Tallberg, M., Hämmäinen, H., Töyli, J., Kamppari, S., & Kivi, A. (2007). Impacts of handset bundling on mobile data usage: The case of Finland. *Telecommunications Policy*, *31*, 648–659. doi:10.1016/j.telpol.2007.08.007

Tan, P.-H. (1999). *Statistical analysis of long-range dependent processes via a stochastic intensity approach, with applications in networking*. Unpublished PhD thesises, North-Carolina State University.

Tapscott, D., Lowy, A., & Ticoll, D. (2000). *Digital Capital - Harnessing the power of business webs*. Boston, MA: Harvard Business School Press.

Taudes, A. (1998). Software growth options. *Journal of Management Information Systems*, *15*(1), 165–185.

Taudes, A., Feurstein, M., & Mild, A. (2000). Options Analysis of Software Platform Decisions: A Case Study. *Management Information Systems Quarterly*, *24*(2), 227–243. doi:10.2307/3250937

TDG. (2007). *GSM continues to have Greatest Gain in U.* S. Cellular Market. TDG Press Releases.

Telefonica and Telefónica Moviles. (2002-2007). *Annual and form 20-F reports*. Madrid, Spain: Author.

Telenor. (2002-2007). *Annual and form 20-F reports, 2002-2007*. Fornebu, Norway: Author.

TeliaSonera. (2002-2007). *Annual and form 20-F reports, 2002-2007*. Stockholm, Sweden: Author.

Theodorakopoulos, G., & Baras, J. S. (2006). On Trust Models and Trust Evaluation Metrics for Ad Hoc Networks. *IEEE Journal on Selected Areas in Communications*, *24*, 318–328. doi:10.1109/JSAC.2005.861390

Theodorakopoulous, G., & Baras, J. S. (2004, October). Trust evaluation in ad-hoc networks. In *Proceedings of the ACM Workshop on Wireless Security,* Philadelphia, (pp.1-10). ACM Publishing.

Tian, Z., Yang, L., & Giannakis, G. B. (2002). *Symbol timing estimation in ultra wideband communications* (pp. 1924–1928). Signals, Systems and Computers.

Timmmers, P. (1998). Business models for electronic markets. *Journal on Electronic Markets*, *8*(2), 3–8. doi:10.1080/10196789800000016

Trigeorgis, L. (1993). The Nature of Option Interactions and the Valuation of Investments with Multiple Real Options. *Journal of Financial and Quantitative Analysis*, *28*(1), 1–20. doi:10.2307/2331148

Trigeorgis, L. (1996). *Real Options: Managerial Flexibility and Strategy in Resource Allocation*. Cambridge, Mass.: The MIT Press.

Trivedi, K. (2001). *Probability and statistics with reliability, queueing, and computer science applications*. New York: John Willey & Sons.

Tsirigos, Z. H. A. (2001, November). Multipath routing in the presence of frequent topological changes. In *IEEE Communication Magazine*.

Tung, T., & Jamalipour, A. (2003). A Kalman-Filter Based Paging Strategy for Cellular Networks. In *36th Annual Hawaii International Conference on System Sciences (HICSS'03)*. Vint project (n.d). http://www.isi.edu/nsnam/ns/.

Tuttlebee, W. (2002). *Software defined radio – enabling technologies*. Wiley.

UCINET. 2007. Received September 25, 2007 from http://www.analytictech.com/

Ulrichsweb.com. (2008). Proquest. Retrieved July 9, 2008, from http://www.ulrichsweb.com

Van Liere, D. W., Hagdorn, L., Hoogeweegen, M. R., & Vervest, P. H. M. (2004). *Organizational Performance of a Firm in a Modular Business Network*. ECIS 2004.

Verkasalo, H. (2007a). *A Cross-Country Comparison of Mobile Service and Handset Usage*. Licentiate's thesis. Helsinki University of Technology, Networking Laboratory, Finland.

Verkasalo, H. (2007b). Empirical Findings on the Mobile Internet and E-Commerce. *Presented at 20th Bled e Conference: E-Mergence: Merging and Emerging Technologies, Processes, and Institutions*. Bled, Slovenia, June 4 - 6, 2007.

Verkasalo, H. (2007c). Measurement of Smartphone Service Evolution in Finland. Forthcoming in *Journal of Targeting, Measurement and Analysis in Marketing; Special Issue; December 2007*.

Verkasalo, H. (2008a). Dynamics of Mobile Service Adoption. Forthcoming in *Journal of E-Business Research, 4(3) July 2008*.

Verkasalo, H. (2008b). Handset-Based Measurement of Mobile Service Demand and Value. Forthcoming in *INFO: The Journal of Policy, Regulation and Strategy, 10*(3); May 2008.

Verkasalo, H., & Hämmäinen, H. (2007). A Handset-Based Platform for Measuring Mobile Service Usage. *INFO: The Journal of Policy, Regulation and Strategy, 9*(1).

Vesa, J. (2005). *Mobile Services in the Networked Economy*. Hershey, PA, USA: IRM Press.

Villen-Altamirano, M., & Villen-Altamirano, J. (2006). On the efficiency of RESTART for multidimensional systems. *ACM Transactions on Modeling and Computer Simulation, 16*(3), 251–279. doi:10.1145/1147224.1147227

Viswanathan, V. (2007). Information Hiding in Wave Files through Frequency Domain. *Applied Mathematics and Computation*.

Vodafone. (2002/2003-2007/2008). *Annual and form 20-F reports*. Newbury, UK: Author.

Wang, H., Jiang, S., Wei, G., & Ji, F. (2006). A Rapid Synchronization Acquisition Scheme for UWB Communications. In *International Symposium on Systems and Control in Aerospace and Astronautics*, 844-848.

Wanta, W., & Craft, S. (2004). Women in the newsroom: Influences of female editors and reporters on the news agenda. *Journalism & Mass Communication Quarterly, 81*(1), 124–138.

Watts, D. (1999). *Small Worlds: The Dynamics of Networks between Order and Randomness*. Princeton, NJ: Princeton University Press.

Webb, W. (2007). *Wireless Communications: The Future*. Wiley. doi:10.1002/9780470065129

Webster's Third New International Dictionary (1971). Springfield MA: G. & C. Merriman Company.

Weill, P., & Vitale, M. R. (2001). *Place to Space: Migrating to E-Business Models*. Boston: Harvard Business School Press.

Welch, G., & Bishop, G. (n.d.). *The kalman filter*. http://www.cs.unc.edu/ welch/kalman/.

Wernerfelt, B. (1984). A Resource-Based View of the Firm. *Strategic Management Journal*, *5*, 171–180. doi:10.1002/smj.4250050207

Wettemann, R., & Kelly, T. (1998, March). The Changing International Telecommunications Environment: Country Case Studies. *2nd World Telecommunication Policy Forum* (pp. 16-18), Geneva. Retrieved April 20, 2008, from, http://www.itu.int/osg/csd/wtpf/wtpf98/cases/index.html

White, D. M. (1950). The "Gate-Keeper": A case study in the selection of news. *The Journalism Quarterly*, *27*, 383–390.

Williams, B. A., & Delli Carpini, M. X. (2004). Monica and Bill all the time and everywhere, the collapse of gatekeeping and agenda setting in the new media environment. *The American Behavioral Scientist*, *47*(9), 1208. doi:10.1177/0002764203262344

Williamson, O. E. (1975). *Markets and hierarchies. NewYork.* Free Press.

Williamson, O. E. (1985). *The economic institution of capitalism. NewYork.* Free Press.

Williamson, O. E. (1989). Transaction cost economics. In Shemlenese, R., & Willig, R. D. (Eds.), *Handbook of industrial organisation, 1*. Amsterdam: Elsevier Science Publishers BV.

Willow, C. C. (2007a). Comprehensive Design for Information Systems with Integrated Systems Design Methodology. *International Journal of Taiwan Academy of Business Management Review*, *3*(1), 49–72.

Willow, C. C. (2007b). Qualitative Decision Making with Integrated Systems Design Methodology. *Journal of Engineering Management and Technology*, *24*(3), 262–282. doi:10.1016/j.jengtecman.2007.05.003

Wills, C. E., Brown, D. C., Dunskus, B. V., & Kemble, J. (1998). Evaluating Network Serviceability. *Computer Networks and ISDN Systems*, *30*(24), 2283–2291. doi:10.1016/S0169-7552(98)00154-8

Win, M. Z., & Scholtz. (2000). Ultra-wide bandwidth time-hopping spread spectrum impulse radio for wireless multiple-access communications. *IEEE Transactions on Communications*, (48): 679–691. doi:10.1109/26.843135

Wireless Federation. (2007). *EU to open up mobile spectrum*. Retrieved June 25, 2009, from http://wirelessfederation.com/news/8794-eu-to-open-up-mobile-spectrum

Woo, S.-C. M., & Singh, S. (2001). Scalable routing protocol for ad hoc networks. *Wireless Networks*, *7*(5), 513–529. doi:10.1023/A:1016726711167

World Bank. (2008). *World development indicators database*. Washington, DC: Author.

Wu, B., Chen, J., Wu, J., & Cardei, M. (2006). A survey on attacks and countermeasures in mobile ad hoc networks. In Y. Xiao, X. Shen, & D.-Z. Du (Eds.), *Wireless mobile networks security* (Ch.12). New York: Springer.

Wu, J. H. K. (2001). Performance study of a multipath routing method for wireless mobile ad hoc networks. In *the 9th international symposium on modeling, analysis and simulation of computer and telecommunication system.*

Xilinx Inc. (2007). Distributed arithmetic filter v8.0, http://www.xilinx.com.

Xiong, L., & Liu, L. (2004). PeerTrust: Supporting Reputation-Based Trust for Peer-to-Peer Electronic Communities. *IEEE Transactions on Knowledge and Data Engineering*, *16*, 843–857. doi:10.1109/TKDE.2004.1318566

Xue, Y., Li, B., & Nahrstedt, K. (2001). A scalable location management scheme in mobile ad-hoc networks. In *IEEE Conference on Local Computer Networks.*

Yan, Z., Zhang, P., & Virtanen, T. (2004). *Trust evaluation based security solution in ad hoc networks* (Nokia Tech. Rep.) Retrieved from http://www.nokia.com/library/files/docs/

Yang, H., Shu, J., Meng, X., & Lu, S. (2006). SCAN: Self-Organized Network Layer Security in Mobile Ad Hoc Networks. *IEEE Journal on Selected Areas in Communications*, *24*, 261–273. doi:10.1109/JSAC.2005.861384

Yang, C. L., & Kubat, P. (1989). Efficient computation of most probable states for communication networks with multimode components. *IEEE Transactions on Communications*, 37.

Yang, C.L., & Kubat, P. (1990). An Algorithm for Network Reliability Bounds. *ORSA Journal on Computing, 2.*

Yang, L., Tian, Z., & Giannakis, G. B. (2003). Non-data aided timing acquisition of ultra-wideband transmissions using cyclostationarity. In *Acoustics, Speech and Signal,* 121-124, IEEE.

Yoon, M. G., & Current, J. (2008). The hub location and network design problem with fixed and variable arc costs: formulation and dual- based solution heuristic. *The Journal of the Operational Research Society*, *59*(1), 80–89. doi:10.1057/palgrave.jors.2602307

Yoon, J., Liu, M., & Noble, B. (2003). Random waypoint considered harmful. In *Infocom 2003: Proceedings of the Twenty-Second Annual Joint Conference of the IEEE Computer and Communications Societies*, 1312–1321. IEEE Press.

Young, O., Burris, P., & Reiss, D. (2009). *The enterprise 2.0 buyer profile 2009–the top priorities and needs of firms looking at enterprise 2.0 technology.* Boston: Forrester.

Yu, Y., Lu, G.-H., & Zhang, Z.-L. (2004). Enhancing location service scalability with high-grade.

Zapata, M. G. (2006). Secure ad hoc on-demand distance vector (SAODV) routing. *ACM SIGMOBILE Mobile Computing and Communications Review*, *6*(3), 106–107. doi:10.1145/581291.581312

Zhang, Y., Liu, W., & Fang, Y. (2006). Securing Mobile Ad Hoc Networks with Certificateless Public Keys. *IEEE Transactions on Dependable and Secure Computing*, *3*, 386–399. doi:10.1109/TDSC.2006.58

Zhang, H., & Goeckel, D. L. (2003). Generalized transmitted-reference UWB systems. In *Ultra-wideband Systems and Technology Proc. 2003*, 147–151 IEEE.

Zhang, Y. S. C., & Fang, Y. (2008, April). Modeling secure connectivity of self-organized wireless ad hoc networks. In *27th IEEE International Conference on Computer Communications (INFOCOM 2008)*. AZ, USA.

Zhang, Y., & Lee, W. (2000, August 6-12). Intrusion detection in wireless ad-hoc networks. In *Proceedings of the 6th Annual Conference on Mobile Computing and Networking (Mobicom 2000)*, Boston (pp. 275-283). ACM Publishing.

Zhao, J., & Cao, G. (2006). Vehicle-Assisted Data Delivery in Vehicular Ad Hoc Networks. In *IEEE INFOCOM'06*. VADD.

Zhou, L., Schneider, F. B., & Renesse, R. V. (2002). Coca: A secure distributed online certification authority. *ACM Transactions on Computer Systems*, *20*(4), 329–368. doi:10.1145/571637.571638

Zhou, T., & Harn, L. 2007. Security in User-Assisted Communications. *Proceedings of IEEE Wireless Telecommunications Symposium (WTS).*

Zhou, T., & Harn, L. 2008. Risk Management of Digital Certification in Ad Hoc and P2P Networks. *Proceedings of the 21st Canadian Conference on Electrical and Computer Engineering.*

Zito, D., Pepe, D., Neri, B., (2002). Feasibility Study of a Low-cost System-on-a-Chip UWB Pulse Radar on Silicon for the Heart Monitoring.

About the Contributors

Michael R. Bartolacci is an associate professor of Information Sciences and Technology at Penn State University - Berks. He holds a PhD in Industrial Engineering and an MBA from Lehigh University. His research includes telecommunications modeling, cultural aspects of emerging technologies, and information security education.

Steven R. Powell is professor emeritus at California State Polytechnic University (Pomona, USA), where he led the university's efforts in interdisciplinary telecommunications education and research. He is the founder and General Chair of the Wireless Telecommunications Symposium, a leading interdisciplinary mobile communications and wireless networking conference supported by the IEEE Communications Society. Powell has taught telecommunications, engineering management, electrical engineering, IS, and computer science courses at the California State Polytechnic University - Pomona, Claremont Graduate University, Long Island University – C.W. Post Campus, and University of Southern California. His industrial experience includes positions as Director of Corporate Planning at American Express Company and Member of the Technical Staff at Bell Laboratories. He organized and chairs the IEEE Communications Society Foothill Chapter. Powell's research interests are primarily in the areas of telecommunications strategy, telecommunications management, telecommunications policy, international telecommunications, and wireless communications. He holds an MBA from Columbia University's Graduate School of Business, a PhD and MSc in electrical engineering from the University of Southern California, and a BS in electrical engineering from the Massachusetts Institute of Technology.

* * *

Philemon Bantimaroudis, PhD, is an assistant professor in the Department of Cultural Informatics at the University of the Aegean, Greece. He is holding a PhD from the University of Texas, Austin, USA. He has been a visiting assistant professor of communication and performance studies department at Northern Michigan University, USA. In 2008 he was an academic visitor at the Judge Business School of the University of Cambridge, UK. His research interests include media and culture, political and international communication. He has been published in different journals such as: *The Harvard International Journal of Press/Politics, The Journal of Communication, Communication Research, Media War & Conflict, The Journal of Media and Religion, Mass Communication and Society, Mediterranean Quarterly etc.* He is a contributor and/or editor to several peer - reviewed volumes. At present, Bantimaroudis is on the editorial board of *The Open Communication Journal.*

Vassiliki Cossiavelou, PhD candidate, is currently the communications secretary A' for the Greek Embassy in Brussels, Belgium, after her diplomatic post in Beijing, China. She has extensive industry experience in the area of telecommunications and networking with such companies as IBM, Mobil and Quest in Europe, North Africa and Middle East, and she has worked on ICT related projects with international organizations such as UNESCO, NATO and the European Union. Mrs. Cossiavelou has submitted her PhD at Aegean University in the Department of Cultural Informatics related to the impact of emerging ICT on the gatekeeping model in news media industries. She is an invited visitor lecturer in Pennsylvania State University, USA, and she is awarded with invited keynote speeches in international conferences of high impact factor. Her current research interests include media gatekeeping model, communication theory, business models in global media industries, emerging Asian media wireless markets, radio-spectrum and digital dividend management and communications 2.0. Her papers have appeared in journals and special issues, such as the International Journal of Mobile Network Design and Innovation (IJMNDI) and International Journal of Interdisciplinary Telecommunications and Networking (IJITN) and in conferences such as IEEE (Institute of Electrical and Electronics Engineers) Communications Society & Springer, Association for Computing Machinery (ACM), Institute for Operations Research and the Management Sciences (INFORMS), Wireless Telecommunications Symposium (WTS), Video and Broadband Telecommunications Symposium (VBTS), (International Conference on Telecommunication Systems - Modeling and Analysis (ICTSM). She is on editorial board for the IJITN and VBTS.

François Cosquer is CTO Security and Technology Strategist for the Alcatel-Lucent Enterprise and Verticals Business Group. Over the past 18 years, he has held senior positions with research institutions, equipment vendors and telecommunications operators. He draws on extensive experience in security architecture, networking, operating systems, middleware and multimedia applications. He has been speaker, panelist and chair at key industry events and conferences. Cosquer graduated in electronics and computing and holds an MSc in computer science and a PhD in computer engineering. He currently serves as adjunct professor at the faculty of engineering and computer science, University of Concordia, Montreal.

Annie Ohayon-Dekel is strategy director for the Alcatel-Lucent Enterprise and Verticals Business group. She is in charge of analyzing financial and new technology markets contributing to the overall enterprise strategy. Prior to joining Alcatel-Lucent, Ohayon-Dekel was technology consultant and held several positions in large technology companies. Ohayon-Dekel graduated in mathematics from Tel-Aviv University.

Floriano De Rango, PhD, received the degree in computer science engineering in October 2000, and a PhD in electronics and communications engineering in January 2005, both at the University of Calabria, Italy. From January 2000 to October 2000 he worked in the Telecom Research LAB C.S.E.L.T. in Turin as visiting scholar student. From March 2004 to November 2004 he was visiting researcher at the University of California at Los Angeles (UCLA). From November 2004 until September 2007 he has been a Research fellow in the D.E.I.S. Department, University of Calabria where he is now assistant professor. He was recipient of Young Researcher Award in 2007. He served as reviewer and TPC member for many International Conferences such as IEEE VTC, ICC, WCNC, Globecom, Med Hoc Net, SPECTS, WirelessCOM, WinSys and reviewer for many journals such as IEEE Communication

Letters, JSAC, IEEE Trans.on Vehicular Technology, Computer Communication, Eurasip JWCN, WI-NET etc. His interests include satellite networks, IP QoS architectures, adaptive wireless networks, ad hoc networks and pervasive computing. He has co-authored more than 130 papers among international journal and conferences proceedings.

Mohammed Ghanbari (M'78-SM'97-F01) is best known for the pioneering work on two-layer video coding for ATM networks, now known as SNR scalability in the standard video codecs, which earned him a Fellowship of IEEE in 2001. He has published more than 450 technical papers and four books on various aspects of video networking. His book on Video coding: an introduction to standard codecs, received the Rayleigh prize as the best book of year 2000 by the IET. He is a fellow of IEEE, fellow of IET and charted engineer (CEng).

Lein Harn received his BS degree in electrical engineering from the National Taiwan University in 1977. In 1980, he received his MS in electrical engineering from the State University of New York at Stony Brook and in 1984 he received his doctorate degree in electrical engineering from the University of Minnesota. He joined as an assistant professor in the department of electrical and computer engineering at the University of Missouri-Columbia in 1984 and in 1986, he moved to Computer Science and Telecommunication Program (CSTP) of University of Missouri-Kansas City (UMKC). While at UMKC he went on development leave to work in Racal Data Group in Florida for a year. His research interests are cryptography, network security and wireless communication security. He has published a number of papers on digital signature design and applications, wireless and network security. He has written two books on security. At present he is investigating new ways of using digital signature in various applications.

Xiaoyan Hong is an assistant professor in the Computer Science Department at the University of Alabama. She received her PhD degree in computer science from the University of California, Los Angeles in 2003. Her research interests are in the areas of mobile and wireless networks, future wireless Internet, challenged networks and wireless sensor networks.

Masoumeh Karimi received her master degree in telecommunications and networking in 2007 from Florida International University (FIU). She is currently pursuing her PhD degree in electrical and computer engineering at FIU. Her research interests include wireless communications, computer networking, and high performance switches. She is a member of IEEE and ACM.

Hak Ju Kim is an assistant professor at the School of Business in the University of Houston - Clear Lake. He received his PhD in information sciences (major: telecommunications) at the University of Pittsburgh. Prior to beginning his academic career, he worked for 6 years in SK Telecom which is a world-class wireless network operator that launched the world's first commercial CDMA cellular phone service. His main research areas are network economics (pricing/costing/investment/options) and policy, technology management (i.e., migration strategy) in telecommunications and networking such as wireless network and IP network, Internet accounting & charging, assessment of new emerging network technologies and services (Wi-Fi, Wi-Max, DMB, RFID, etc.).

Izabella Lokshina, PhD, is associate professor of MIS and chair of Management, Marketing and Information Systems Department at SUNY Oneonta, USA. Her main research interests are intelligent information systems and communication networks, queuing systems and performance analysis, complex system modeling and simulation.

Jun Liu received his PhD in computer science at the University of Alabama in 2007. His research interests include protocol design, performance evaluation, security and privacy, routing, and power-aware solutions for wireless mobile networks and sensor networks.

Kia Makki received his PhD in computer science from the University of California, Davis, and two MS degrees from The Ohio State University and West Coast University. He is currently a lucent technology professor with tenure and he had been associate dean of the College of Engineering and co-director of the Telecommunications and Information Technology at the Florida International University. Previously he was a tenured faculty with an O'Krepki Chair Professorship at the University of Louisiana, Lafayette, the director of the Center For Telecommunications Studies, and the co-director of the NOMAD and Advanced Computer Laboratory which is partially funded by NSF and NASA. His technical interests include mobile and wireless communication networks, ad-hoc and sensor networks, netwok security and privacy, multicasting and optical telecommunications. He has numerous publications in refereed journals and conferences and his work has been extensively cited in books and papers. Makki has been associate editor, editorial board member, guest editor of several international journals including editorial board member of several telecommunication journals. Makki has been involved in numerous conferences as a steering committee member, general chair, program chair, program vice-chair and program committee member. His current professional activities include standing committee member and technical program co-chair of the IEEE INFOCOM and the general chair of IEEE Globecom 2010. Makki is the co-author of a well-received book on mobile and wireless internet.

Zhaomin Mo received her PhD in electrical and computer engineering in 2007 from Florida International University (FIU). She is currently a software engineering at Microsoft Corp. Her research interests include wireless communications, vehicular ad hoc networks, and signal processing.

Benoit Morel, did his PhD in theoretical physics at the University of Geneva (Switzerland). His post-doctoral career took him to Harvard, Caltech and Stanford University, before he joined the faculty of Carnegie Mellon in the Department of Engineering and Public Policy. His research and teaching interests have focused at the interface between policy and math (from chaos and complex system to stochastic processes and option theory).

Niki Pissinou received her PhD in computer science from the University of Southern California, her MSc in computer science from the University of California at Riverside, and her BSISE in Industrial and systems engineering from The Ohio State University. She is currently a tenured professor and the director of the Telecommunication & Information Technology Institute at the Florida International University. Previously she was a tenured faculty at the Center For Advanced Computer Studies at the University of Louisiana at Lafayette where she was also the director of the Telecommunication & Information & Technology Laboratory partially funded by NASA, and the co-director of the NOMAD:

A Wireless and Nomadic Laboratory partially funded by NSF, and the Advanced Network Laboratory. Pissinou is active in the fields computer networks, information technology and distributed systems. Pissinou has published over hundred refereed publications and has received best paper awards. She has also co-edited eight volumes and is the author of an upcoming book on wireless internet computing. She has received extensive funding for her research work from such agencies as NSF, NASA, DARPA and ARO including two recent NSF awards on wireless networks. Pissinou has also served as a steering committee, general chair and program committee member of over hundred program and organizational committees of IEEE and ACM sponsored technical conferences. Pissinou has also refereed for a wide variety of conferences and journals, and has served as a reviewer on NSF and NASA panel reviews including the KDI and ITR panels. She has been the editor of eight journals including IEEE TKDE and guest editor of seven journals. She has received eight achievement awards and she has been invited speaker and keynote speaker of conferences.

Dimitar Radev DSc., PhD, is professor at Department of Communication Technique and Technologies at University of Rousse, Bulgaria. His main research interests are teletraffic theory, simulation and modeling of communication networks, rare event simulation and performance analysis of queuing systems.

Svetla Radeva, DSc, PhD is associate professor at Department of Computer Aided Engineering at University of Architecture Civil Engineering and Geodesy, Sofia, Bulgaria. Her main research interests are simulation and modeling of traffic processes, and rare event simulation.

Swadesh Kumar Samanta is on leave from BSNL, India where he worked as Deputy General Manager looking after planning and deployment of telecommunications services for 13 years. He was the recipient of Chevening Scholarships awarded by British Government and currently doing PhD on telecom service pricing under UKIERI Scholarships scheme. He worked with Network Research Centre, BT, UK where he developed a tool to indentify and account the users and their applications which cause congestion in the internet. He is regularly invited to give talks at international conferences.

Lei Tang is a graduate student in the Department of Computer Science at Rice University. His research interests include routing schemes of mobile and wireless systems, network security and privacy. He was a graduate student at the University of Alabama before joining the Rice University.

William Webb joined Ofcom as head of Research and Development and senior technologist in 2003. Here he manages a team providing technical advice and performing research across all areas of Ofcom's regulatory remit. He also leads some of the major reviews conducted by Ofcom including the Spectrum Framework Review and Ultra-Wideband Consultation. Previously, Webb worked for a range of communications consultancies in the UK in the fields of hardware design, computer simulation, propagation modelling, spectrum management and strategy development. Webb also spent three years providing strategic management across Motorola's entire communications portfolio, based in Chicago, Webb has published ten books, sixty papers, and four patents. He is a visiting professor at Surrey University and a fellow of the Royal Academy of Engineering. He is a fellow of the IET where he has served as a vice president and is a fellow of the IEEE where he was General Chair for the DySpan 2007 conference. His biography is included in multiple "Who's Who" publications around the world. He sits on the judging panels of the Wall Street Journal "Annual Innovation Awards". Webb has a first class honours degree in electronics, a PhD and an MBA.

Martin B.H. Weiss is associate dean for academic affairs and research at the School of Information Science at the University of Pittsburgh. He holds a PhD in engineering and public policy from Carnegie Mellon University (1988) where he studied the standards development process under professor Marvin A. Sirbu. He also holds an MSE in computer, control, and information engineering from the University of Michigan (1979) and a BSE in electrical engineering from Northeastern University (1978). His overall research themes are the analysis of situations where competing firms must cooperate, cost modeling of new technologies in telecommunications systems and the evolution of telecommunications industry. His industrial experience includes technical and professional work at several R&D and consulting firms in the United States. From 1978 to 1981, he was a member of the technical staff at Bell Laboratories; from 1983 to 1985, he was a member of the technical staff at the MITRE Corp; and from 1985 to 1987 he was a senior consultant with Deloitte, Haskins, and Sells. His current research projects include technical and cost studies of IP and ATM telephony, spectrum markets, secondary spectrum usage, bandwidth markets, interconnection of packet networks that support quality of service (QoS), and economics of new technologies in telecommunications. Together with Christopher Sterling and Phyllis Bernt, he has written the book Shaping American Telecommunications, published by Erlbaum.

Charles Willow, PhD, is a professor of management information systems (MIS) and Management of Technology (MOT) in the School of Business Administration at the Monmouth University, West Long Branch, New Jersey, USA. As an engineer-and-computer-scientist-turned-management-faculty, Dr. Willow's research agenda has been extensive, ranging from information systems development to case-driven strategic management issues, systems engineering, and operations research. His current research interests include information systems and network security, neural network applications, intelligent software agents, databases, computer-generated graphics, cost-model analysis of Internet-based business, and strategic management of technology. Dr. Willow's papers have appeared in journals such as ACM Transactions on Information Systems, Journal of Intelligent Manufacturing, IEEE Transactions on Systems, Man, and Cybernetics, Springer Lecture Notes on Computer Science, International Journal of Intelligent Information Technologies, and the Journal of Engineering and Technology Management, among others. He is a member of the IEEE Computer Society, ACM, INFORMS Computing Society, Association of Information Systems, and the National Society of Professional Engineers (NSPE). At present, Dr. Willow is on the editorial board for the Journal of Management Information Systems (JMIS), MIS Quarterly (MISQ), and Information Systems Research (ISR). Outside academia, he has been an active member of the International Simultaneous Interpreters Society (ISIS) and the International Judo Federation (IJF).

John Woods is a senior lecturer and his interests include image processing, autonomous robotics, intelligent power control, networks and network pricing. During his brief career he has accumulated over 60 journal and conference publications accompanied by grants in these areas. He is a member of the IEE and regularly attends and presents at national and international conferences.

Hannu Verkasalo is an acting project manager at the national research project MoMI (modeling the usage and business of the mobile Internet) at the Helsinki University of Technology in Finland. Hannu Verkasalo has actively published on the evolution of the mobile services market. He recently finished his doctoral dissertation on networking business. Verkasalo has focused particularly on the empirical modeling of mobile service usage, utilizing the recently developed handset-based research method in his analysis work. In addition to research work and consulting of the mobile industry, Verkasalo gives lectures on network economics, operator business and telecommunications regulation.

Qunwei Zheng received his PhD in computer science from the University of Alabama in 2007. His research interests include wireless networks and mobile computing, especially on mobility modeling, analysis and applications.

Tong Zhou is a PhD candidate in interdisciplinary studies at University of Missouri - Kansas City, USA and a principal engineer at Comcast Cable, USA. He received his BEng in communications engineering from Beijing Information Science & Technology University, P. R. China in 1991 and MS in electrical and computer engineering from State University of New York - Stony Brook, USA in 1998. He worked at Sprint Nextel, USA as a principal engineer from 1998 to 2007 and Motorola China Electronics Ltd. as a cellular field engineer from 1993 to 1997. His research interests are in the areas of information security, next generation networks and services, peer-to-peer trust management, social networks and ad hoc networks. He has over twenty patents issued/pending in USA.

Hao Zhu received his BS degree in computer science and engineering from Xian Jiaotong University, Xian, China in 1997. He received the MS degree in computer science from the Institute of Software, Chinese Academy of Sciences, Beijing, China in 2000. After receiving his PhD in computer science and engineering at the Pennsylvania State University in August 2004, he joined the Department of Electrical and Computer Engineering at Florida International University as an assistant professor. His research interests focus on resource management in wireless networks, cross-layer optimization of wireless networking protocols and cognitive radio networks.

Index

E

F

G

H

I

N

O

P

Q

R

S

T

U

V

W

Y

Breinigsville, PA USA
31 March 2011
258903BV00007B/2/P

9 781609 605056